SUSTAINABLE HYBRID COOLING AND DEHUMIDIFICATION STRATEGIES FOR CONTROLLED ENVIRONMENTS

EMERGING TECHNOLOGIES AND MATERIALS IN THERMAL ENGINEERING

SUSTAINABLE HYBRID COOLING AND DEHUMIDIFICATION STRATEGIES FOR CONTROLLED ENVIRONMENTS

Edited by

UZAIR SAJJAD

Interdisciplinary Research Center of Sustainable Energy Systems (IRC-SES), King Fahd University of Petroleum and Minerals, Dhahran, Saudi Arabia

HAFIZ MUHAMMAD ALI

Interdisciplinary Research Center for Sustainable Energy Systems (IRC-SES), King Fahd University of Petroleum and Minerals; Department of Mechanical Engineering, King Fahd University of Petroleum and Minerals, Dhahran, Saudi Arabia

MUHAMMAD SULTAN

Department of Agricultural Engineering, Bahauddin Zakariya University, Pakistan

Series Editor

HAFIZ MUHAMMAD ALI

ELSEVIER

Elsevier
Radarweg 29, PO Box 211, 1000 AE Amsterdam, Netherlands
125 London Wall, London EC2Y 5AS, United Kingdom
50 Hampshire Street, 5th Floor, Cambridge, MA 02139, United States

ISBN: 978-0-443-48555-8

For information on all Elsevier publications visit our website at https://www.elsevier.com/books-and-journals

Publisher: Priscilla Braglia
Acquisitions Editor: Fran Kennedy-Ellis
Editorial Project Manager: Harshita Manchanda
Production Project Manager: Gomathi Sugumar
Cover Designer: Arumugam Kothandan

Typeset by Integra Software Services

Contents

Contributors

Fakhar Abbas
Department of Agricultural Engineering, Bahauddin Zakariya University, Multan, Pakistan

Muhammad Adil
Department of Agricultural Engineering, Bahauddin Zakariya University, Multan, Pakistan

Md Shamim Ahamed
Department of Biological and Agricultural Engineering, University of California, Davis, CA, United States

Faheem Hassan Akhtar
Department of Chemistry and Chemical Engineering, Lahore University of Management Sciences (LUMS), Lahore, Pakistan

Muhammad Aleem
Department of Agricultural Engineering, Bahauddin Zakariya University, Multan, Pakistan

Hafiz Muhammad Ali
Interdisciplinary Research Center for Sustainable Energy Systems (IRC-SES); Mechanical Engineering Department, King Fahd University of Petroleum and Minerals, Dhahran, Saudi Arabia

Hadeed Ashraf
Department of Agricultural Engineering, Bahauddin Zakariya University, Multan, Pakistan
Agricultural Engineering Institute, National Agricultural Research Centre, Islamabad, Pakistan

Muhammad Nauman Ashraf
Department of Agricultural Engineering, Bahauddin Zakariya University, Multan, Pakistan

Muhammad Asif
Department of Agricultural Engineering, Bahauddin Zakariya University, Multan, Pakistan

Muhammad Waheed Azam
Interdisciplinary Research Center for Industrial Nuclear Energy, King Fahd University of Petroleum & Minerals, Dhahran, Saudi Arabia

Muhammad U.B.U. Butt
Department of Agricultural Engineering, Bahauddin Zakariya University, Multan, Pakistan

Ghulam Qadir Chaudhary
Mechanical Engineering Department, MUST, Mirpur, Pakistan

Muhammad Farooq
Department of Mechanical Engineering, University of Engineering and Technology, Lahore, Pakistan

Ghulam Abbas Gohar
Department of Mechanical Engineering, King Fahd University of Petroleum & Minerals, Dhahran, Saudi Arabia

Naveed Husnain
Department of Mechanical Engineering, Bahauddin Zakariya University, Multan, Pakistan

Muhammad Ali Imran
Department of Agricultural Engineering, Bahauddin Zakariya University, Multan, Pakistan

Nida Imtiaz
Faculty of Mechanical Engineering, Universiti Teknologi Malaysia, Johor Bahru, Malaysia

Muhammad Irfan
Department of Agricultural Engineering, Bahauddin Zakariya University, Multan, Pakistan

Muhammad Ahmad Jamil
Department of Mechanical and Construction Engineering, Northumbria University, Newcastle Upon Tyne, United Kingdom

Muhammad Yasar Javaid
Department of Mechanical Engineering and Technology, Government College University Faisalabad, Faisalabad, Pakistan

Muhammad Khalid
Department of Agricultural Engineering, Bahauddin Zakariya University, Multan, Pakistan

Zahid Mahmood Khan
Department of Agricultural Engineering, Bahauddin Zakariya University, Multan, Pakistan

Jie Liu
College of Engineering, Huazhong Agricultural University, Wuhan, China

Muhammad Hamid Mahmood
Department of Agricultural Engineering, Bahauddin Zakariya University, Multan, Pakistan

Takahiko Miyazaki
Department of Advanced Environmental Science and Engineering, Faculty of Engineering Sciences, Kyushu University, Kasuga, Fukuoka, Japan

Saliha Qamar
Herff College of Engineering, The University of Memphis, Memphis, TN, United States

Junaid Ramzan
Department of Agricultural Engineering, Bahauddin Zakariya University, Multan, Pakistan

Hafiz Muhammad Umar Raza
Department of Agricultural Engineering, Bahauddin Zakariya University, Multan, Pakistan

Nadia Riaz
Department of Agricultural Engineering, Bahauddin Zakariya University, Multan, Pakistan

Uzair Sajjad
Interdisciplinary Research Center for Sustainable Energy Systems (IRC-SES), King Fahd University of Petroleum and Minerals, Dhahran, Saudi Arabia

Aneela Sarwar
Department of Agricultural Engineering, Bahauddin Zakariya University, Multan, Pakistan

Giovani Sempirini
Department of Industrial Engineering (DIN), Alma Mater Studiorum - University of Bologna, Bologna, Italy

Muhammad Wakil Shahzad
Department of Mechanical and Construction Engineering, Northumbria University, Newcastle Upon Tyne, United Kingdom

Muhammad Sultan
Department of Agricultural Engineering, Bahauddin Zakariya University, Multan, Pakistan

Mazlan Abdul Wahid
Faculty of Mechanical Engineering, Universiti Teknologi Malaysia, Johor Bahru, Malaysia

Haseeb Yaqoob
Department of Mechanical and Construction Engineering, Northumbria University, Newcastle Upon Tyne, United Kingdom

Muhammad Zeshan
Department of Agricultural Engineering, Bahauddin Zakariya University, Multan, Pakistan

Zhaoli Zhang
School of Mechanical Engineering, Southwest Jiaotong University, Chengdu, China

Preface

Cooling and dehumidification are now a key part of many of the most pressing challenges of this century. The need for controlled environments with precise control of temperature and humidity has grown significantly due to rapid urbanization, population growth, climate change, and the growing demand for high-quality food production. Whether in greenhouses, postharvest storage areas, industrial buildings, indoor agriculture, or specialized manufacturing settings, effective climate control is crucial for boosting productivity, maintaining quality, and promoting human health and well-being. Meanwhile, traditional vapor-compression air-conditioning systems remain a significant contributor to global energy use and greenhouse gas emissions, highlighting the critical need for sustainable and energy-efficient solutions.

In recent years, hybrid cooling and dehumidification technologies have become promising technologies that can solve these problems. These systems can incorporate evaporative cooling, desiccant dehumidification, membrane-based energy recovery, renewable energy, advanced heat recovery, and intelligent control strategies, providing substantial potential for energy savings while providing excellent thermal and humidity control. The development of next-generation cooling technologies is further propelled by ongoing developments in computational modeling, artificial intelligence, optimization techniques, and new materials, all of which are more efficient, environmentally friendly, and flexible to different climatic conditions.

While there has been extensive research on each of the cooling technologies, there are few comprehensive references that combine these various technologies in the context of controlled environments. This book has been developed to fill this gap, providing a comprehensive overview of sustainable hybrid cooling and dehumidification systems. It compiles the most recent research advances, engineering techniques, numerical analysis, experimental research, integration of renewable energy technologies, and future technology trends in one place. The combined knowledge of the contributing authors gives readers a solid foundation of knowledge and practical experience in the design, analysis, optimization, and implementation of sustainable cooling solutions.

The chapters are well structured and arranged in a logical sequence from basic to advanced applications. The first chapters present the fundamentals of hybrid cooling technologies and provide the scientific background of evaporative cooling, indirect cooling systems, and desiccant-assisted air-conditioning. The advanced cooling and dehumidification methods for greenhouse climate control, numerical modeling with computational fluid dynamics, membrane-based energy recovery ventilators, integration of renewable energy, and solar-assisted cooling methods are discussed in the following chapters. The volume also emphasizes the increasing importance of intelligent modeling and artificial neural networks in predicting system performance and design optimization. Finally, the book addresses emerging technologies, future research directions, and practical avenues for large-scale deployment of sustainable cooling systems. The chapters collectively offer a thorough overview of the state of the art and future potential of hybrid cooling and dehumidification technologies.

One of the unique aspects of this volume is the interdisciplinary approach. The contributors cover a range of expertise from thermal engineering, mechanical engineering, agricultural engineering, sustainable energy systems, computational modeling, and building technologies. This diversity is indicative of the interdisciplinary character of sustainable cooling research and the fact that progress is increasingly being made through interdisciplinary collaboration. The book is designed to give balanced coverage of the subject, both theoretical and experimental, and with practical engineering applications, useful for both research and industry.

The book will be used as a reference book for researchers, graduate students, practicing engineers, HVAC professionals, agricultural engineers, policymakers, and industry specialists in the fields of sustainable cooling technologies and controlled-environment systems. We hope that readers will not only gain a comprehensive and current overview of the state-of-the-art but also practical insights into how to tackle real-world engineering problems and inspiration for creating new solutions to enhance energy efficiency and environmental sustainability.

This volume has been completed with the help and dedication of many people. We gratefully acknowledge the contributions of all the contributing authors for their knowledge and dedication to creating quality scholarly chapters. Their contributions are the result of years of research and engineering innovation and are the basis of this book. Finally, we would like to thank the anonymous reviewers for their insightful comments and helpful suggestions that greatly enhanced the quality and clarity of the individual chapters. We would also like to thank the editorial and production staff at Elsevier for their professionalism, encouragement, and ongoing support during the development and publication of this volume.

We hope this book will not only be a useful scientific reference but also a springboard for future innovation. With the transition to low-carbon energy systems, climate-resilient agriculture, and sustainable built environments, the development of efficient cooling and dehumidification technologies will gain more significance. We hope that the information contained in this book will inspire interdisciplinary cooperation, inspire new research, and help to translate scientific findings into practical engineering solutions that will help build a more sustainable future.

Uzair Sajjad
Interdisciplinary Research Center for Sustainable Energy Systems,
King Fahd University of Petroleum and Minerals, Dhahran, Saudi Arabia

Hafiz Muhammad Ali
Interdisciplinary Research Center for Sustainable Energy Systems (IRC-SES),
King Fahd University of Petroleum and Minerals; Department of Mechanical Engineering,
King Fahd University of Petroleum and Minerals,
Dhahran, Saudi Arabia

Muhammad Sultan
Department of Agricultural Engineering,
Bahauddin Zakariya University Multan, Pakistan

Acknowledgements

The editors gratefully acknowledge the support of the International UK Research and Innovation (UKRI) Project "Super-efficient Sustainable Cooling Solution for All Applications (UKRI312-S2Cool)," funded by UKRI under the Ayrton Challenge Programme. This initiative, led by Northumbria University (United Kingdom) with the collaboration of Bahauddin Zakariya University (Pakistan) and other partners, has provided valuable insights into energy-efficient HVAC technologies and smart evaporative cooling systems, contributing to several themes discussed in this volume.

The editors further acknowledge the support of King Fahd University of Petroleum and Minerals (KFUPM), Saudi Arabia, for providing an enabling research environment and institutional support that contributed to the successful completion of this volume.

The editors sincerely thank all chapter authors for their valuable contributions and cooperation in the preparation of this book. We also appreciate the constructive comments of the anonymous reviewers, which helped improve the quality of the chapters. Finally, we acknowledge the guidance and support of the editorial and production teams at Elsevier throughout the development and publication of this volume.

CHAPTER 1

Introduction: Hybrid cooling and dehumidification for controlled environments

Uzair Sajjad[a], Ghulam Abbas Gohar[b], Hafiz Muhammad Ali[a,b], and Muhammad Sultan[c]
[a]Interdisciplinary Research Center for Sustainable Energy Systems (IRC-SES), King Fahd University of Petroleum and Minerals, Dhahran, Saudi Arabia
[b]Department of Mechanical Engineering, King Fahd University of Petroleum and Minerals, Dhahran, Saudi Arabia
[c]Department of Agricultural Engineering, Bahauddin Zakariya University, Multan, Pakistan

1.1 Background, significance, and scope

Reliable operation of controlled environments such as greenhouses, postharvest storage facilities, and climate-sensitive buildings requires cooling and dehumidification. Mechanical vapor compression-based traditional cooling systems are slowly becoming limited due to their high energy requirements, restricted ability to control humidity, and environmental concerns, especially in hot and humid climates. These restrictions have led to the creation of hybrid cooling and dehumidification systems that combine different cooling systems such as evaporative cooling and desiccant-based dehumidification, energy recovery, and renewable heat use to decouple sensible and latent loads and enhance overall efficiency.

In this book, the system-level and application-oriented perspective on sustainable hybrid cooling technologies for controlled environments is presented. By combining basic principles with the integration of hybrid systems, optimization, and deployment pathways, this book shows that hybrid cooling systems are feasible and scalable ways of attaining energy-efficient and climate-adaptive cooling. Fig. 1.1 presents an overview of the four sections of the book structure and theme development of hybrid cooling and dehumidification systems. Section-wise summaries are presented in Tables 1.1–1.4.

Sustainable Hybrid Cooling and Dehumidification Strategies for Controlled Environments
https://doi.org/10.1016/B978-0-443-48555-8.00013-X

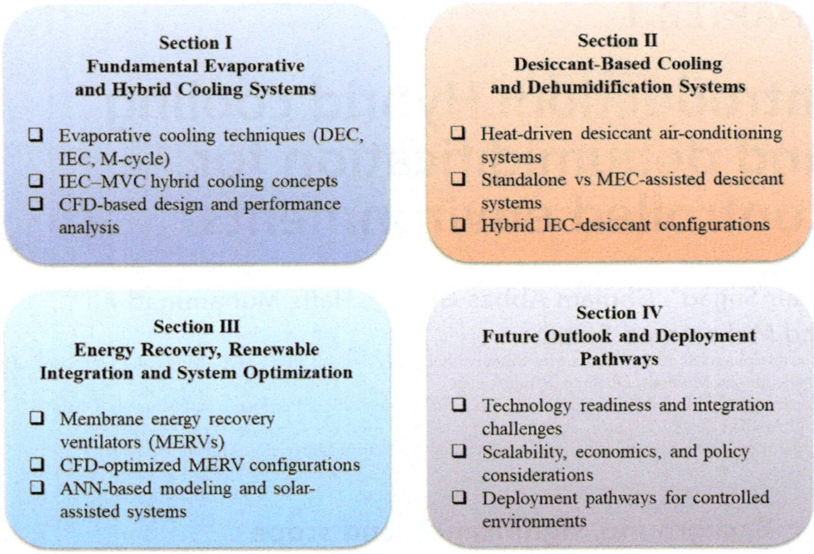

Fig. 1.1 Overview of the four sections of the book structure and thematic progression of hybrid cooling and dehumidification systems.

1.2 Section I: Fundamental evaporative and hybrid cooling systems

Section I introduces the thermodynamic and heat-mass transfer principles of evaporative cooling, including direct, indirect, and Maisotsenko-cycle systems, and examines their performance through analytical and CFD-based studies (see Table 1.1).

1.3 Section II: Desiccant-based cooling and dehumidification systems

Section II focuses on latent load control using solid and liquid desiccant technologies, highlighting hybrid evaporative-desiccant configurations for independent temperature and humidity regulation in controlled environments (see Table 1.2).

1.4 Section III: Energy recovery, renewable integration, and system optimization

Section III explores membrane energy recovery, renewable-assisted cooling, passive strategies, and system optimization methods to enhance energy efficiency, reliability, and scalability of hybrid cooling systems (see Table 1.3).

Table 1.1 Section I—Fundamental evaporative and hybrid cooling systems.

Chapter	Core focus	Perspective	Key contribution	Relevance
Energy-efficient evaporative cooling techniques for greenhouse temperature and humidity control in Pakistan	Direct, indirect, and Maisotsenko-cycle evaporative cooling	Application-oriented analysis under climatic conditions	Evaluates the feasibility and performance of evaporative cooling strategies for greenhouse environments	Establishes sensible cooling foundations for agricultural controlled environments
Engineering insights into indirect evaporative cooling and mechanical vapor compression air-conditioning systems	Hybrid IEC–MVC cooling systems	Comparative engineering assessment	Demonstrates energy-saving potential of IEC as a precooling stage for conventional HVAC systems	Supports hybridization of evaporative cooling with mechanical systems
Study on CFD simulation of a novel indirect evaporative cooler	Indirect evaporative cooler design	CFD-based design and performance analysis	Identifies effects of geometry and airflow on cooling effectiveness and pressure drop	Guides optimized design of evaporative cooling units

Table 1.2 Section II—Desiccant-based cooling and dehumidification systems.

Chapter	Core focus	Perspective	Key contribution	Relevance
Investigating heat-driven desiccant air conditioning systems for greenhouse farming in Pakistan	Solid desiccant dehumidification	Experimental and system-level evaluation	Demonstrates effectiveness of thermally driven desiccant systems for greenhouse humidity control	Addresses latent load challenges in humid agricultural environments
Comparative thermal modeling of Standalone desiccant and Maisotsenko-Cycle assisted desiccant air conditioning systems for greenhouse climate control	DAC vs. M-cycle-assisted DAC systems	Comparative thermal modeling	Quantifies performance enhancement of desiccant systems assisted by evaporative cooling	Supports integration of sensible and latent cooling techniques
Advanced indirect evaporative cooling and desiccant dehumidification systems for greenhouse farming in arid and semi-arid Climates	Hybrid IEC–desiccant systems	Climate-specific system analysis	Demonstrates independent temperature-humidity control under arid and semi-arid conditions	Enables climate-adaptive greenhouse cooling solutions

Table 1.3 Section III—Energy recovery, renewable integration, and system optimization.

Chapter	Core focus	Perspective	Key contribution	Relevance
Recent developments in membrane energy recovery ventilators for optimum channel geometry, operating parameters, and membrane materials	Membrane energy recovery ventilators (MERVs)	Review and design-oriented synthesis	Identifies optimal membrane properties and geometric configurations for energy recovery	Reduces ventilation-related energy losses in buildings
Investigating energy-efficient configurations of membrane energy recovery ventilators using CFD	MERV performance optimization	CFD-based parametric analysis	Quantifies effects of geometry and operating parameters on heat and moisture recovery	Supports design optimization of energy recovery systems
Investigating membrane energy recovery ventilator-assisted evaporative cooling system in subtropical and tropical climates	Hybrid MERV-evaporative cooling systems	Climate-based performance evaluation	Demonstrates energy efficiency improvements in humid climates	Enhances hybrid cooling performance in ventilation-intensive buildings
Artificial neural network-based analysis of dehumidification performance of desiccant air-conditioning system	Data-driven modeling of desiccant systems	Predictive modeling and optimization	Enables accurate prediction of dehumidification performance	Facilitates system control and performance optimization
Application of solar chimney-assisted passive cooling in postharvest storage	Passive cooling using a solar chimney	Application-focused case study	Demonstrates passive temperature control in postharvest storage	Reduces energy use in agricultural storage facilities
Sustainable greenhouse farming through integration of solar energy and desiccant dehumidification system using design-builder	Solar-assisted desiccant systems	Simulation-based system integration	Evaluates energy and environmental benefits of solar-desiccant integration	Supports sustainable greenhouse operation
Integration of solar thermal collectors with desiccant cooling systems for controlled environments	Solar thermal-desiccant cooling	System integration and performance analysis	Demonstrates use of low-grade solar heat for desiccant regeneration	Enhances sustainability of hybrid cooling systems

Table 1.4 Section IV—Future outlook and deployment pathways.

Chapter	Core focus	Perspective	Key contribution	Relevance
Future perspectives on sustainable Hybrid cooling and dehumidification systems	Technology readiness and future pathways	Strategic synthesis and outlook	Identifies research gaps, integration challenges, and deployment barriers	Guides transition from research to real-world implementation

1.5 Section IV: Future outlook and deployment pathways

Section IV synthesizes technology readiness, remaining challenges, and future research directions, outlining pathways for scaling hybrid cooling and dehumidification systems toward real-world deployment (see Table 1.4).

AI disclosure

During the preparation of this work, the author(s) used Paper Pal and Research Pal to improve the readability and coherence. The tools were used with proper supervision and care. After using this tool/service, the author(s) reviewed and edited the content as needed and take(s) full responsibility for the content of the publication.

CHAPTER 2

Energy-efficient evaporative cooling techniques for greenhouse temperature and humidity control in Pakistan

Muhammad Adil[a], Muhammad Sultan[a], Muhammad Wakil Shahzad[b], Muhammad Ahmad Jamil[b], Muhammad Aleem[a], Muhammad Hamid Mahmood[a], Muhammad Ali Imran[a], Naveed Husnain[c], Uzair Sajjad[d], and Muhammad Farooq[e]

[a]Department of Agricultural Engineering, Bahauddin Zakariya University, Multan, Pakistan
[b]Department of Mechanical and Construction Engineering, Northumbria University, Newcastle Upon Tyne, United Kingdom
[c]Department of Mechanical Engineering, Bahauddin Zakariya University, Multan, Pakistan
[d]Interdisciplinary Research Center for Sustainable Energy Systems (IRC-SES), King Fahd University of Petroleum and Minerals, Dhahran, Saudi Arabia
[e]Department of Mechanical Engineering, University of Engineering and Technology, Lahore, Pakistan

2.1 Introduction

Evaporative cooling (EC) techniques were first applied in greenhouses thousands of years ago. Farmers in ancient Egypt, for example, used to cool the air inside their homes and storage facilities by covering windows and doors with moist mats. This technique was also used in greenhouses to help regulate temperature and humidity (Austin Refrigeration & Air Conditioning, 2023). To boost plant growth and production, EC systems in greenhouses became more and more common in the 19th century. One of the oldest references to this that is known to exist occurred in a greenhouse in London in 1839, where a "cooling wall" was used (Buchholz, n.d.). The water in the cooling wall's network of pipes was used to chill the air within the greenhouse. Advanced materials, science, and technology around the turn of the 20th century produced sophisticated EC systems for greenhouses. These pads are made of a porous substance that has been wetted; as air travels through the material, the water evaporation cools the air. A greenhouse is a contained space where all external elements are managed. It is also appropriate for

Sustainable Hybrid Cooling and Dehumidification Strategies for Controlled Environments
https://doi.org/10.1016/B978-0-443-48555-8.00004-9

off-season and constrained crop production. The greenhouse is a building with controlled humidity and temperature. It was discovered that greenhouses were an effective way to keep the crops' temperature, soil moisture, humidity, and light levels consistent. The ideal climate is maintained via greenhouse buildings. To regulate the conditions needed for agricultural growth, including temperature, light intensity, soil moisture, and humidity, human tracking is required regularly (Soussi et al., 2022).

Photosynthesis and evapotranspiration are fundamental processes that contribute to plant growth by releasing water vapor into the atmosphere, thereby continuously raising the humidity levels within a greenhouse. This elevated humidity can lead to potential issues such as insect infestations, fungal growth, and condensation. Photosynthesis is a crucial process in which plants convert carbon dioxide (CO_2) into carbohydrates using light energy from the sun. As part of this process, plants expel a significant portion of the water they absorb through their roots into the air, as given by Eq. (1) (Sultan et al., 2016).

$$2nCO_2 + 4n(H_2O)\text{from roots} \xrightarrow{\text{Photons}} 2(CH_2O)_n + 2nO_2 + 2n(H_2O)\text{ into air} \tag{1}$$

Evapotranspiration encompasses both the evaporation occurring from the ground and the transpiration originating from plants. For various plant species, the typical pace of transpiration can vary between 150 and 450 g per day per square meter of leaf area, depending on specific conditions (Trimble S., n.d.). While photosynthesis primarily takes place during daylight hours, evapotranspiration persists continuously, both day and night. Adjusting the flow rate of CO_2-enriched outdoor air and solar radiation intensity can affect the efficiency of these processes, which in turn introduces fluctuations in the sensible and latent cooling in an air conditioning system requirement (Sultan et al., 2016). Given that plants are extremely sensitive to variations in humidity and temperature, it is clear from the description above that successful greenhouse farming depends on maintaining optimal temperature and humidity levels. Factors such as photosynthesis, evapotranspiration, and the greenhouse effect result in the rise of temperature and humidity inside the greenhouse, which leads to substantial yield and quality loss (Amani et al., 2020). Fig. 2.1 shows possible impacts of low and high humidity levels on a plant inside the greenhouse. The humidity level increases in the greenhouse owing to continuous transpiration by plants and

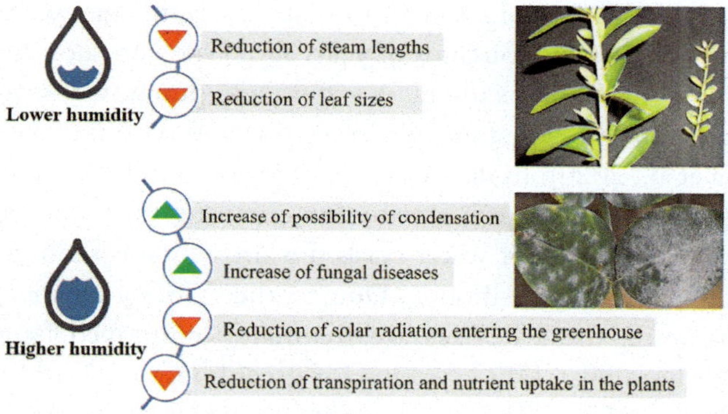

Fig. 2.1 Possible impacts of low and high humidity levels on a plants inside the greenhouse. *(From Amani, M., Foroushani, S., Sultan, M., Bahrami, M., 2020. Comprehensive review on dehumidification strategies for agricultural greenhouse applications. Appl. Therm. Eng. 1359-4311. 181, 115979. doi:10.1016/j.applthermaleng.2020.115979.)*

water vapor condensation or soaking through the greenhouse walls, which results in pest/fungus/disease attack, thereby affecting overall yield (Sultan et al., 2021). Keeping in view the above scenario, ideal management of humidity and temperature units is principally needed for developing a greenhouse farming culture in Pakistan. Conventional technologies and practices adopted in Pakistan to regulate humidity and temperature in greenhouses are either expensive or inefficient. In the developed nations, systems and technologies based on vapor compression are employed; however, they consume huge amounts of electricity, which makes such systems economically unfeasible at least in the Pakistani situation (Sultan et al., 2015). The concept of EC pad (swamp cooler) systems is not new in Pakistan. There are many companies and products available in Pakistan that control the air temperature via water vapor evaporation, and these devices are common for air-conditioning in the summer season. Utilizing the Maisotsenko cycle, indirect evaporative cooling (IEC) and desiccant dehumidification are advanced and emerging temperature and humidity control technologies that can be utilized for greenhouse farming. The distinction of the advanced IEC concept is that it can control the temperature without increasing the absolute humidity, which will make it suitable for the monsoon (rainy) season too.

Cool the air by turning sensible heat into latent heat using this oldest and most basic form ofEC, which involves bringing outdoor air into direct contact with water. Several permutations of clever strategies were employed by ancient civilizations thousands of years ago. One common ECS is direct ECS, whose design is an example. A water tank, pump, wet-table evaporative media, fan or blower, and a water distribution system make up a direct ECS (Raza et al., 2020). In direct evaporative cooling (DEC) systems, water is pumped directly into the airstream that circulates throughout the greenhouse. The evaporation of water cools the air, which is then circulated by fans throughout the greenhouse. Due to their ability to increase atmospheric moisture, these devices are frequently employed in places with low relative humidity. DEC systems for greenhouse cooling can be more or less effective depending on a variety of variables, encompassing the outdoor temperature and humidity, the airflow rate, and cooling pad design.

In IEC, two airstreams exchange energy while being kept apart by a heat-exchanging apparatus with two sides: a wet side that cools the air and the water, along with a dry side that cools the air on its own. The heat is cooled by delivering the hot main air exchanger's dry-side cooling system. These methods involve using a heat exchanger that is cooled by a water stream to circulate hot air from the greenhouse. The heat exchanger's surface evaporating water cools the air going through it. The greenhouse is then filled with the cooled air again. Since these systems don't add moisture to the air, they are usually utilized in locations with high relative humidity. The efficiency by use of IEC systems for greenhouse cooling can also differ based on a number of variables, including airflow rate, heat exchanger configuration, and ambient temperature and humidity. IEC systems can typically reduce ambient air temperature by 5.5°C–8.3°C (Raza et al., 2020; Shahzad et al., 2021).

The Maisotsenko cycle evaporative cooling (MEC) system is the advanced thermodynamic concept of IEC known as MEC, which allows product air to be cooled to the ambient air's dew-point temperature (Shahzad et al., 2021). This method chills the warm air in the greenhouse without adding moisture by first circulating it through a heat exchanger. After the air has been cooled, a second heat exchanger humidifies it with a thin mist of water. The humidified air is then pumped back into the greenhouse. Since M-cycle technology cools the air without adding too much moisture, it is more effective than conventional EC systems (Ashraf et al., 2021; Sultan et al., 2015).

2.2 Review of literature

2.2.1 History

The earlier research on EC is presented in this chapter. In this chapter, we will also reviews the various categories of EAC systems. Based on the Maisotsenko cycle, there are divided in to three types: DEC, IEC, and Maisotsenko evaporative cooling (MEC). This chapter also discusses factors that affect the rate at which water evaporates, as well as the benefits and drawbacks of EC systems. The MEC, IEC, and DEC systems are the three types of EC systems. The application of MEC systems in HVAC (heating, ventilation, and air conditioning) systems has been thoroughly studied by the scientific community. A freestanding Maisotsenko-cycle mass exchanger and evaporative heat source with eight distinct configurations was quantified and compared by Pandalids and Anisimov (Mahmood et al., 2016). Energy studies on the independent MEC system for building air conditioning applications were carried out by Caliskan. Zhan looked into standalone MEC systems with counterflow for applications involving building air conditioning.

2.2.2 History of EC techniques for greenhouse temperature and humidity control in Pakistan

2.2.2.1 Traditional cooling methods

In the past, passive cooling methods were used in Pakistan to control greenhouse temperatures. To produce a cooling effect, farmers employed manual watering, straw mat shading, and natural ventilation. Despite their simplicity, these methods were inexpensive and frequently used in small-scale farming. Mud walls, clay pots, and water-soaked jute cloth were used to keep the humidity levels somewhat constant. These conventional techniques were labor-intensive and inconsistent, which limited their applicability for large-scale commercial farming (Sultan et al., 2016).

2.2.2.2 Introduction of mechanical cooling systems

As agricultural technology advanced in the latter half of the 20th century, organized EC systems were introduced. One of the first technological solutions was the wet pad and fan system, which became well-liked in the 1980s and 1990s. This technique used water-soaked cellulose or aspen pads, through which exhaust blowers were used to suck air. Internal temperatures were considerably lowered as air moved through the damp pads and was cooled by water evaporation before entering the greenhouse (Soussi et al., 2022). In Punjab and Sindh, where greenhouse farming was growing as a result of rising demand for high-value crops and off-season vegetables, this technology was frequently used.

2.2.2.3 Advancements in high-pressure fogging and misting systems

In Pakistan, high-pressure fogging and misting devices were being promoted by research institutes and agricultural extension organizations by the early 2000s. By dispersing tiny water droplets into the atmosphere, these sophisticated EC techniques produced a cooling effect and preserved optimal humidity levels. Misting systems were especially useful for growing crops that are sensitive to moisture, such as tomatoes, bell peppers, cucumbers, and flowers, because they allowed for more accurate management of greenhouse microclimates than the pad-and-fan system. These techniques were incorporated by numerous commercial greenhouse operators to increase crop yields and improve product quality (Ghoulem et al., 2019).

2.2.3 History of greenhouse in Pakistan

The growth of Pakistan's agricultural industry has been greatly aided by the introduction of greenhouse farming. Farmers may grow crops under controlled circumstances in greenhouses, which boosts output, improves quality, and lessens reliance on seasonal fluctuations. In Pakistan, greenhouse technology has developed over time from simple buildings to contemporary, climate-controlled spaces that improve food security and long-term economic viability.

2.2.3.1 Early development and traditional practices

In Pakistan, the idea of greenhouse farming dates back to the middle of the 20th century, when conventional techniques for sheltered cultivation were being used. To protect crops from inclement weather, farmers first erected crude buildings out of bamboo, plastic sheets, and jute mats. To prolong the growing season and increase harvests, these early greenhouses were mainly utilized for the wintertime cultivation of food and ornamental plants. However, because these constructions were unable to regulate humidity and temperature, they were susceptible to drastic changes in the weather.

2.2.3.2 *Introduction of modern greenhouse farming*

Pakistan saw a transition to contemporary greenhouse farming in the 1980s and 1990s when government organizations and research institutes started to support protected agriculture. To increase crop protection and yield stability, farmers began implementing shade-net homes and tunnels wrapped in polyethylene with assistance from agricultural extension programs. Vegetable farming was transformed with the advent of polyethylene tunnels, which allowed high-value crops such as bell peppers, tomatoes, and cucumbers to be grown year-round (Khan and Khan, 2020).

2.2.3.3 *Expansion of commercial greenhouses*

The growing need for high-quality food for export markets and off-season veggies in Pakistan propelled commercial greenhouse farming by the early 2000s. As a result of the implementation of controlled-environment agriculture (CEA), sophisticated technology such as automatic climate control, heating systems, and EC was introduced. Large-scale greenhouse facilities were built by entrepreneurs and agribusiness investors to produce fruits, vegetables, and floriculture products for both domestic and international markets. To promote the use of greenhouses, the government and foreign organizations offered financial incentives as well as technical assistance. Training programs and the dissemination of information about contemporary greenhouse management practices were conducted in large part by the Pakistan Agricultural Research Council and other research organizations. In addition, partnerships with Dutch and Chinese agricultural specialists made it easier to implement soilless and hydroponic farming methods, which increased greenhouse productivity even more.

2.3 Materials and methods

2.3.1 Evaporative cooling systems

A cheaper, more environmentally friendly, and alternative kind of air-conditioning for both human and nonhuman uses is EC. The EC systems come in three categories: DEC, IEC, and MEC as shown in Fig. 2.2. Applications of EC are a draft of air over water to create a cooling effect. Because water vapor is being released from the lake's surface, there is not enough water vapor in the air. This "evaporation" mechanism gives water a cooling sensation. When air comes into direct contact (DEC) or indirect contact (IEC, MEC) with cooled water, it cools (Kapilan et al., 2023). A working schematic and psychrometric representation of DEC, IEC, and

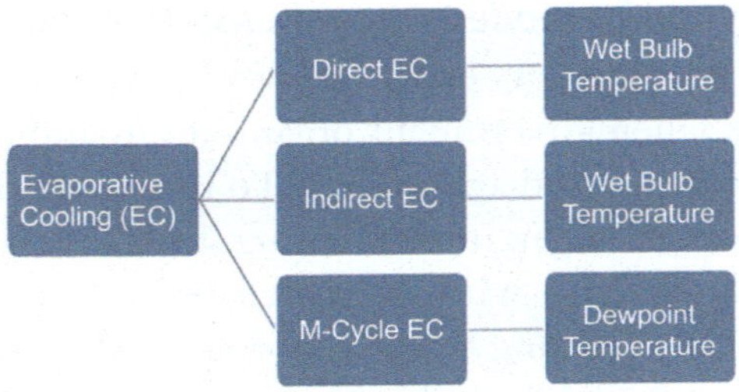

Fig. 2.2 Classification of evaporative cooling (EC) systems.

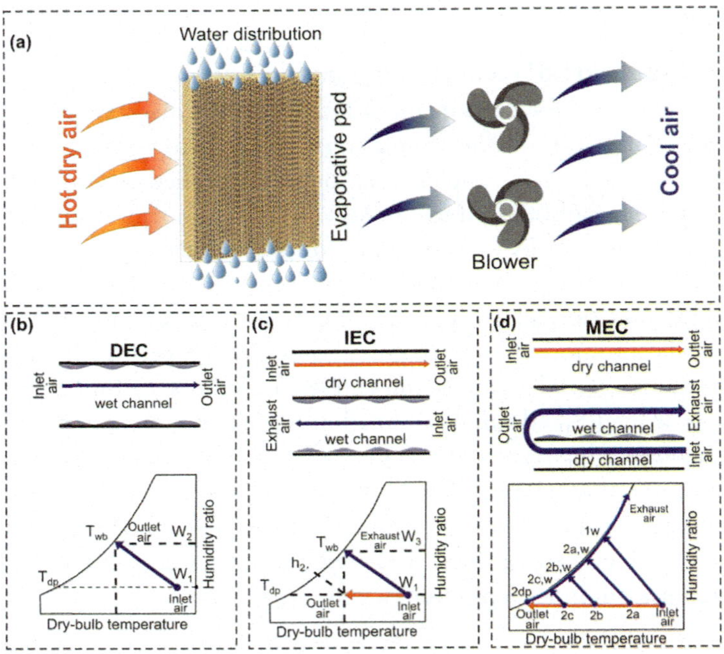

Fig. 2.3 (A) Working schematic of evaporative cooling (B-D) working schematic and psychrometric representation of DEC, IEC, and MEC systems. *(With permission from Shahzad, K., Sultan, M., Bilal, M., Ashraf, H., Farooq, M., Miyazaki, T., Sajjad, U., Ali, I., Hussain, M.I., 2021. Experiments on energy-efficient evaporative cooling systems for poultry farm application in Multan (Pakistan). Sustainability. 2071-1050. 13 (5), 2836. doi:10.3390/su13052836.)*

MEC systems are shown in Fig. 2.3. The humidity ratio of the air, which relates to how much moisture is in the air overall, directly determines how much water vapor it can contain. This proves the theory that EC works best in hotter air. Therefore, forced or natural ventilation can prevent over-heating within the greenhouse.

2.3.1.1 Direct evaporative cooling

Water vapors continuously drain up into the sky during cooling by direct evaporation (DEC) until air saturation happens; this is the fundamental principle underlying ordinary EC. Within the DEC system, air and water come into direct contact. The cooling pad is one of the moist channels that make up this structure. Product air moves from the left to the right, and water flows from upper to lower via the cooling pad. Water evaporation causes the air that passes over a cooling pad to become both colder and more humid. The DEC experimental setup includes a water pump, water tank, fan, and cooling pad. For optimal functioning and water evaporation, a constant supply of water is needed. Performance of the DEC system is estimated by the effectiveness method as given by Eq. (2) (Raza et al., 2020, 2021).

$$\varepsilon_{wb} = \frac{T_{in} - T_{out}}{T_{in} - T_{wb}} \tag{2}$$

2.3.1.2 Indirect evaporative cooling

The primary disadvantage of a typical DEC system is that it raises the product air's humidity. Consequently, an indirect potential use for an IEC system is maintaining a constant absolute humidity. IEC allows for hygienic air quality because it does not raise the air moisture content. Process (1) sensible heat transfer from the dry-channel, (2) isenthalpic cooling through the DEC process (wet-channel), which entails evaporating water vapor into the atmosphere; through these two thermodynamic processes are combined, the IEC cooling is achieved. Water in the IEC system does not come into direct touch with the product air. There are moist and dry passages in it. In wet channels, water flows vertically along the breadth of the sheet, whereas in dry channels, the cross-flow between the two media is created by air moving horizontally along the sheet's length. Wet channels allow process air to flow, evaporating water. The 30 aluminum sheets that make up the indirect heat mass exchanger are covered in felt, an absorbent medium for water. Approximately 280 g/m^2 of water may be retained and held by felt. Using liquid Samad bond, it is applied to aluminum sheets. Heat transmission between wet and dry channels is dependent on the thickness of the felt-coated aluminum sheet wall; a thin wall will transfer heat between wet and dry channels at a high rate, while a thick wall will transmit heat at a low rate. Rubber is used to make channels, both wet and dry, with a 4 mm thickness. Two channels are separated by one inch.

Because the thickness of the channel guides can be changed to change the highs of the dry and wet channels, guide thickness is important. IHMX channel guides travel down the length to generate dry channels and down the width to create wet channels to create a cross-flow (Raza et al., 2020, 2021). Process air travels vertically through the wet channel, and product air travels horizontally down the dry channels, arriving at the exit as cool product air due to heat transfer between the wet and dry channels. The IHMX sheet has eight dry channels on the dry side and twenty wet channels on the west side. Each channel's height and gap measure 4 mm, which is the same as the rubber guides' thickness. The effectiveness of the IEC system is estimated as given by Eq. (3) (Raza et al., 2020, 2021).

$$(\varepsilon_{wb})_{IEC} = \frac{T_{in} - T_{out}}{T_{in} - T_{wb}} \tag{3}$$

2.3.1.3 Maisotsenko cycle (M-Cycle) evaporative cooling

The IEC developed the Maisotsenko cycle (M-Cycle), an advanced thermodynamic idea that allows the product air to be cooled to the dew point temperature of the ambient air (Sultan et al., 2016). Wet-bulb effectiveness has resulted in MEC systems substantially greater than unity, based on research findings that have been documented in the literature. This suggests that MEC can reasonably lower the wet-bulb temperature of the product air below that of the surrounding air. Thus, the effectiveness of the MEC system is estimated as given by Eq. (4) (Raza et al., 2020, 2021). The optimum temperature and humidity zone for greenhouse farming with hourly based climatic conditions of Multan is illustrated in Fig. 2.4.

$$\varepsilon_{wb} = \frac{T_{in} - T_{out}}{T_{in} - T_{dp}} \tag{4}$$

For $\varepsilon_{dp} = 0.95$, the MEC unit's outlet air temperature is estimated for the climates of Fukuoka and Multan.

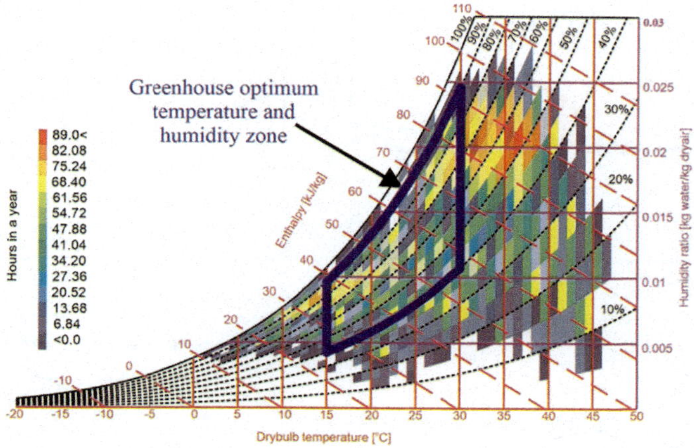

Fig. 2.4 The psychrometric illustration of optimum temperature and humidity zones for greenhouse farming with hourly based climatic conditions of Multan. *(From Raza, H.M.U., Sultan, M., Aleem, M., Shahzad, M.W., Jamil, M.A., Miyazaki, T., Sajjad, U., Farooq, M., Zhang, Z., 2026. Experimental study on advanced indirect evaporative cooling and desiccant dehumidification systems for agricultural greenhouses. Int. Commun. Heat Mass Transf. 0735-1933. 172, 110321. doi:10.1016/j.icheatmasstransfer.2025.110321.)*

2.4 Results and discussion

The experimental results of temperature and relative humidity for the studied DEC, IEC, and MEC systems for May, June, July, and August on daily bases are presented in Fig. 2.5. Vapor pressure deficit (VPD) profiles for tomatoes and chilies with optimum VPD range at different greenhouse conditions are presented in Fig. 2.6. Meanwhile, profiles of VPD for studied DEC, IEC, and MEC systems for May, June, July, and August are described in Fig. 2.7. Results in terms of psychrometric representation of ambient conditions and outlet conditions of the studied DEC, IEC, and MEC systems are described in Figs. 2.8 – 2.11. Relative humidity, wet bulb temperature, and dew point temperature are measured hourly, daily, and monthly to assess the effectiveness of EC systems. This study looks at EC systems' dew point and wet bulb efficacy (Raza et al., 2020, 2021). Systems for air conditioning, ventilation, and heating use a lot of energy. Around the world, buildings use between 30% and 40% of their total energy for primary purposes, of which 50% is used for air conditioning. Vapor compression air

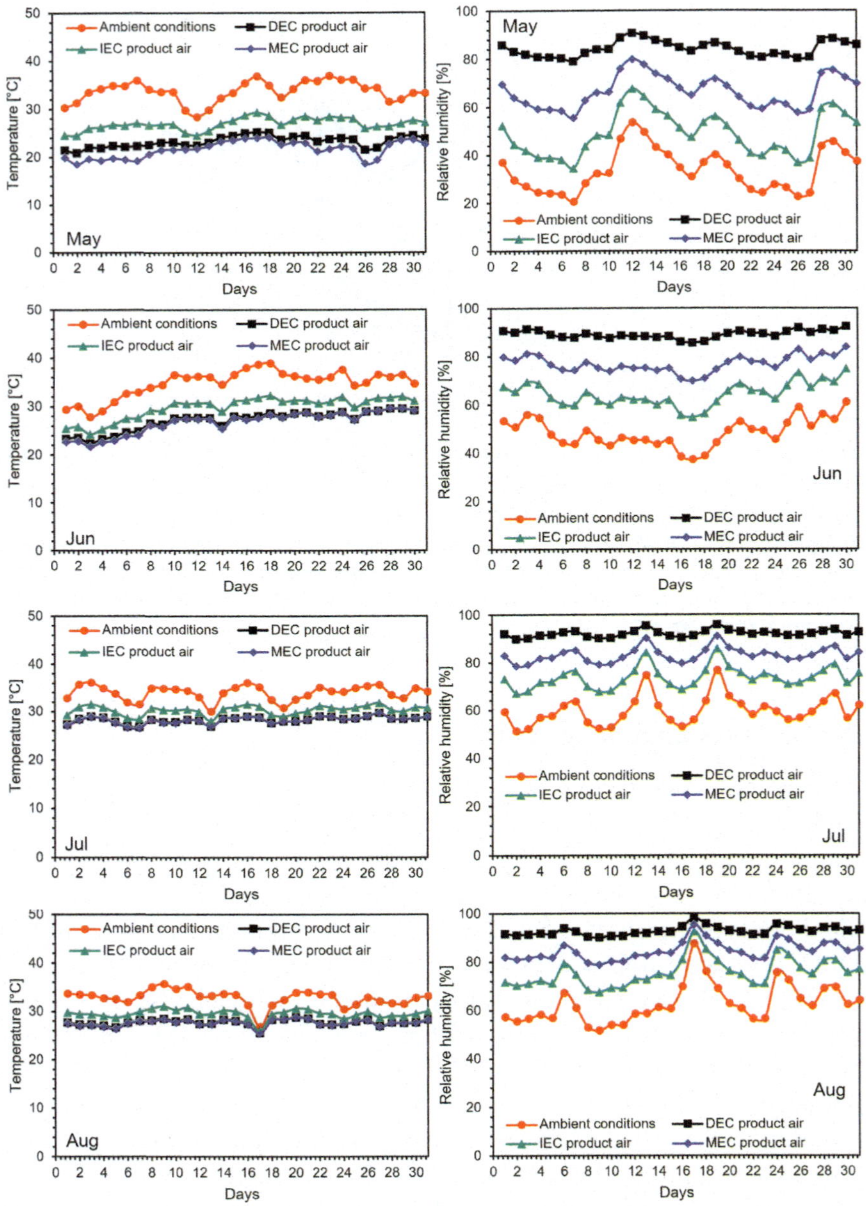

Fig. 2.5 Daily-based profiles of temperature and relative humidity for studied DEC, IEC, and MEC systems for May, June, July, and August.

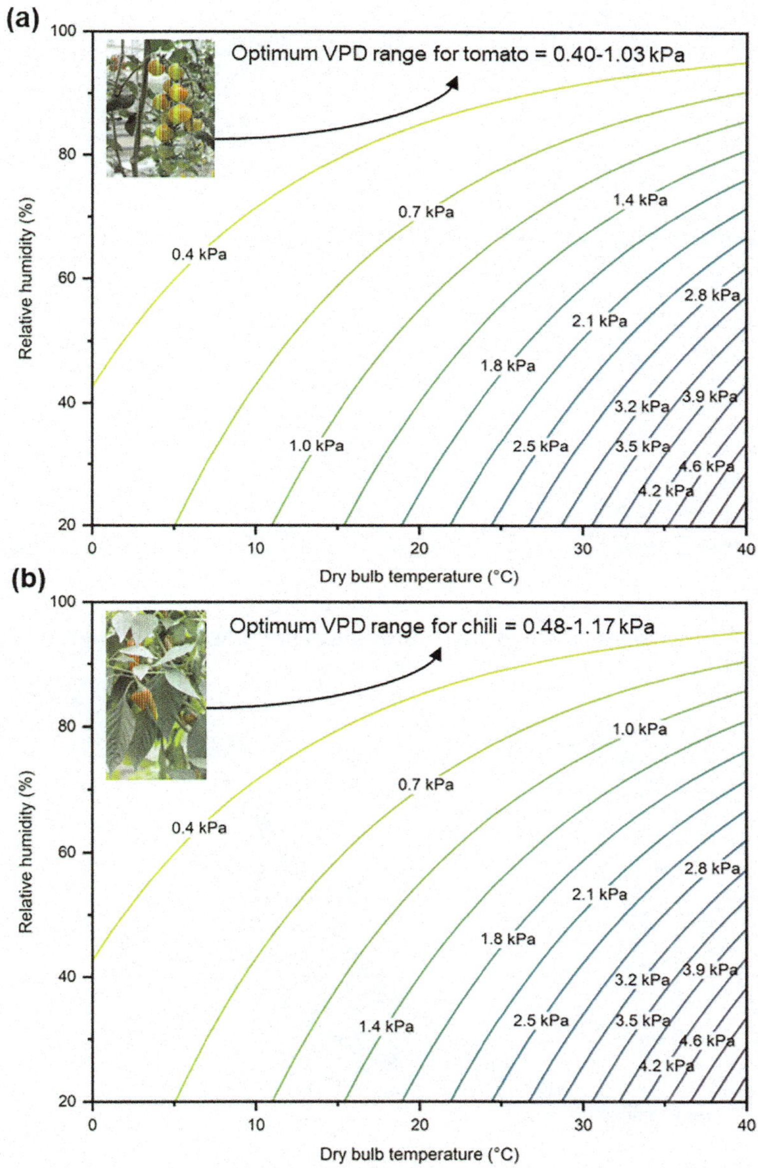

Fig. 2.6 Vapor pressure deficit (VPD) profiles for (A) tomatoes with an optimum VPD range, and (B) chilies with an optimum VPD range at different greenhouse conditions.

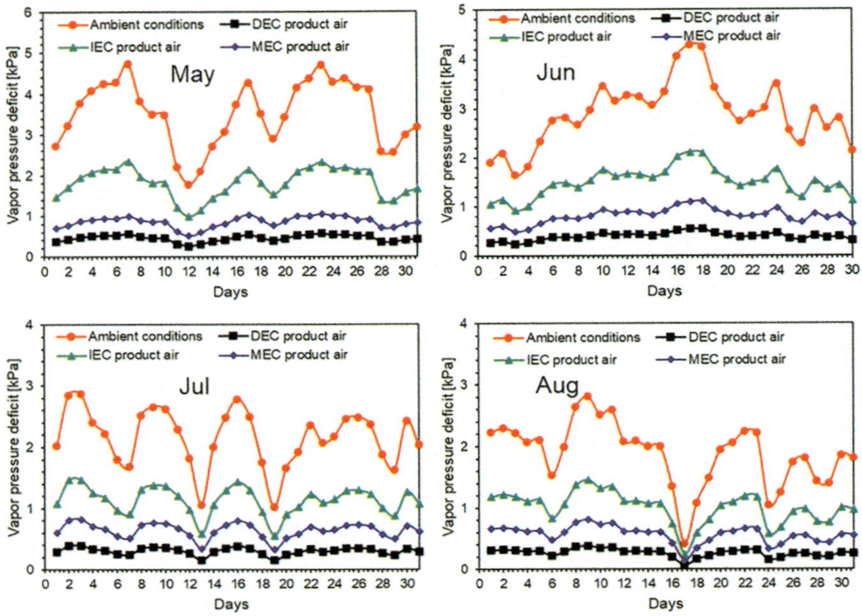

Fig. 2.7 Daily-based profiles of vapor pressure deficit (VPD) for the studied DEC, IEC, and MEC systems for May, June, July, and August.

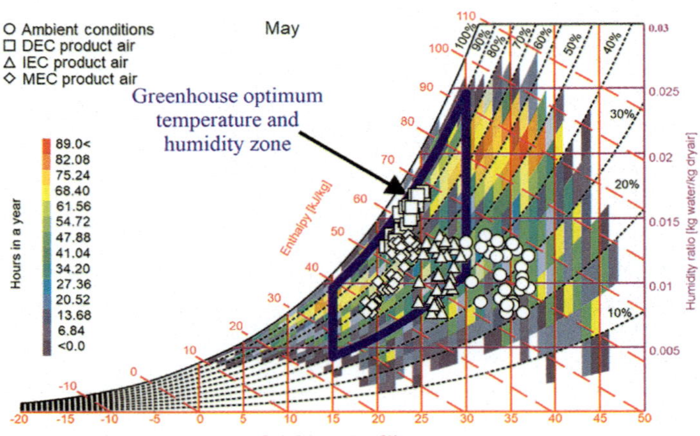

Fig. 2.8 Psychrometric representation of ambient conditions and outlet conditions of studied DEC, IEC, and MEC systems for May.

conditioning (VCAC) is the basis for over 95% of air conditioning appliances (Aleem et al., 2020). They contribute to global warming in addition to consuming large amounts of basic energy. A cheap, energy-efficient, and ecologically benign method isEC.

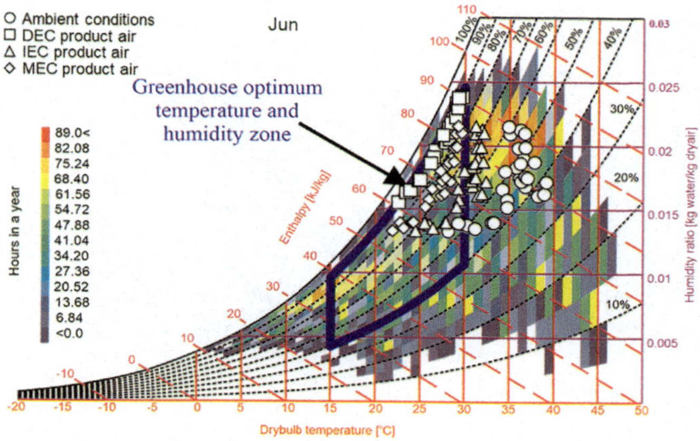

Fig. 2.9 Psychrometric representation of ambient conditions and outlet conditions of studied DEC, IEC, and MEC systems for June.

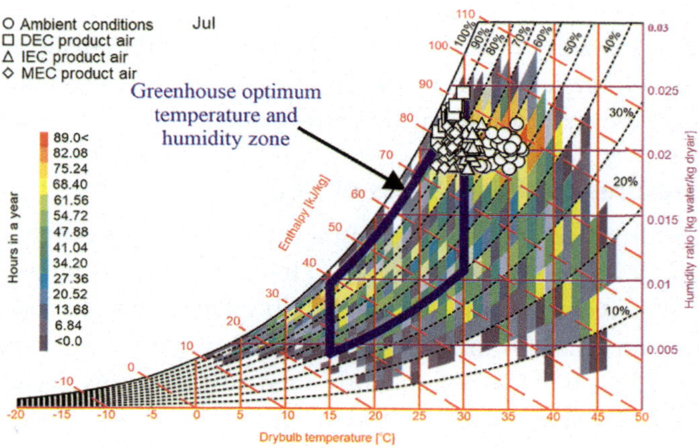

Fig. 2.10 Psychrometric representation of ambient conditions and outlet conditions of studied DEC, IEC, and MEC systems for July.

2.4.1 Assessment of the designed experimental configuration's performance

To assess their performance, DEC, IEC, and MEC systems are constructed at a lab scale. The lab scale experimental sets are created in Multan City. The developed systems have undergone performance and applicability checks for this city. The state of the surrounding air affects how well EC systems function. When sensors were employed to detect the temperature and humidity of the air at the intake and outflow, the following findings were drawn. Hygrometer sensors are used to measure changes in the temperature

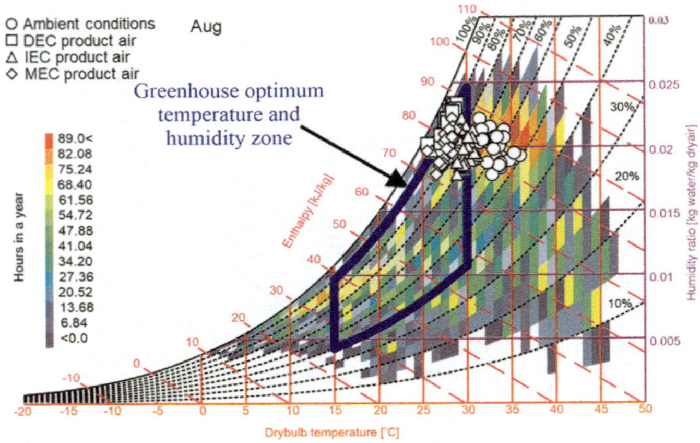

Fig. 2.11 Psychrometric representation of ambient conditions and outlet conditions of studied DEC, IEC, and MEC systems for August.

and relative humidity of ambient and product air at the intake and exit of DEC and IEC systems. The trial data in Multan was promptly assessed every day and every month for the entire year.

2.4.2 Measures of effectiveness of DEC

DEC's performance was evaluated using two metrics: Effectiveness of wet bulb (ε_{WBT}) and dew point (ε_{DPT}). The outcomes showed that DEC systems' effectiveness during the wet bulb effectiveness period was only moderate. Since Eq. (5) makes it possible to lower the temperature to the level of the wet bulb, it is frequently used to assess how well the DEC system cools (Raza et al., 2020, 2021).

$$\epsilon_{wb} = \frac{T_{in} - T_{out}}{T_{in} - T_{wb}} \tag{5}$$

2.4.3 Measures of effectiveness of IEC

Wet bulb effectiveness and dew point effectiveness were used to assess IEC's performance. The data showed that IEC systems were only slightly effective throughout the day, particularly when contrasted with M-Cycle systems, which proved to be more effective. The wet bulb effectiveness equation is typically utilized to ascertain the IEC system's cooling efficacy because it can potentially bring the temperature down to the wet bulb temperature (Raza et al., 2020, 2021).

$$\varepsilon_{wb} = \frac{T_{in} - T_{out}}{T_{in} - T_{wb}} \tag{6}$$

2.4.4 Measures of effectiveness of MEC

The dew point effectiveness Eq. (7) is widely used to evaluate the cooling effectiveness of MEC systems since they have the potential to reduce the temperature to the point of dew. It is the ratio of the temperature difference between the air entering and leaving the dry bulb to the temperature difference between the dry bulb and the corresponding dew point temperature (Raza et al., 2020, 2021).

$$\varepsilon_{db} = \frac{T_{in} - T_{out}}{T_{in} - T_{db}} \tag{7}$$

2.4.5 Feasibility analyses of EC systems

The ambient conditions are suitable for 6 months throughout the year, specifically from January to March and from October to December, during which no additional cooling systems are required. The direct cooling (DEC) system can be effectively implemented from April to June and again from October to December. IEC systems are feasible from February to June and from September to November. In addition, the M cycle, can be utilized year-round except during the peak summer months of July and August.

2.5 Conclusions

In Pakistan, greenhouse agriculture is becoming more and more important for raising crop yields and guaranteeing food security, especially in areas with harsh weather. To encourage healthy plant growth and maximize productivity, these structures must have effective temperature and humidity control. EC techniques are one of the more promising approaches to controlling the microclimate in greenhouses. The main objectives of this study were developing a key understanding of direct, indirect, and MEC techniques for greenhouse air-conditioning and thermodynamic investigation of the EC techniques for greenhouse climate control applications. The optimum VPD for tomato and chili crops ranges between 0.40–1.03 kPa and 0.48–1.17 kPa, respectively. EC systems could provide better results during summer conditions. Ambient conditions from January to March and October to December are in the acceptable range for greenhouse farming so, there is no need for EC systems. A DEC system is feasible from April to June and October to December. An IEC system is not feasible from June to August. MEC is feasible throughout the year except from July to August and December to January. According to the study, the methods of DEC, IEC, and MEC are very successful at controlling the humidity and temperature in greenhouses in the climate of Pakistan. Every method has unique benefits for greenhouse settings, helping to improve control over the microclimates that are essential for the best possible crop growth. The three methods—DEC, IEC, and MEC—all turned out to be more energy-efficient than traditional cooling approaches. But when compared to DEC alone, IEC and MEC systems provide better cooling performance with lower energy use. MEC, a blend of DEC and IEC, balances energy efficiency and cooling efficacy, making it especially appropriate for a variety of greenhouse applications. IEC provides the benefit of cooling without appreciably raising humidity. In Pakistan's arid and hot locations, DEC, IEC, and MEC methods work incredibly well since low humidity increases their ability to cool. While DEC and MEC can be utilized efficiently in locations with moderate humidity, making them adaptable across diverse geographic zones in Pakistan, IEC is especially helpful in situations where increasing humidity may not be appropriate for crop production. By preserving steady and ideal growing conditions, the application of DEC, IEC, and MEC cooling systems has positively impacted crop output and quality. These techniques led to longer growing seasons, less stress from high temperatures, and improved plant development in greenhouses, which increased agricultural productivity in Pakistan's harsh climate.

Acknowledgment

This research work is completed with the financial support from the Higher Education Commission (HEC) of Pakistan through the research project NRPU-17029 awarded to Bahauddin Zakariya University. The United Kingdom government funds this research (and/or) pilot activity through the Project: APP47457, titled "Super-efficient Sustainable Cooling Solution for All Applications (S2Cool)," under the Ayrton Challenge Programme of the UK Research and Innovation (UKRI); however, the views expressed do not necessarily reflect the United Kingdom government's official policies. The Project: APP47457 is implemented/led by Northumbria University, UK together with partners to develop sustainable cooling solutions for several applications.

Funding statement

This research work is completed with the financial support from the HEC of Pakistan through the research project NRPU-17029 awarded to Bahauddin Zakariya University. The UK government funds this research (and/or) pilot activity through the Project: APP47457, titled "Super-efficient Sustainable Cooling Solution for All Applications (S2Cool)," under the Ayrton Challenge Programme of the UKRI; however, the views expressed do not necessarily reflect the UK Government's official policies. The Project: APP47457 is implemented/led by Northumbria University UK together with partners to develop sustainable cooling solution for several applications.

CRediT authorship contribution statement

Muhammad Adil was involved in the conceptualization, methodology, software, formal analysis, investigation, writing of the original draft. **Muhammad Sultan and Muhammad Wakil Shahzad** were involved in the conceptualization, methodology, validation, resources, visualization, supervision, project administration, writing (review and editing), funding acquisition. **Muhammad Ahmad Jamil, Muhammad Aleem, Muhammad Hamid Mahmood, Muhammad Ali Imran, Naveed Husnain, Uzair Sajjad, and Muhammad Farooq:** Conceptualization, methodology, software, formal analysis, investigation, writing (review and editing).

Data availability statement

The datasets used and/or analyzed during the current study are available from the corresponding authors upon request.

Conflicts of interest

The authors declare no conflict of interest.

Declaration of generative AI and AI-assisted technologies in the writing process

The author declares that generative AI was used to support language refinement and editorial revision of the manuscript. Specifically, OpenAI's ChatGPT (GPT-5 Mini) was employed to improve clarity, coherence, and academic style in existing draft text. The conceptual framework, analysis, interpretations, and conclusions were developed independently by the author. All AI-assisted revisions were critically reviewed, verified, and edited by the author, who assumes full responsibility for the content of the final manuscript.

References

Aleem, M., Hussain, G., Sultan, M., Miyazaki, T., Mahmood, M.H., Sabir, M.I., Nasir, A., Shabir, F., Khan, Z.M., 2020. Experimental investigation of desiccant dehumidification cooling system for climatic conditions of multan (Pakistan). Energies. 19961073. 13 (21). Available from: https://www.mdpi.com/1996-1073/13/21/5530. doi:10.3390/en13215530.

Amani, M., Foroushani, S., Sultan, M., Bahrami, M., 2020. Comprehensive review on dehumidification strategies for agricultural greenhouse applications. Appl. Therm. Eng. 13594311. 181, 115979. doi:10.1016/j.applthermaleng.2020.115979.

Ashraf, H., Sultan, M., Shamshiri, R.R., Abbas, F., Farooq, M., Sajjad, U., Md-Tahir, H., Mahmood, M.H., Ahmad, F., Taseer, Y.R., Shahzad, A., Niazi, B.M.K., 2021. Dynamic evaluation of desiccant dehumidification evaporative cooling options for greenhouse air-conditioning application in multan (Pakistan). Energies. 19961073. 14 (4). Available from: https://www.mdpi.com/1996-1073/14/4/1097/pdf. doi:10.3390/en14041097.

Austin Refrigeration & Air Conditioning, 2023. Egyptian evaporative cooling: Austin refrigeration's ancient cooling techniques 101. Available from: https://medium.com/@austinrefrigeration/egyptian-evaporative-cooling-austin-refrigerations-ancient-cooling-techniques-101-1d9fd944b20e (Accessed 5 June 2024).

Buchholz, M., 2010. Overcoming drought: a scenario for the future development of the agricultural and water sector in arid and hyper arid areas, based on recent technologies and scientific results.

Ghoulem, M., Moueddeb, K.E., Nehdi, E., Boukhanouf, R., Calautit, J.K., 2019. Greenhouse design and cooling technologies for sustainable food cultivation in hot climates: review of current practice and future status. Biosyst. Eng. 15375110. 183, 121–150. doi:10.1016/j.biosystemseng.2019.04.016.

Kapilan, N., Isloor, A.M., Karinka, S., 2023. A comprehensive review on evaporative cooling systems. Results Eng. 25901230. 18, 101059. doi:10.1016/j.rineng.2023.101059.

Khan, M.B., Khan, J., 2020. An economic analysis of tunnel farming in enhancing productivity of off-season vegetables in district Peshawar. Sarhad J. Agric. 22245383. 36 (1), 153–160. Available from: http://researcherslinks.com/base/downloads.php?jid=14&aid=2715&acid=1&path=pdf&file=1581890503SJA_36_1_153-160. doi:10.17582/journal.sja/2020/36.1.153.160.

Mahmood, M.H., Sultan, M., Miyazaki, T., Koyama, S., Maisotsenko, V.S., 2016. Overview of the Maisotsenko cycle – a way towards dew point evaporative cooling. Renew. Sustain. Energy Rev. 18790690. 66, 537–555. Available from: https://www.journals.elsevier.com/renewable-and-sustainable-energy-reviews. doi:10.1016/j.rser.2016.08.022.

Raza, H.M.U., Ashraf, H., Shahzad, K., Sultan, M., Miyazaki, T., Usman, M., Shamshiri, R.R., Zhou, Y., Ahmad, R., 2020. Investigating applicability of evaporative cooling systems for thermal comfort of poultry birds in Pakistan. Appl. Sci. (Switz.) 20763417. 10 (13). Available from: https://res.mdpi.com/d_attachment/applsci/applsci-10-04445/article_deploy/applsci-10-04445.pdf. doi:10.3390/app10134445.

Raza, H.M.U., Sultan, M., Bahrami, M., Khan, A.A., 2021. Experimental investigation of evaporative cooling systems for agricultural storage and livestock air-conditioning in Pakistan. Build. Simul. 19968744. 14 (3), 617–631. Available from: http://www.springer.com/engineering/journal/12273?detailsPage=aboutThis. doi:10.1007/s12273-020-0678-2.

Shahzad, K., Sultan, M., Bilal, M., Ashraf, H., Farooq, M., Miyazaki, T., Sajjad, U., Ali, I., Hussain, M.I., 2021. Experiments on energy-efficient evaporative cooling systems for poultry farm application in Multan (Pakistan). Sustainability. 2071-1050. 13 (5), 2836. doi:10.3390/su13052836.

Soussi, M., Chaibi, M.T., Buchholz, M., Saghrouni, Z., 2022. Comprehensive review on climate control and cooling systems in greenhouses under hot and arid conditions. Agronomy. 20734395. 12 (3). Available from: https://www.mdpi.com/2073-4395/12/3/626/pdf. doi:10.3390/agronomy12030626.

Sultan, M., Ashraf, H., Miyazaki, T., Shamshiri, R.R., Hameed, I.A., 2021. Temperature and humidity control for the next generation greenhouses: overview of desiccant and evaporative cooling systems. IntechOpen. doi:10.5772/intechopen.97273.

Sultan, M., Miyazaki, T., Saha, B.B., Koyama, S., 2016. Steady-state investigation of water vapor adsorption for thermally driven adsorption based greenhouse air-conditioning system. Renew. Energy. 18790682. 86, 785–795. Available from: http://www.journals.elsevier.com/renewable-and-sustainable-energy-reviews/. doi:10.1016/j.renene.2015.09.015.

Sultan, M., Miyazaki, T., Saha, B.B., Koyama, S., Maisotsenko, V.S., 2015. Steady-state analysis on thermally driven adsorption air-conditioning system for agricultural greenhouses. Procedia Eng. 18777058. 118, 185–192. http://www.sciencedirect.com/science/journal/18777058. doi:10.1016/j.proeng.2015.08.417.

Trimble, S. Transpiration in plants: its importance and applications. 2021. https://cid-inc.com/blog/transpiration-in-plants-its-importance-and-applications/.

CHAPTER 3

Engineering insights into indirect evaporative cooling and mechanical vapor compression air-conditioning systems

Nida Imtiaz[a], Mazlan Abdul Wahid[a], Muhammad Ali Imran[b], Muhammad Sultan[b], and Muhammad Wakil Shahzad[c]
[a]Faculty of Mechanical Engineering, Universiti Teknologi Malaysia, Johor Bahru, Malaysia
[b]Department of Agricultural Engineering, Bahauddin Zakariya University, Multan, Pakistan
[c]Department of Mechanical and Construction Engineering, Northumbria University, Newcastle Upon Tyne, United Kingdom

3.1. Introduction

Sustainable and energy-efficient cooling technologies have become necessary to meet growing global energy demands under severe climatic conditions and climate change (Ahmad and Zhang, 2020; Ghasemian et al., 2024). A considerable amount of electricity is consumed worldwide for air conditioning and refrigeration systems, which are mainly driven by mechanical vapor compression (MVC) systems. These systems are widely used due to their thermodynamic efficiency. To enhance the performance and energy efficiency, MVC systems have been incorporated with different technologies such as desiccant dehumidification, solar energy, and subcooling assistance as discussed in the following sections. However, MVC systems rely on toxic refrigerants and are recognized as potential contributors to ozone depletion and global warming (Andersen et al., 2013) due to greenhouse gas emissions and carbon footprints (Sultan et al., 2018; Rafique et al., 2015). Thus, alternative sustainable cooling solutions such as evaporative cooling are gaining attention because they can deliver comparable comfort and performance with minimal environmental impact.

Indirect evaporative cooling (IEC) is emerging as an efficient alternative with the advantages of energy efficiency, environmental sustainability, and cost-effectiveness. The IEC system drops the air temperature without

Sustainable Hybrid Cooling and Dehumidification Strategies for Controlled Environments
https://doi.org/10.1016/B978-0-443-48555-8.00010-4

humidifying the supply air by transferring sensible heat through a heat and mass exchanger (Ali et al., 2021). This simple process, along with low operational cost and the absence of chemical refrigerants, makes IEC an attractive solution for a wide range of applications (Sultan and Miyazaki, 2017; Xuan et al., 2012). However, IEC systems can experience inconsistent cooling performance, which limits the adoption and application of the standalone IEC systems under variable ambient climatic conditions (Kapilan et al., 2023; Raza et al., 2021). In this regard, various design configurations have been studied, as discussed in the following sections.

The performance of the IEC system is greatly affected by the geometry of the IEC system as well as the configuration of the heat and mass exchanger. In this regard, the arrangement of dry and wet channels, plate materials, and airflow direction play critical roles in determining thermal performance (Alam et al., 2024; Castillo-González et al., 2023). Over the years, various designs such as cross–flow, counter–flow, parallel–flow, and counter-cross–flow configurations have been explored to balance cooling effectiveness, pressure drop, and structural simplicity (Kim, 2022). Each design has distinct advantages and limitations. For example, cross–flow systems provide high cooling capacity but require larger installation space, while counter–flow designs deliver higher effectiveness at the expense of increased pressure losses (Coolerado, 2006; Patunkar and Dingare, 2022). Besides these configurations, other approaches have also been adopted to improve the performance of IEC systems, including improvements in structural design, operation, and geometrical aspects of systems. A wide variety of materials, heat exchanger types, and plate geometries have been used to enhance performance, reliability, and manufacturability (Alam et al., 2024; Zhao et al. (2008)). In addition, advanced assembly and fabrication technologies such as fused filament fabrication and 3D printing have enabled the development of compact, lightweight, and modular exchanger systems. These improvements contribute not only to enhancing the system performance but also to simplified manufacturing and maintenance.

In addition, limitations associated with the IEC performance can be managed by integrating IEC with MVC in a hybrid system (Xue et al., 2023). In such systems, IEC is responsible to precool the air, resulting in higher effectiveness, while the MVC system provides consistent performance. This integration combines the energy-efficient IEC system for precooling of air with the consistent cooling performance of the MVC system, which combats the limitations of both systems. Several studies have shown that the performance and applicability are improved with the hybrid

IEC–MVC system (Chen et al., 2022). Thus, several configurations have been evaluated as feasible pathways toward consistent performance and versatile applications under variable ambient climatic conditions, as described in detail in the section 3.4. Accordingly, a novel IEC integrated with MVC (NIEC–MVC) system is a fresh development. The NIEC–MVC introduces an improved precooling of working air, simplified the heat exchanger design, and overall compactness and scalability of system design (Jamil et al., 2023). Overall, these progressions are paving the way towards low-energy, high-performance, and widely adoptable cooling solutions. Although numerous studies have investigated IEC and hybrid IEC–MVC systems, a consolidated perspective emphasizing geometry-driven compactness, material selection, and scalable system design remains limited.

This book chapter addresses this gap by systematically reviewing the working principles, material considerations, and geometrical designs, configurations, and scalable structural designs of IEC, MVC, and hybrid IEC–MVC systems. It also presents a compact computer-aided design (CAD) of a hybrid IEC–MVC system with distinct design and features, case studies of different systems and configurations, and strengths, weaknesses, opportunities, and threats (SWOT) analysis among the different cooling options. These insights underscore the critical role of innovation in material selection, design optimization, and hybridization in advancing next-generation cooling technologies that align with global energy and environmental goals. The following sections systematically examine the working principles, geometric configurations, material considerations, and hybridization strategies of IEC-based cooling systems.

3.2. Working principle of the systems

3.2.1 IEC systems

IEC is a low-cost, energy-efficient, and environmentally sustainable cooling system that has the potential to decrease the temperature without humidification of the air (Mahmood et al., 2016). The IEC system consists of wet and dry channels, as shown in Fig. 3.1. The system reduces the temperature of ambient air by two thermodynamic processes: (1) isenthalpic cooling in wet channels through water vapor evaporation, and (2) sensible heat exchange between primary and secondary air streams to cool the air in dry channels (Sultan and Miyazaki, 2017; Raza et al., 2020). An IEC heat exchanger is an integral part that transfers heat between two or more fluids of different temperatures without allowing them to mix (Zohuri, 2018).

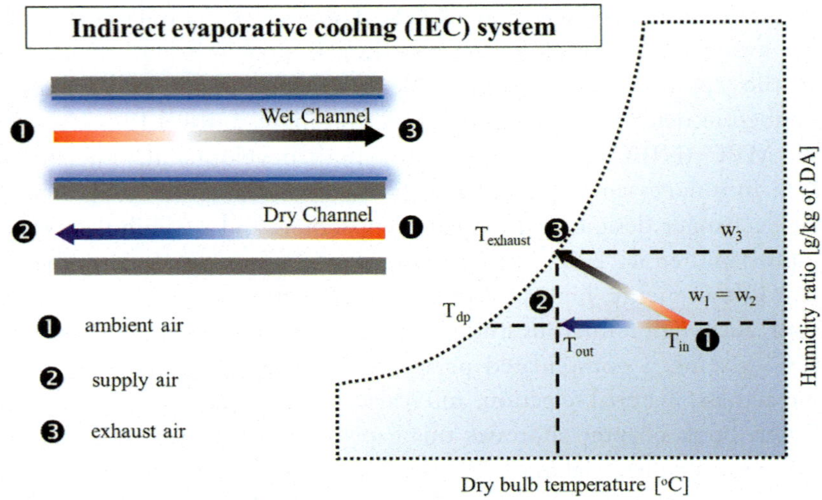

Fig. 3.1 Schematic and psychrometric illustration of the IEC system.

The transfer of cooling, heating, and energy recovery processes occurs in distinct tubes separated by conducting material, typically metal, which continues until thermal equilibrium is reached. The efficiency of heat exchangers depends on the available surface for heat transmission, flow types, and the thermal conductivity of the separating material. The IEC cooling potential is the available margin between the dry-bulb temperature and wet-bulb temperature of the ambient air, wherein the wet-bulb temperature is the upper cooling limit. Thus, the effectiveness of the IEC system can be calculated in terms of wet–bulb effectiveness and can be calculated by the equation (Fan et al., 2021):

$$\varepsilon_{\mathrm{wb}} = \frac{T_{\mathrm{in}} - T_{\mathrm{out}}}{T_{\mathrm{in}} - T_{\mathrm{in,wb}}}$$

where, $\varepsilon_{\mathrm{wb}}$ is wet-bulb effectiveness of the IEC system, T_{in} is ambient temperature, T_{out} is oultlet temperature, and $T_{\mathrm{in,wb}}$ is wet-bulb temperature of outdoor air.

3.2.2 MVC system

The MVC system operates on a basic thermodynamic refrigeration cycle. It is a thermodynamic process in which chemical refrigerants such as hydrochlorofluorocarboncarbons/hydrofluorocarbons/chlorofluorocarbons are

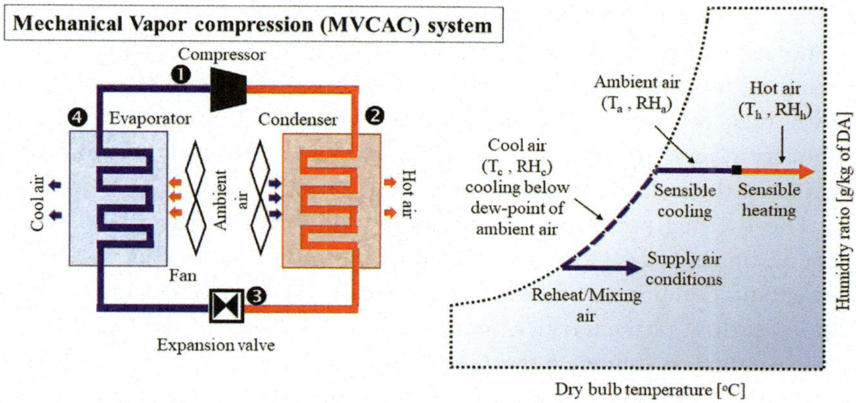

Fig. 3.2 Schematic and psychrometric illustration of MVC system.

used to achieve desired indoor environment conditions (Jamil et al., 2023; Sajjad et al., 2021). The system consumes high energy and is not environmentally sustainable because of refrigerants, resulting in ozone depletion and global warming potential and subsequently contributing to climate change (Khalid et al., 2016; Jamil et al., 2024). However, these systems are most commonly used across the world due to consistent performance across varying climatic conditions (Sultan et al., 2015). Fig. 3.2 shows the basic refrigeration cycle and psychrometric behavior of the MVC system.

In the MVC system, the chemical refrigerants circulate through four basic operations, such as compression, condensation, expansion, and evaporation. In the first step, refrigerant that is in a gaseous state enters into the compressor at low temperature and pressure. The function of a compressor is to compress the refrigerants and increase the temperature and pressure to produce a high-pressure superheated vapor. In the second step, the high–pressure refrigerants leave the compressor and enter into the condenser. Within the condenser, the refrigerant releases its thermal energy to the surroundings by convective heat transfer, condensing to a liquid state from vapor to liquid with almost constant pressure. Then, this liquid refrigerant with high-pressure moves toward the expansion valve, where it expands and releases pressure through throttling. This irreversible expansion process results in a significant decrease in temperature, yielding a low–pressure, low–temperature two-phase mixture of liquid (~75%) and vapor (~25%) refrigerant. In fourth step, the cold refrigerant mixture then flows into the evaporator coil, located within the conditioned space. After that, the refrigerant transfers cooling to indoors while air passes over the evaporator.

During this heat absorption process, the refrigerant undergoes complete evaporation, thereby lowering the air temperature and, in many cases, removing moisture via the condensation process as the coil temperature is lower than the dew point temperature. The low-pressure refrigerant in vapor form recirculates to the compressor, enabling continuous removal of heat from the indoor environment.

The thermal performance and energy efficiency of the MVC system depend on several factors, including the thermophysical properties of the refrigerant, compressor efficiency, heat exchanger surface area and configuration, airflow characteristics, and operating pressure levels. Collectively, these parameters influence the Coefficient of Performance (COP), system stability, and overall energy consumption in practical applications.

3.2.3 IEC–MVC systems

To address the limitations of both IEC and MVC systems, the MVC system is coupled to the IEC system that provides the combined advantages of the energy efficiency of IEC and the consistent performance of MVC under variable climates (Chen et al., 2021). Fig. 3.3 presents the schematic and thermodynamic representation of a hybrid IEC–MVC system, where the IEC system is dedicated to precool the high-temperature outdoor air initially, and the MVC system subsequently achieves the required temperature and humidity conditions as described in the above sections. To augment energy efficiency, relatively cool exhaust air from the supply area is directed into the wet channel as the working air for precooling the ambient air in the dry channel. This process allows the system to recover latent and sensible energy from the recirculating exhaust air and improves the cooling

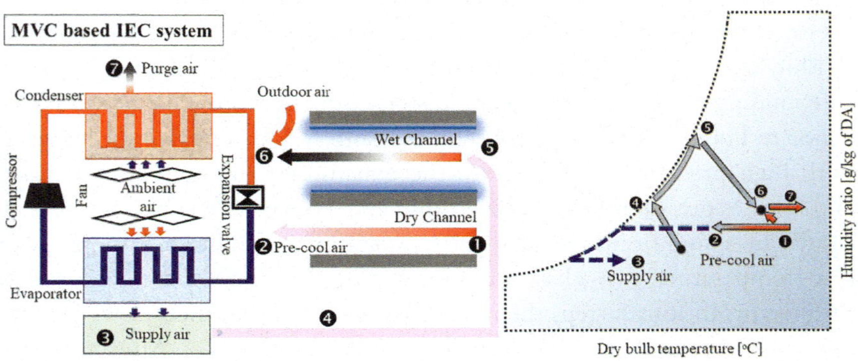

Fig. 3.3 Schematic and psychrometric illustration of IEC–MVC system.

performance. Afterward, the humidified and heated working air is mixed with outdoor air and directed to the MVC condenser. The hybrid IEC–MVC system has demonstrated promising results in the form of consistent performance and energy efficiency (Chen et al., 2022). Consequently, researchers have developed various hybrid IEC–MVC configurations for different applications, which are discussed below.

3.2.4 NIEC–MVC systems

The NIEC–MVC system mainly focuses on precooling ambient air through the NIEC system. In the NIEC system, a humidifier is supplemented to dedicatedly humidify and cool the working air before delivering it to the heat exchanger (Jamil et al., 2023). In this setup, the working air first flows into a humidification system that generates a fine water mist using a series of high-pressure nozzles and a pump. The working air cools down to the wet-bulb temperature and becomes fully humidified by interacting with the water mist. This cooled and saturated working air is then directed to the wet channel, where it absorbs heat from the hot air flowing in the dry channels and cools the supply air. The precooled air is further processed by the MVC system to achieve the required conditioning, as illustrated in Fig. 3.4. Meanwhile, the working air experiences a temperature increase due to sensible and latent heat transfer, and a slight reduction in moisture occurs as the mist evaporates before being exhausted from the heat exchanger.

In NIEC, heat exchanger, aluminum sheets are used to simplify the structure and to enhance the heat transfer rate. This arrangement eliminates the need for wetted surfaces such as hydrophobic or hydrophilic coatings and avoids issues related to durability and fouling since saturated working air

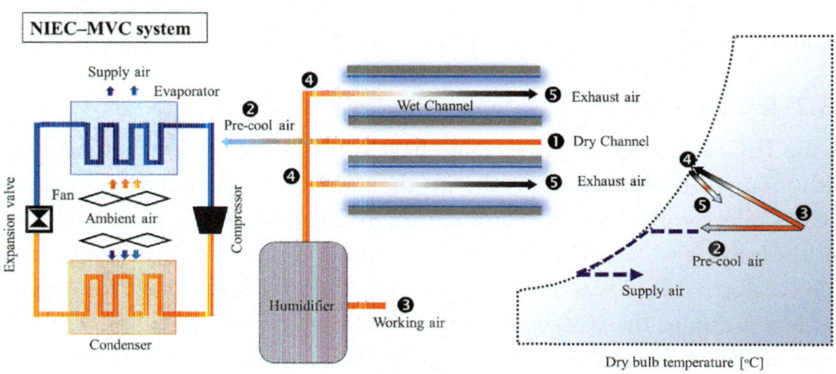

Fig. 3.4 Schematic and psychrometric illustration of NIEC–MVC system.

with wet-bulb temperature is produced during the additional humidification process. This NIEC system improves the precooling performance, which further supports the MVC system in enhancing overall performance and energy efficiency.

3.3. Geometry and design configuration of the systems

3.3.1 Material used for the systems

The heat exchanger of an IEC system consists of many sheets for the wet and dry channels. Several materials are used to fabricate these sheets, as material properties have a direct impact on the cooling performance, and a hydrophobic coating to avoid the accumulation of moisture in product air and a hydrophilic material to improve the saturation of working air are used in dry and wet channels, respectively (Castillo-González et al., 2023). The properties such as thermal conductivity, resistance to corrosion and bacterial growth, shape formation, compatibility with coating, and inexpensiveness are important for considering the material (Alam et al., 2024). Particularly, thermal conductivity and thickness of materials are directly related to heat transfer; water retention and absorption are associated with mass transfer, and the strength of materials is important for construction shape formation of heat exchangers. In this regard, metals such as aluminum, copper, and their alloys are commonly employed for fabricating sheets of indirect heat exchangers (Chen et al., 2021; Comino et al., 2018; Fatahian et al., 2020). Other materials, including zeolite, fiber, carbon, and ceramic, are used in the literature (Alam et al., 2024; Al-Zubaydi and Hong, 2019; Gómez et al., 2005; Xu et al., 2016). Although fibrous materials are affordable, thin-structured, and strong in wettability, they are mechanically weak. Therefore, suitable material should contain a reasonable thermal conductivity, water-holding capacity, and mechanical strength.

3.3.2 IEC systems

In the IEC heat exchanger, sensible heat is transferred from the supply air without any humidification as the product air passes through dry channels. Meanwhile, working air flowing through adjacent wet channels absorbs both sensible and latent heat that are separated by heat-exchanging plates. The concept of IEC was first introduced in 1935 by Ray (1935) for cooling ambient air up to the wet-bulb temperature. Fig. 3.5 illustrates the different types of IEC heat and mass exchangers: (1) cross–flow, (2) counter–flow, and (3) parallel–flow, depending upon the flow directions. Table 3.1

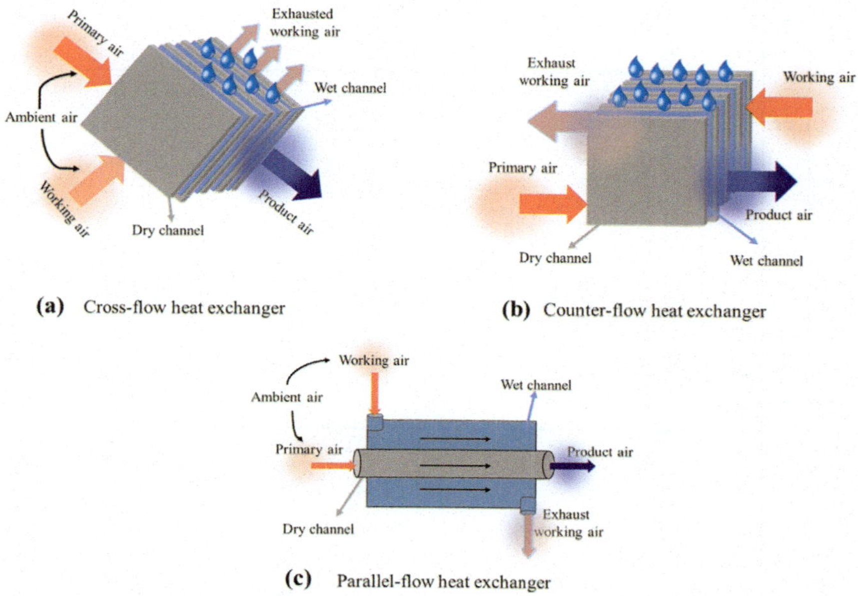

Fig. 3.5 Types of heat exchangers: (A) cross–flow, (B) counter–flow, and (C) parallel–flow.

summarizes the various geometrical designs and outlines the airflow direction for the IEC heat exchangers and systems. The heat transfer, cooling capacity, and design structure are greatly associated with the structural arrangement of channels and airflow directions of product and working air. The perpendicular interaction in the cross–flow IEC system improves the cooling capacity (Coolerado, 2006), while counter–flow IEC has the advantage of temperature gradient increases across the channel length (Patunkar and Dingare, 2022).

From geometrical aspects, the cross–flow arrangement requires more installation space and manufacturing costs, and the counter–flow design lacks in pressure drop and complex structural design. To overcome these issues, the counter-cross–flow heat exchanger is also explored to combine the advantages of both designs. The effectiveness of this system is comparable to the counter–flow and more than the cross–flow arrangements (Jia et al., 2019). Nevertheless, uneven air and water distribution, a drop in air pressure, complex geometry, and flow arrangement are concerns with the system.

With recent advancements, studies paid attention to increasing the heat transfer and simplifying the design and fabrication. For instance, a modified

Table 3.1 Summary of the geometrical design and layouts of the IEC systems.

Title	System design	Remark	References
Cross–flow dew point heat exchanger	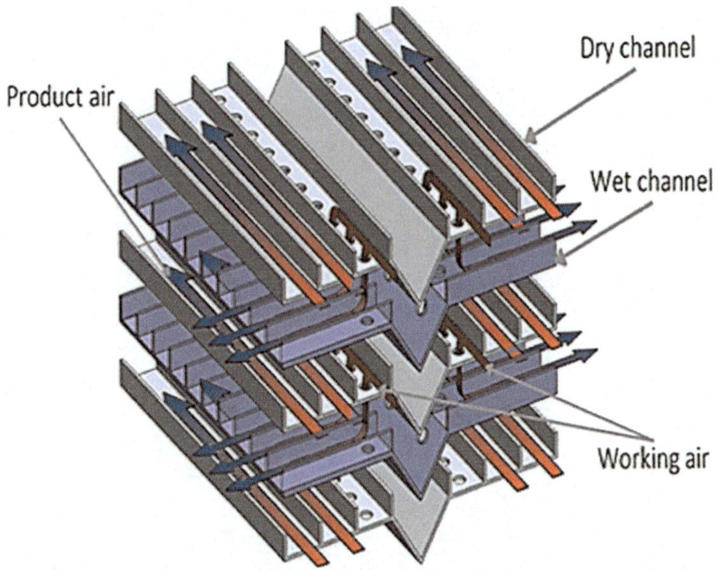 Coolerado, C., 2006. Coolerado HMX (Heat and Mass Exchanger) Brochure. Coolerado Corporation Arvada, CO, USA.	• Working air flow direction is perpendicular to product air flow. • More cooling capacity than counter–flow design. • Design size and cost are challenge.	Coolerado (2006)

<table>
<tr><td>Counter–flow indirect evaporative heat exchanger</td><td>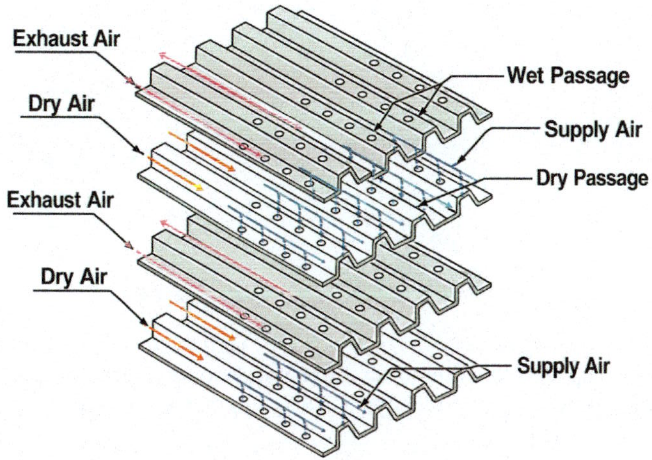
Patunkar, P., Dingare, S., 2022. Study of the experimental performance of dew point evaporative cooling arrangement for modified fluid flow passages. Front. Mech. Eng. 2297-3079. 8. doi:10.3389/fmech.2022.957898.</td><td>• Working air and product air flow in opposite directions in parallel channels
• Relatively simple fabrication design and high cooling effectiveness.
• The design faces a high pressure drop.</td><td>Patunkar and Dingare (2022)</td></tr>
</table>

Continued

Table 3.1 Summary of the geometrical design and layouts of the IEC systems—cont'd.

Title	System design	Remark	References
Counter-cross–flow indirect evaporative heat exchanger	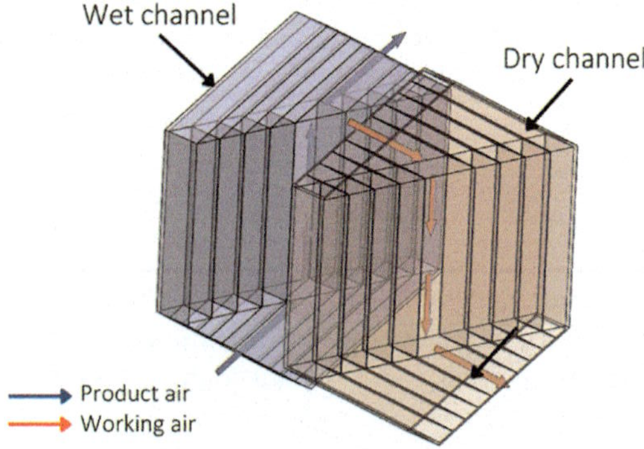 Jia, L., Liu, J., Wang, C., Cao, X., Zhang, Z., 2019. Study of the thermal performance of a novel dew point evaporative cooler. Appl. Therm. Eng. 13594311. 160, 114069. doi:10.1016/j.applthermaleng.2019.114069.	• Counter-cross–flow design combines advantages of both counter and cross air movements. The design showed higher effectiveness than cross flow design.	Jia et al. (2019)
Counter–flow indirect evaporative heat exchanger	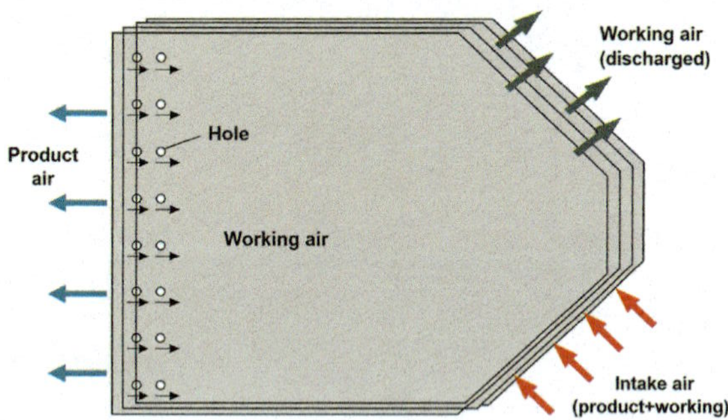 Zhao, X., Li, J.M., Riffat, S.B., 2008. Numerical study of a novel counter-flow heat and mass exchanger for dew point evaporative cooling. Appl. Therm. Eng. 13594311. 28 (14–15), 1942–1951. doi:10.1016/j.applthermaleng.2007.12.006.	• Design is modified to polygonal shape. • Dry and channel option is different than simple counter flow design. • Dry channel received both the product and working air and cooled down by transfer heat to adjacent wet channel.	Zhao et al. (2008)

Baffled indirect evaporative heat exchanger	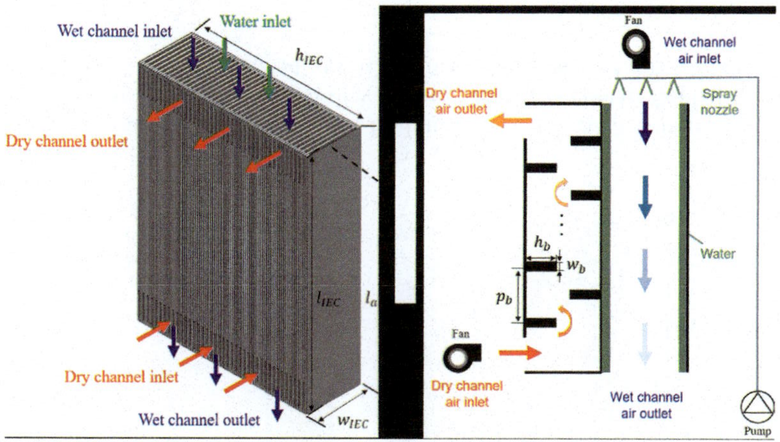 Kim, S.B., Moon, K.A., Choi, H.U., Choi, K.H., 2024. Experimental study on the performance of an air conditioning unit with a baffled indirect evaporative cooler. Energies. 19961073. 17 (13). Available from: http://www.mdpi.com/journal/energies/. doi:10.3390/en17133231.	• Wet channel is straightly moving towards downward. • Surface area is improved with baffles in dry channel to promote turbulence and heat transfer.	Kim et al. (2024)

Continued

Table 3.1 Summary of the geometrical design and layouts of the IEC systems—cont'd.

Title	System design	Remark	References
Advanced evaporative heat exchanger in a tubular design	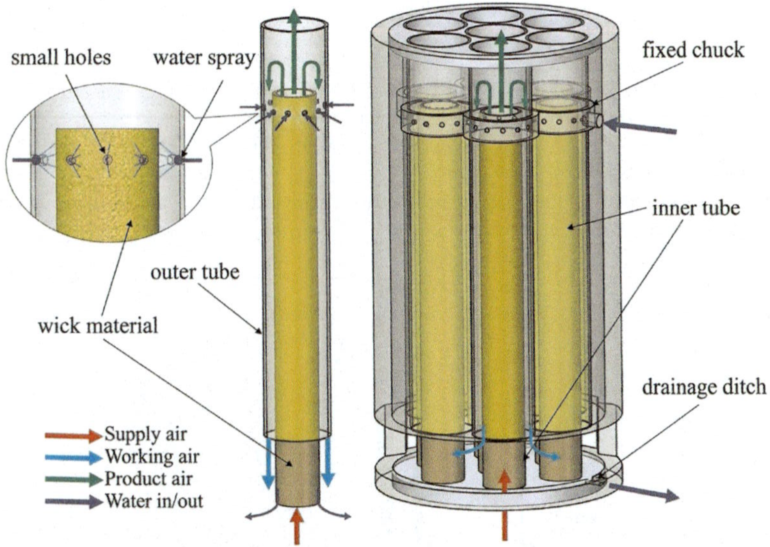 Gao, F., Thu, K., Wang, S., Zhao, F., Lin, J., Wu, K., 2023. Numerical investigation of a novel tubular dew-point evaporative cooler. Appl. Therm. Eng. 13594311. 223, 120064. doi:10.1016/j.applthermaleng.2023.120064.	• A tube–shaped counter–flow heat and mass exchanger supports the uniform water distribution. • Supply air flow in the inner tubes • Working airflow in the outer tubes in opposite directions.	Gao et al. (2023)

Simplified indirect evaporative cooler	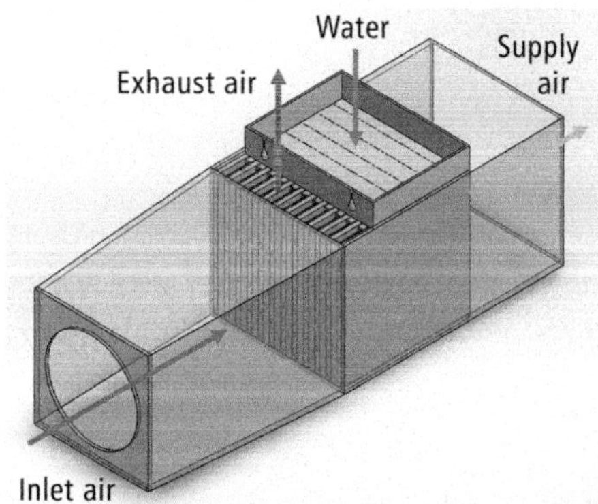 Comino, F., Castillo-González, J., Navas-Martos, F.J., Romero, P.E., de Adana, M.R., n.d. Design of highly compact indirect evaporative coolers. In: Proceedings of CLIMA 2022 Conference.	• A compact designed of IEC system is proposed with counter–flow design of heat exchanger. • Geometry and operating parameters of IEC were optimized with using a mathematical modeling. • Results depicted a cooling capacity of 177 kW/m^3 per system volume and 10.9 $kW/(m^3/s)$ per system airflow rate.	Comino et al. (n.d.)

Continued

Table 3.1 Summary of the geometrical design and layouts of the IEC systems—cont'd.

Title	System design	Remark	References
Indirect evaporative cooling heat exchanger (patent)	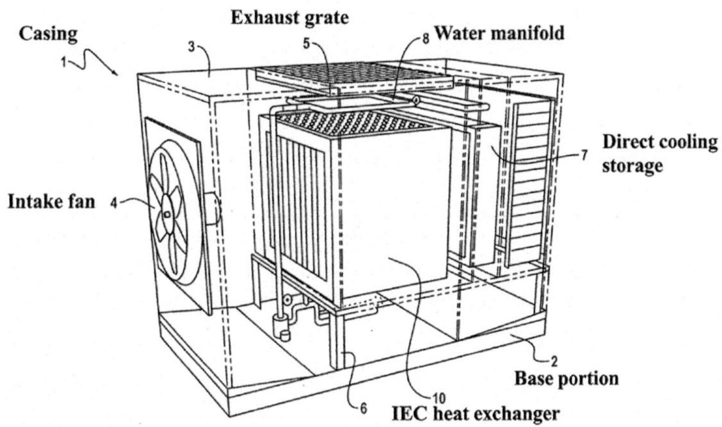 Des Champs, N.H., 2006. Indirect evaporative cooling heat exchanger.	• Nonmetallic material is used to improve the durability and reduce the risk of scaling and mineral deposition. • Integrated cooling tower functionality within IEC heat exchanger. • Design based on low manufacturing cost and ease of assembly. • Target to reduce the maintenance cost and improve the mechanical integrity for long term–performance.	Des Champs (2006)

<table>
<tr>
<td>Indirect evaporating cooler (patent)</td>
<td>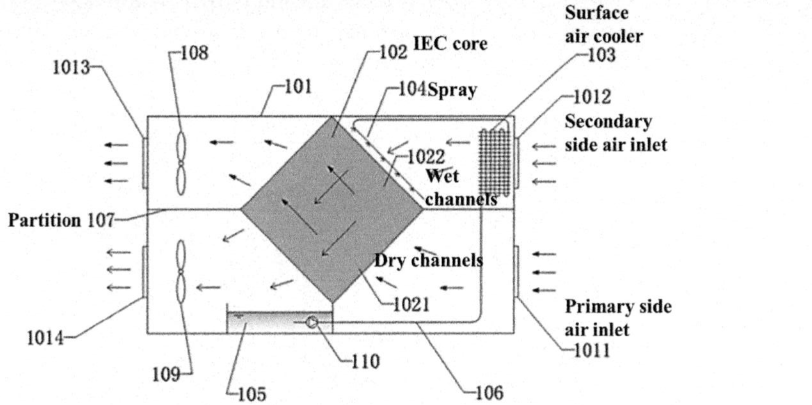

Chengbin, W., 2019. Produce the indirect evaporating-cooling equipment and air-conditioning system of dew-point temperature cold wind.</td>
<td>

• Proposed a designed of an IEC system-based system with separated dry and wet channels.
• Inexpensive and resistive to corrosion panel are made of plastic sheets with fiber cloth and acrylic dividers.
• Advanced IEC with cross–flow achieved close to dew–point temperature of ambient air.
• Structural design is found to be suited for arid regions.

</td>
<td>Chengbin (2019)</td>
</tr>
<tr>
<td>Indirect evaporative cooling-based heat recovery ventilation system</td>
<td>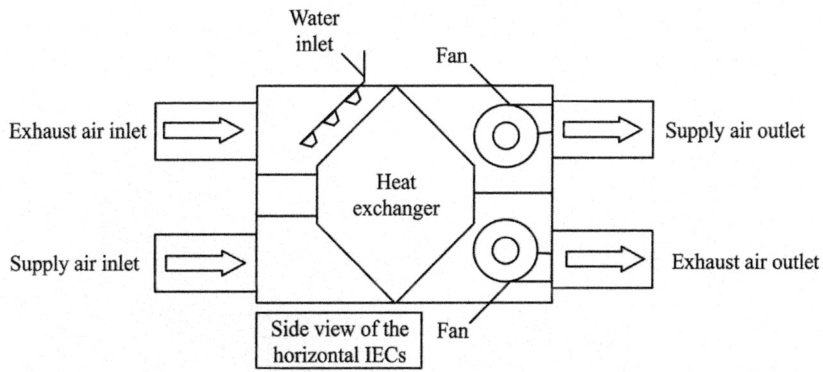

Huang, S., Li, W., Lu, J., Li, Y., 2017. Experimental study on two type of indirect evaporative cooling heat recovery ventilator. Procedia Eng. 18777058. 205, 4105–4110. doi:10.1016/j.proeng.2017.09.910.</td>
<td>

• Study is performed on IEC-based heat recovery ventilation system.
• Counter-cross–flow design arrangements with flat aluminum plates (0.1 mm thick) was used.
• Compared the horizontal and vertical arrangements of heat exchanger under different summer conditions.
• Vertical placement showed good results due to uniform water distribution.

</td>
<td>Huang et al. (2017)</td>
</tr>
</table>

Continued

Table 3.1 Summary of the geometrical design and layouts of the IEC systems—cont'd.

Title	System design	Remark	References
3D printing-based fabrication of an advanced indirect evaporative cooler	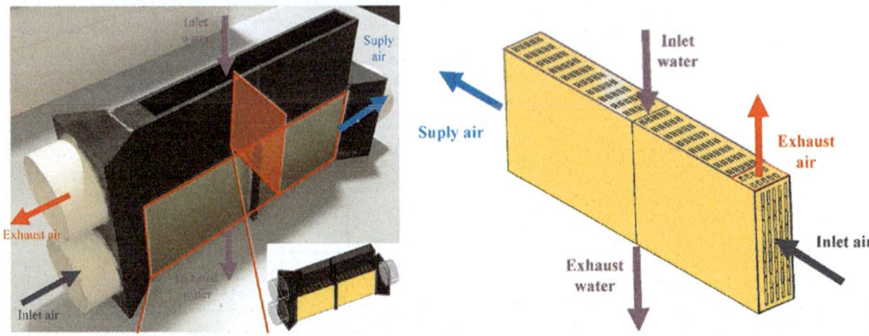 Castillo-González, J., Comino, F., Navas-Martos, F.J., Ruiz de Adana, M., 2023. Manufacturing and experimental analysis of a dew-point indirect evaporative cooler using fused deposition modeling 3D printing and polymeric materials. Appl. Therm. Eng. 13594311. 230, 120683. doi:10.1016/j.applthermaleng.2023.120683.	• Advanced IEC system was designed and fabricated with using fused filament fabrication technology. • Plate thickness is optimized to 0.5 mm. • Hydrophilic material for dry channel and hydrophobic material wet channel is used. • Results showed an improvement in cooling performance and reduction in energy consumption.	Castillo-González et al. (2023)
Counter–flow based regenerative evaporative cooler	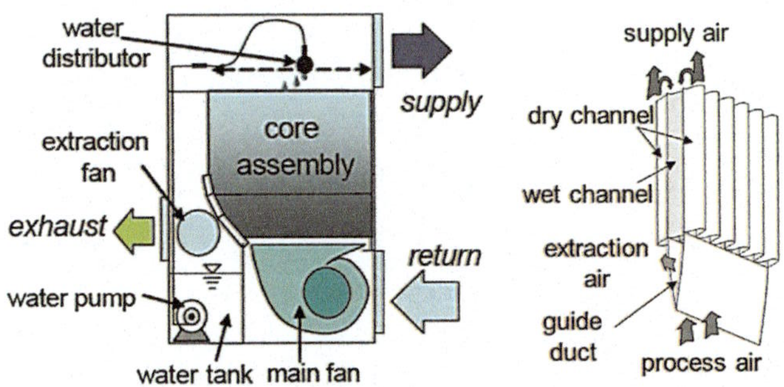 Lee, J., Lee, D.Y., 2013. Experimental study of a counter flow regenerative evaporative cooler with finned channels. Int. J. Heat Mass Transf. 00179310. 65, 173–179. doi:10.1016/j.ijheatmasstransfer.2013.05.069.	• IEC with finned counter–flow heat exchanger is developed. • Aluminum is used as a fabrication material for plates and associated fins. • Finned channel improved the heat transfer area. • A 100 mm long fin is arranged in the flow direction.	Lee and Lee (2013)

polygonal counter–flow heat exchanger was developed to optimize the contact between dry and wet channels, increasing the energy efficiency and uniform water distribution (Zhao et al., 2008). Similarly, the tubular advanced IEC heat exchanger is employed to support water distribution as well as stable interaction between air and water, which improves operational reliability (Gao et al., 2023). In addition, the IEC performance is critically related to the plate geometry, which determines the heat transfer area. Several plate designs such as flat plate, corrugated, finned or circular–finned, protrusion and pitched, and those with multiple spacers are used (Ali et al., 2021; Alam et al., 2024; Khalid et al., 2016; Shahzad et al., 2021). For instance, aluminum fins are incorporated in the counter–flow design to enhance the surface area (Lee and Lee, 2013), and the baffled-shaped dry channel is used to promote turbulence for improving air mixing and thermal exchange (Kim et al., 2024).

Besides, the overall design and geometry of the system are critical to flexible airflow that helps to reduce pressure drop, improve performance, and applicability of the system (Alam et al., 2024; Park et al., 2025). A typical system is composed of a heat exchanger, fans for inlet and outlet of air, a water distribution system, a water collecting arrangement, and a small recirculating pump (Coolerado, 2006). According to the operational setup, fans draw ambient air into and out of the system via the heat exchanger and cool it down before it is supplied to the closed space for air conditioning. Proper water distribution with a recirculating pump is connected to a water storage tank with circulating tubes. Water distribution is required to pursue the evaporative mechanism within the heat exchanger, and saturated air is exhausted out of the system. The external casing provides insulation from the surroundings to minimize heat exchange and is usually made of metals (e.g., aluminum) for the stability of the system. Several design approaches and configurations of IEC-based systems in terms of patented innovations and experimental developments are summarized in Table 3.1, aiming to develop understanding of design compactness associated with cooling effectiveness, applicability, and energy efficiency. In general, the system geometry is considered as simply described in the simulation-based study (Fatahian et al., 2020), and a simplified compact design of an IEC system with a counter–flow heat exchanger is developed and optimized with mathematical modeling (Comino et al., n.d.). As explained above, the numerical and simulation-based designing is underscoring the importance of compact geometries for high-performance cooling.

As a sustainable and energy-efficient alternative, IEC technology is critically important to combat the toxic refrigerant release into the environment and growing energy demands for air conditioning requirements. Thus, several designs and modified IEC systems have been developed as patent-based innovations, which have also contributed to system advancement. For example, a nonmetallic IEC heat exchanger design was introduced to overcome the issues of corrosion, scaling, and maintenance by using plastic as a fabrication material (Des Champs, 2006). Likewise, the advanced IEC system in a crossflow configuration was constructed with plastic and fiber composite panels (Chengbin, 2019). This system was featured to maintain a design with a simple structure and reduce the manufacturing design for arid climatic conditions. Keeping in view the adaptation of IEC for diverse applications, a heat recovery ventilator coupled with the counter-cross–flow aluminum plate exchanger is employed (Huang et al., 2017). This design performed well in a vertical arrangement due to improvement in the water distribution mechanism.

In addition to fundamental designs, advanced fabrication technologies are also emphasized in recent research. A 3D–printed advanced IEC model was developed by predetermining CAD based on experimental and simulation investigation. And polymeric materials with coated hydrophilic and hydrophobic materials were employed to improve the evaporative process and overall efficiency (Castillo-González et al., 2023). Meanwhile, efforts were made to develop a simple IEC with aluminum plates and semispherical dimples (Comino et al., 2018). This simplified system offers high applicability potential to data centers, desiccant cycles, and residential cooling. However, geometrical design lacks compactness. Another study focused on the compact configuration of a regenerative evaporative cooler, wherein channels of a heat exchanger are embraced with fins to improve the heat transfer and wettability (Lee and Lee, 2013).

In summary, these case studies illustrate the evolution of IEC-based systems from various aspects such as simple design, specific application, improved efficiency and effectiveness, and even water distribution. Also, it is considering the design compactness, corrosion and bacterial resistive material, improvement in manufacturability, and integration with conventional cooling systems. Along with the development of compact IEC-based systems, the adaptation will become easy to achieve high efficiency and environmental sustainability across residential, healthcare, and commercial and other sectors.

3.3.3 MVC systems

Table 3.2 summarizes the different configurations of MVC-integrated systems to enhance cooling performance, energy efficiency, and adaptability for various applications. In case of very hot ambient conditions, the heat rejection of the MVC condenser is assisted with a water spray system, resulting in improved performance and energy efficiency (Youbi-Idrissi et al., 2007). The MVC system is combined with a subcooling system and organic Rankine cycle in different studies to improve the cooling capacity, exergy efficiency, and thermal efficiency (Alsouda et al., 2023; Alshammari et al., 2023). Likewise, a desiccant dehumidification setup is coupled with the MVC system in a study that used condenser waste heat as input energy to regenerate the desiccant material (Jani et al., 2017). In addition, a sustainable solar–powered integrated MVC system is particularly designed for the cold storage of agricultural goods. This system exhibited consistent cooling performance under varying solar radiation, providing a suitable alternative solution for remote agricultural demands (Hussain et al., 2025). Furthermore, an experimental investigation on R–718 MVC cycles showed that integrating a condensing wave rotor can overcome the limits of high pressure ratios for water as a refrigerant. Results exhibited that thermal performance and operational stability were increased with these arrangements (Alsouda et al., 2023). Overall, these advancements validate the continued growth of MVC technology via hybridization and thermal optimization for high-efficiency cooling applications. However, enivronmental sustainability is being compromised with the use of toxic refrigerants, which are facing very high pressure from aspects of environmental protection.

3.3.4 IEC–MVC systems

In addition to the standalone IEC systems, various studies have shown that MVC-integratded IEC systems allow performance stability along with improved cooling capacity and energy efficiency compared to the MVC system. In MVC-assisted IEC systems, evaporative precooling reduces compressor load and increases the overall performance and applicability under a wide range of climatic conditions. Table 3.3 highlights the compact and versatile IEC systems for different applications. For instance, for the hybrid IEC–MVC system, a compact design is demonstrated, which shows a good COP, implying high energy-saving efficiency (Chen et al., 2021). A household exhaust air heat pump is integrated with the IEC system that

Table 3.2 Summary of design and configuration MVC integrated systems.

Title	System design	Remark	References
MVC system with water sprayer	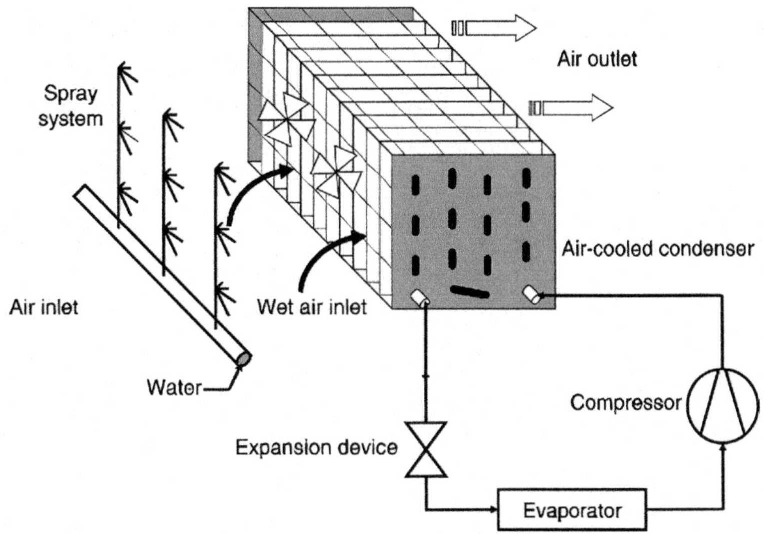 Youbi-Idrissi, M., Macchi-Tejeda, H., Fournaison, L., Guilpart, J., 2007. Numerical model of sprayed air cooled condenser coupled to refrigerating system. Energy Convers. Manag. 01968904. 48 (7), 1943–1951. doi:10.1016/j.enconman.2007.01.025.	• Development of MVC system with water sprayer for hot area. • Study the performance of water sprayer-integrated condenser. • Result showed 55% more COP with using water spray than the air–cooled condenser.	Youbi-Idrissi et al. (2007)

MVC coupled with organic Rankine cycle	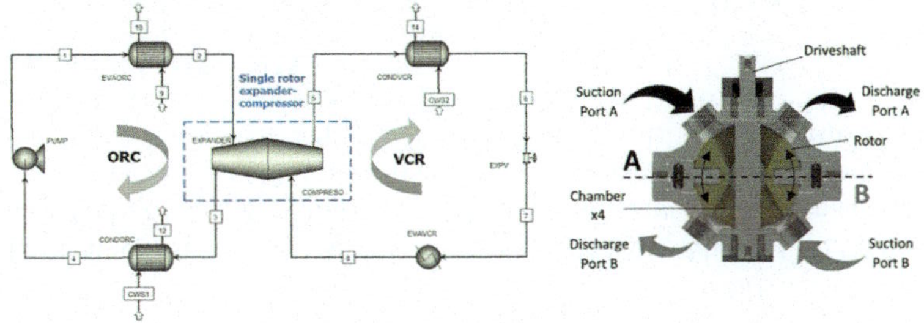 Alshammari, S., Kadam, S.T., Yu, Z., 2023. Assessment of single rotor expander-compressor device in combined organic Rankine cycle (ORC) and vapor compression refrigeration cycle (VCR). Energy. 03605442. 282, 128763. doi:10.1016/j.energy.2023.128763.	• An MVC system is coupled with single-rotor compressor and organic Rankine cycle system. • Numerical model is used to validate this system. • This system converted 56% heat–to–cooling efficiency and obtained 63% efficiency of exergy. • Rotor speed linearly related with cooling effect, but not with cooling efficiency.	Alshammari et al. (2023)
Mechanical vapor compression integrated with subcooling system	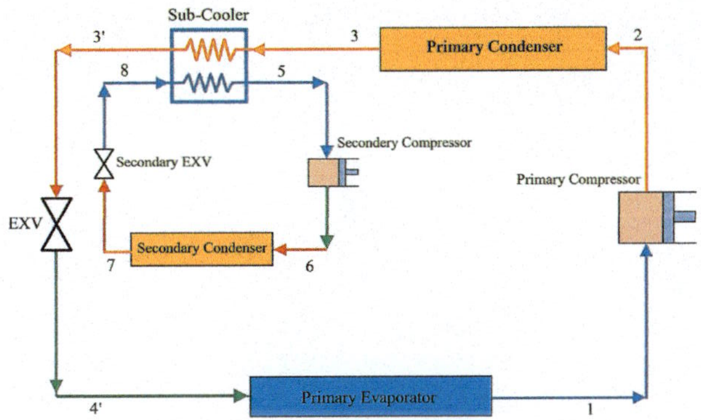 Alsouda, F., Bennett, N.S., Saha, S.C., Salehi, F., Islam, M.S., 2023. Vapor compression cycle: a state-of-the-art review on cycle improvements, water and other natural refrigerants. Clean Technol. 25718797. 5 (2), 584–608. Available from: www.mdpi.com/journal/cleantechnol. doi:10.3390/cleantechnol5020030.	• General configuration for subcooling assisted MVC system. • Subcooling system is incorporated to enhance the overall cooling capacity • Different refrigerants used for primary and subcooling cooling system.	Alsouda et al. (2023)

Continued

Table 3.2 Summary of design and configuration MVC integrated systems—cont'd.

Title	System design	Remark	References
Solar–power assisted cold storage system	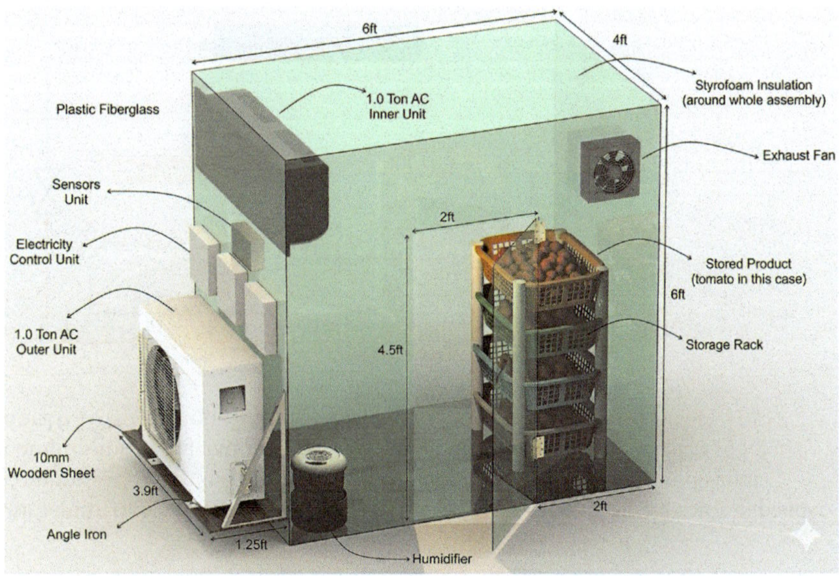 Hussain, G., Sultan, M., Aleem, M., Shahzad, M.W., Ahamed, M.S., Farooq, M., Sajjad, U., Zhang, Z., Ahmad, S., 2025. Development and performance evaluation of solar-powered cold storage system for perishable agricultural products. Int. Commun. Heat Mass Transf. 07351933. 168. Available from: https://www.sciencedirect.com/science/journal/07351933. doi:10.1016/j.icheatmasstransfer.2025.109408.	• Solar powered MVC based cold storage system was developed. • Temperature between 16°C–29°C and relative humidity between 75%–99% were achieved to store different agriculture products. • Storage system is developed with mild steel and Styrofoam is used for insulation purpose.	Hussain et al. (2025)

R–718 cycle couples with condensing wave rotor	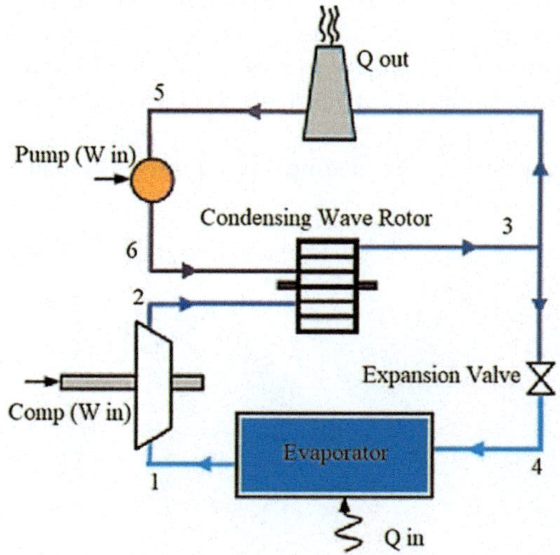 Alsouda, F., Bennett, N.S., Saha, S.C., Salehi, F., Islam, M.S., 2023. Vapor compression cycle: a state-of-the-art review on cycle improvements, water and other natural refrigerants. Clean Technol. 25718797. 5 (2), 584–608. Available from: www.mdpi.com/journal/cleantechnol. doi:10.3390/cleantechnol5020030.	• Combined centrifugal compressor and condensing wave rotor for R–718 cycle • Design is dedicated to overcome the issues related to high pressure ratios and compressor discharge temperatures. • System improved the COP considerably.	Alsouda et al. (2023)

Table 3.3 Summary of compact design of hybrid IEC–MVC systems.

Title	System design	Remark	References
Hybrid IEC–MVC system with cross–flow heat exchanger	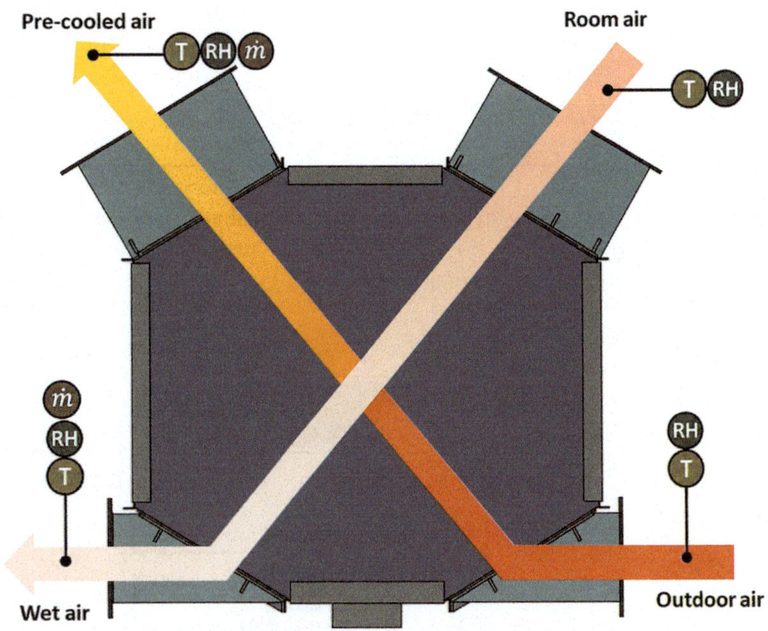 Chen, Q., Kum Ja, M., Burhan, M., Akhtar, F.H., Shahzad, M.W., Ybyraiymkul, D., Ng, K.C., 2021. A hybrid indirect evaporative cooling-mechanical vapor compression process for energy-efficient air conditioning. Energy Convers. Manag. 01968904. 248. Available from: https://www.journals.elsevier.com/energy-conversion-and-management. doi:10.1016/j.enconman.2021.114798.	• Designed a 1–Rton IEC system with cross–flow heat exchanger. • Channel walls of heat exchanger were constructed with 0.3 mm thick aluminum plates. • Air flow channels are designed by cutting corner and connected with acrylic ducts.COP ranged from 6–16 for IEC, and~5–6 for IEC–MVC system.	Chen et al. (2021)

IEC-based air heat pump for exhaust system	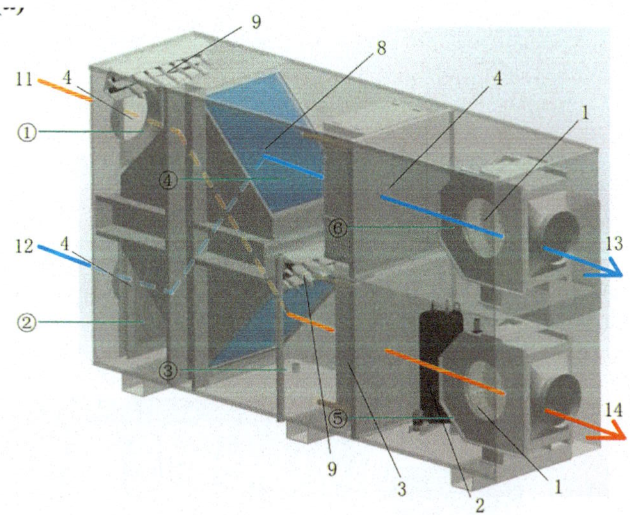 Li, W., Shi, W., Wang, J., Li, Y., Lu, J., 2021. Experimental study of a novel household exhaust air heat pump enhanced by indirect evaporative cooling. Energy Build. 03787788. 236, 110808. doi:10.1016/j.enbuild.2021.110808. Key parts: 1. Centrifugal fans; 3. Condenser of heat pump; 4. Evaporator; 8. Heat exchanger; 9. water spray systems.	• Air heat pump coupled with IEC is designed for household exhausting system. • Counter–cross heat exchanger with flat plates is made of aluminum of 0.1 mm thickness • System showed efficient energy efficiency ratio (4.6–5.65) under different summer conditions. • This compact design and efficient system are suggested for various applications.	Li et al. (2021)

Continued

Table 3.3 Summary of compact design of hybrid IEC?–?MVC systems—cont'd.

Title	System design	Remark	References
IEC chiller	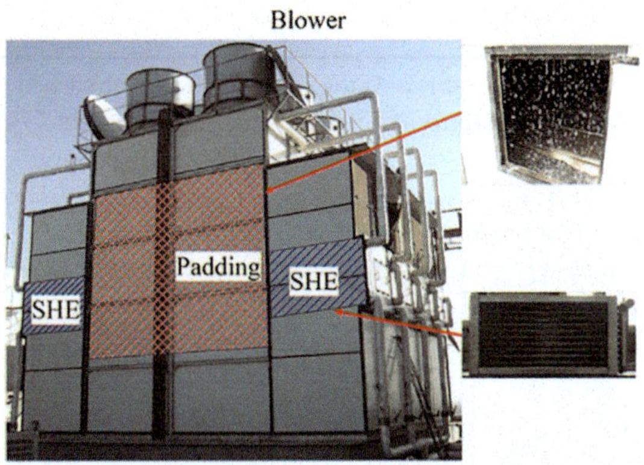 Jing, Y., Zhao, C., Xie, X., Jiang, Y., 2022. Optimization and performance analysis of water-mediated series indirect evaporative chillers: experimental and simulated investigation. Energy Convers. Manag. 01968904. 268, 115990. doi:10.1016/j.enconman.2022.115990.	• IEC chiller is fabricated to supply chilled water for hospital use. • The system is based on one or more sensible heat exchanger along with direct EC system. • Wet-bulb efficiency reached up to 106.57% in or a hospital in Xinjiang. • The system showed better performance in dry and hot regions.	Jing et al. (2022)

Advanced IEC with heat recovery ventilation system	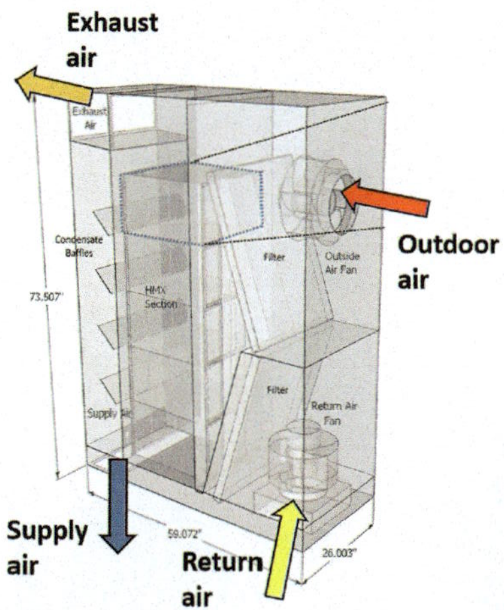 Chakraborty, S., Vernon, D., Jha, A., Narayanan, V., 2023. Performance characterization of M-cycle indirect evaporative cooler and heat recovery ventilator for commercial buildings—experiments and model. Energy Build. 03787788. 281, 112762. doi:10.1016/j.enbuild.2022.112762.	• A dew–point IEC system is used for heat-recovery ventilation in winters and switch to IEC during summers. • The system contains a series of five heat exchanger. • Integrated heat exchangers are made of polyethylene. • Results showed advantage of using IEC with a COP up to 20.6 in summer and 10.3 in winter.	Chakraborty et al. (2023)

uses a counter-cross–flow aluminum plate exchanger to enhance the energy efficiency (Li et al., 2021). This is an advanced geometrical compact design for residential use. Similarly, a series water–mediated indirect evaporative chiller was developed for hospital cooling and optimized with a numerical model for different climatic conditions (Jing et al., 2022). Results showed a satisfying applicability for arid regions. Moreover, a multiexchanger advanced IEC system was tested for commercial buildings, functioning as a heat recovery ventilator in winter and an IEC in summer (Chakraborty et al., 2023). However, these systems are designed for a particular region or specific applications, which limits the applicability and adoptability in comparison to the MVC systems.

3.4. NIEC–MVC system

Fig. 3.6 illustrates an innovative and compact design that integrates IEC and MVC in a single system developed under the UKRI–funded project "Super–Efficient Sustainable Cooling Solutions for All Applications (S2COOL)."

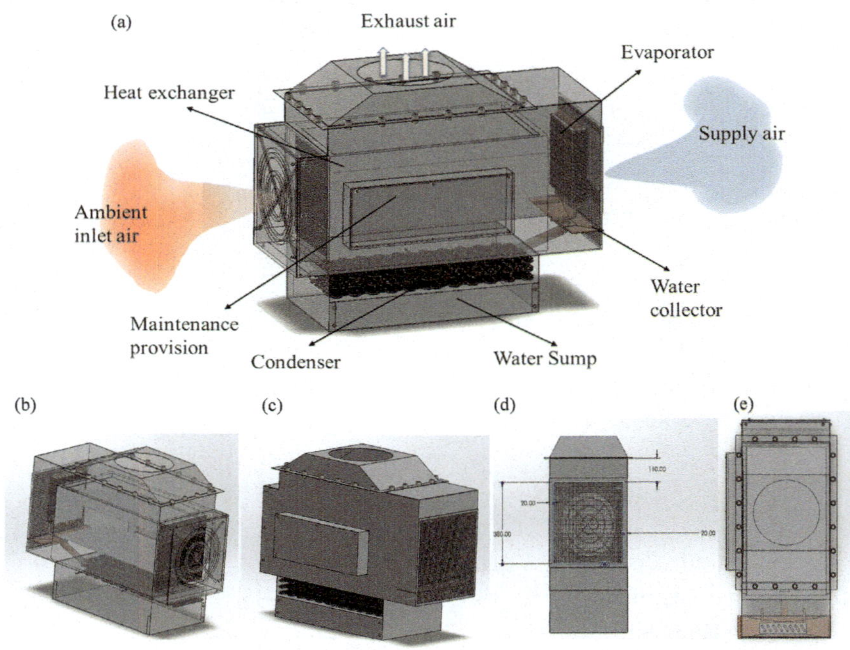

Fig. 3.6 CAD design of hybrid NIEC–MVC system (A) main transparent assembly design, (B) transparent orthogonal view, (C) solid view, (D) front view, and (E) top view of the model.

The key assembly design includes a heat exchanger, evaporator, condenser, compressor, and water circulation system with a pump, all embodied in a metallic casing (Fig. 3.6A). Ambient air is precooled through a cross–flow heat exchanger, and then the MVC system further processes it as required for conditioning. The vertical structure minimizes the system footprint by optimizing airflow and thermal exchange efficiency. Fig. 3.6B–E shows the dimensional and structural details of the compact system. The overall dimensions of the system are 900 mm in height and 500 mm in width, emphasizing easy installation and modular scalability. The top mounted outlet for exhaust air effectively improves thermodynamic performance. This compact and robust design exemplifies durability, scalability, and applicability for diverse applications. The system is designed to ensure safety and provide access for maintenance and cleaning operations. In addition, this compact geometry is equipped with an Internet of Things (IoT) panel, integration with sustainable solar power energy, and proper placement of sensors for research and monitoring purposes. Further details of this system will be patented and published in the future.

3.5. SWOT analysis

Table 3.4 presents the SWOT analysis for the discussed IEC, MVC, hybrid IEC–MVC, and NIEC–MVC systems to outline the design geometry, layout, system compactness, applicability, and their impact on performance, efficiency, and sustainability. The key attraction of IEC systems is their simple geometry and construction, sustainable operation, and energy efficiency. However, dependency on ambient temperature and humidity, water-related issues, and space requirements remain concerning factors. The mature MVC technology offers consistent cooling performance under varying conditions but requires intensive energy and is harmful to the environment.

On the other hand, IEC integrated MVC systems address these problems by combining energy-efficient and sustainable precooling with the constant performance of the MVC. Still, configuration requirements, airflow design, water utilization, and the relatively bulky and complex layout limit its adoptability and versatile applicability. In the NIEC–MVC system, the simplified configuration and improved performance of the NIEC support better system operation and design flexibility. The proposed compact design of NIEC–MVC promotes applicability and adoptability across a wide range of applications. This system is emphasizing the sustainable nature of evaporative cooling technology with the strength of the MVC systems, benchmarking

Table 3.4 SWOT analysis of IEC, MVC, hybrid IEC–MVC, and NIEC–MVC cooling systems.

Category	Strengths	Weaknesses	Opportunities	Threats
IEC Systems	• Energy efficient and sustainable to environment • Simple design and fabrication • Easy to scale • Effective in dry climatic conditions	• Performance dependency on ambient conditions • Sensitive to blockage and water circulation • Footprints depend on design and capacity	• Optimization of design and geometries. • Integration and commercialization in different sectors • Material exploration in terms of corrosion, fouling and wettability and applications	• Water availability and quality requirements • Blockage with scaling and fouling • Limited climatic compatibility • Competition with hybrid systems
MVC Systems	• Consistent performance across climatic conditions • Reliable control for varying temperature and humidity • Mature technology and compact design for different applications	• Energy inefficient • Refrigerants contribute to carbon footprints and greenhouse gases • Higher load and operational cost • Limited design geometry for system performance	• Exploration for relatively sustainable refrigerants • Integration with renewable energy and remote applications • Improvement in compressor design and technology.	• Refrigerants regulations • High energy prices and demands • Pressure to environmental sustainability

IEC-based MVC hybrid systems	• IEC-based precooling to reduce the MVC load • Reduces operational cost, energy efficient and stabile performance • Customized adaptability with applications	• Structural and design complexity • System layout and air flow directions required more considerations • Geometry and configuration of IEC influence the performance and design of system • High initial cost	• Hybrid model development for different applications • Integration with sustainable energy systems • Advanced controls and IoT optimization to reduce the complexity. • Improvement in design and configurations	• Adoptability restriction for various sectors • Design complexity and maintenance • Water availability and quality requitements • Completion with NIEC–MVC system due to improved effectiveness
NIEC–MVC systems	• No need for coated media • Simple plate geometry of heat exchanger and fabrication • High precooling performance to decrease the MVC load • Improves performance stability with humidification system • Compact design for versatile applicability	• Bulky structure • Maintenance of nozzle performance • Complex system compared to the simple IEC system	• Future patent potential • Integration with solar energy and IoT optimization • Application in portable/mobile cooling • Potential for scalability	• Water availability • Competition from ultrahigh-efficiency heat pumps

progression toward the more mature sustainable cooling approach. However, the bulkiness with MVC integration remains the consistent for both hybrid systems. The SWOT analysis outlines strong opportunities for future aspects of development, including optimization of compact system, material selection, integration with renewable solar energy and IoT, and effective water management, which can further expand system applicability.

3.6. Conclusions

IEC systems have the ability to meet worldwide demands for sustainable and energy-efficient air conditioning. The geometrical design and parametric optimization of different cooling systems are fundamentally important in terms of system performance and efficiency, along with applicability. The IEC-based systems have progressed from simple parallel flow to complicated channel arrangements, plate geometry, designs, and water distribution. A range of materials for plates and casings is explored based on the design structure, intended application, and prospective technical requirements. With advancements, digital design tools, numerical and simulation models, and experimental investigation continue to improve the IEC system in terms of durability, water management, compactness, energy savings, and operating flexibility of the IEC systems. Case studies across various applications show that MVC integration with different technologies has the potential to enhance performance and energy efficiency. The hybrid IEC–MVC systems addressed the limitations of standalone systems, improving sustainability and stabilizing performance by combining the strengths of both IEC and MVC systems. The NIEC system presents a simple and effective precooling configuration coupled with the MVC system, which offers high cooling performance and applicability and removes the concerns related to coated materials. Although NIEC improves the design compactness, applicability, and performance, bulkiness is still common for both hybrid IEC–MVC systems. The SWOT analysis underscores future opportunities, including geometry optimization, water management, digital integration, and policy-driven adoption to speed up the global shift to clean and efficient cooling systems.

Acknowledgment

The United Kingdom Government funds this research (and/or) pilot activity through the Project, APP47457, titled "Super–efficient Sustainable Cooling Solution for All Applications (S2Cool)," under the Ayrton Challenge Programme of the UK Research and

Innovation (UKRI); however, the views expressed do not necessarily reflect the UK Government's official policies. The Project: APP47457 is implemented/led by Northumbria University UK together with partners to develop sustainable cooling solution for several applications.

Funding statement

The United Kingdom Government funds this research (and/or) pilot activity through the Project, APP47457, titled "Super–efficient Sustainable Cooling Solution for All Applications (S2Cool)," under the Ayrton Challenge Programme of the UK Research and Innovation (UKRI); however, the views expressed do not necessarily reflect the UK Government's official policies. The Project: APP47457 is implemented/led by Northumbria University UK together with partners to develop sustainable cooling solution for several applications.

Conflicts of interest

The authors declare no conflict of interest.

Declaration of generative AI and AI-assisted technologies in the writing process

No generative AI/AI-assisted technology is used in this paper.

CRediT authorship contribution statement

Nida Imtiaz was involved in the conceptualization, methodology, software, formal analysis, investigation, writing of the original draft. **Mazlan Abdul Wahid, Muhammad Sultan, and Muhammad Wakil Shahzad** were involved in the conceptualization, methodology, validation, resources, visualization, supervision, project administration, writing (review and editing), funding acquisition. **Muhammad Ali Imran** was involved in the conceptualization, methodology, software, formal analysis, investigation, writing (review and editing).

References

Ahmad, T., Zhang, D., 2020. A critical review of comparative global historical energy consumption and future demand: the story told so far. Energy Rep. 23524847. 6, 1973–1991. doi:10.1016/j.egyr.2020.07.020.

Alam, M.S., Mohd Zubir, M.N.B., Muhamad, M.R.B., Kazi, S.N., Öztop, H.F., Abdullah, S., Shaikh, K., 2024. A technological review of dew point evaporative cooling: experimental, analytical, numerical and optimization perspectives. J. Build. Eng. 23527102. 91. Available from: https://www.sciencedirect.com/science/journal/23527102. doi:10.1016/j.jobe.2024.109544.

Ali, M., Ahmad, W., Sheikh, N.A., Ali, H., Kousar, R., Rashid, T.U., 2021. Performance enhancement of a cross flow dew point indirect evaporative cooler with circular finned channel geometry. J. Build. Eng. 23527102. 35. Available from: http://www.journals.elsevier.com/journal-of-building-engineering/. doi:10.1016/j.jobe.2020.101980.

Alshammari, S., Kadam, S.T., Yu, Z., 2023. Assessment of single rotor expander-compressor device in combined organic Rankine cycle (ORC) and vapor compression refrigeration cycle (VCR). Energy. 03605442. 282, 128763. doi:10.1016/j.energy.2023.128763.

Alsouda, F., Bennett, N.S., Saha, S.C., Salehi, F., Islam, M.S., 2023. Vapor compression cycle: a state-of-the-art review on cycle improvements, water and other natural refrigerants. Clean Technol. 25718797. 5 (2), 584–608. Available from: www.mdpi.com/journal/cleantechnol. doi:10.3390/cleantechnol5020030.

Al-Zubaydi, A.Y.T., Hong, G., 2019. Experimental study of a novel water-spraying configuration in indirect evaporative cooling. Appl. Therm. Eng. 13594311. 151, 283–293. Available from: http://www.journals.elsevier.com/applied-thermal-engineering/. doi:10.1016/j.applthermaleng.2019.02.019.

Andersen, S.O., Halberstadt, M.L., Borgford-Parnell, N., 2013. Stratospheric ozone, global warming, and the principle of unintended consequences-an ongoing science and policy success story. J. Air Waste Manag. Assoc. 21622906. 63 (6), 607–647. Available from: http://www.tandfonline.com/loi/uawm20?. doi:10.1080/10962247.2013.791349.

Castillo-González, J., Comino, F., Navas-Martos, F.J., Ruiz de Adana, M., 2023. Manufacturing and experimental analysis of a dew-point indirect evaporative cooler using fused deposition modelling 3D printing and polymeric materials. Appl. Therm. Eng. 13594311. 230, 120683. doi:10.1016/j.applthermaleng.2023.120683.

Chakraborty, S., Vernon, D., Jha, A., Narayanan, V., 2023. Performance characterization of M-cycle indirect evaporative cooler and heat recovery ventilator for commercial buildings—experiments and model. Energy Build. 03787788. 281, 112762. doi:10.1016/j.enbuild.2022.112762.

Chen, Q., Kum Ja, M., Burhan, M., Akhtar, F.H., Shahzad, M.W., Ybyraiymkul, D., Ng, K. C., 2021. A hybrid indirect evaporative cooling-mechanical vapor compression process for energy-efficient air conditioning. Energy Convers. Manag. 01968904. 248. Available from: https://www.journals.elsevier.com/energy-conversion-and-management. doi:10.1016/j.enconman.2021.114798.

Chen, Q., Burhan, M., Ja, M.K., Shahzad, M.W., Ybyraiymkul, D., Zheng, H., Cui, X., Ng, K.C., 2022. Hybrid Indirect evaporative cooling-mechanical vapor compression system: a mini-review. Energies. 19961073. 15 (20). Available from: http://www.mdpi.com/journal/energies/. doi:10.3390/en15207810.

Wu, C., 2019. Produce the indirect evaporating-cooling equipment and air-conditioning system of dew-point temperature cold wind. CN110469929A, Available from: https://patents.google.com/patent/CN110469929A/en?oq=CN110469929A.

Comino, F., Milani, S., De Antonellis, S., Joppolo, C.M., Ruiz de Adana, M., 2018. Simplified performance correlation of an indirect evaporative cooling system: development and validation. Int. J. Refrig. 01407007. 88, 307–317. doi:10.1016/j.ijrefrig.2018.02.002.

Comino, F., Castillo-González, J., Navas-Martos, F.J., Romero, P.E., de Adana, M.R., n.d. Design of highly compact indirect evaporative coolers. In: Proceedings of CLIMA 2022 Conference.

Coolerado, C., 2006. Coolerado HMX (Heat and Mass Exchanger) Brochure. Coolerado Corporation Arvada, CO, USA.

Des Champs, N.H., 2006. Indirect evaporative cooling heat exchanger, U.S. Patent USRE46343E1 . The Munters Corporation (Accessed 23 March 2017). Available from: https://worldwide.espacenet.com/patent/search/family/035423685/publication/USRE46343E?q=pn%3DUSRE46343E.

Fan, X., Lu, X., Wang, J., Li, Z., Wang, Q., Dong, Z., Zhang, R., 2021. Performance evaluation of a maisotsenko cycle cooling tower with uneven length of dry and wet channels in hot and humid conditions. Energies. 1996-1073. 14 (24), 8249. doi:10.3390/en14248249.

Fatahian, E., Salarian, H., Fatahian, H., 2020. A parametric study of the heat exchanger copper coils used in an indirect evaporative cooling system. SN Appl. Sci. 2523-3963. 2 (1). doi:10.1007/s42452-019-1915-0.

Gao, F., Thu, K., Wang, S., Zhao, F., Lin, J., Wu, K., 2023. Numerical investigation of a novel tubular dew-point evaporative cooler. Appl. Therm. Eng. 13594311. 223, 120064. doi:10.1016/j.applthermaleng.2023.120064.

Ghasemian, S., Faridzad, A., Abbaszadeh, P., Taklif, A., Ghasemi, A., Hafezi, R., 2024. An overview of global energy scenarios by 2040: identifying the driving forces using cross-impact analysis method. Int. J. Environ. Sci. Technol. 1735-1472. 21 (11), 7749–7772. doi:10.1007/s13762-020-02738-5.

Gómez, E.V., Martínez, F.J.R., Diez, F.V., Leyva, M.J.M., Martín, R.H., 2005. Description and experimental results of a semi-indirect ceramic evaporative cooler. Int. J. Refrig. 01407007. 28 (5), 654–662. doi:10.1016/j.ijrefrig.2005.01.004.

Huang, S., Li, W., Lu, J., Li, Y., 2017. Experimental study on two type of indirect evaporative cooling heat recovery ventilator. Procedia Eng. 18777058. 205, 4105–4110. doi:10.1016/j.proeng.2017.09.910.

Hussain, G., Sultan, M., Aleem, M., Shahzad, M.W., Ahamed, M.S., Farooq, M., Sajjad, U., Zhang, Z., Ahmad, S., 2025. Development and performance evaluation of solar-powered cold storage system for perishable agricultural products. Int. Commun. Heat Mass Transf. 07351933. 168. Available from: https://www.sciencedirect.com/science/journal/07351933. doi:10.1016/j.icheatmasstransfer.2025.109408.

Jamil, M.A., Imtiaz, N., Ng, K.C., Xu, B.B., Yaqoob, H., Sultan, M., Shahzad, M.W., 2023. Experimental and parametric sensitivity analysis of a novel indirect evaporative cooler for greener cooling. Therm. Sci. Eng. Prog. 24519049. 42. Available from: https://www.journals.elsevier.com/thermal-science-and-engineering-progress. doi:10.1016/j.tsep.2023.101887.

Jamil, M.A., Ashraf, W.M., Imtiaz, N., Xu, B.B., Zubair, S.M., Yaqoob, H., Imran, M., Shahzad, M.W., 2024. Machine learning-based process design of a novel sustainable cooling system. Energy Convers. Manag. 01968904. 319. Available from: https://www.sciencedirect.com/science/journal/01968904. doi:10.1016/j.enconman.2024.118941.

Jani, D.B., Mishra, M., Sahoo, P.K., 2017. A critical review on solid desiccant-based hybrid cooling systems. Int. J. Air-Cond. Refrig. 2010-1325. 25 (03), 1730002. doi:10.1142/s2010132517300026.

Jia, L., Liu, J., Wang, C., Cao, X., Zhang, Z., 2019. Study of the thermal performance of a novel dew point evaporative cooler. Appl. Therm. Eng. 13594311. 160, 114069. doi:10.1016/j.applthermaleng.2019.114069.

Jing, Y., Zhao, C., Xie, X., Jiang, Y., 2022. Optimization and performance analysis of water-mediated series indirect evaporative chillers: experimental and simulated investigation. Energy Convers. Manag. 01968904. 268, 115990. doi:10.1016/j.enconman.2022.115990.

Kapilan, N., Isloor, A.M., Karinka, S., 2023. A comprehensive review on evaporative cooling systems. Results Eng. 25901230. 18, 101059. doi:10.1016/j.rineng.2023.101059.

Khalid, O., Ali, M., Sheikh, N.A., Ali, H.M., Shehryar, M., 2016. Experimental analysis of an improved Maisotsenko cycle design under low velocity conditions. Appl. Therm. Eng. 13594311. 95, 288–295. Available from: http://www.journals.elsevier.com/applied-thermal-engineering/. doi:10.1016/j.applthermaleng.2015.11.030.

Kim, S.B., Moon, K.A., Choi, H.U., Choi, K.H., 2024. Experimental study on the performance of an air conditioning unit with a baffled indirect evaporative cooler. Energies.

19961073. 17 (13). Available from: http://www.mdpi.com/journal/energies/. doi:10.3390/en17133231.

Kim, W.S., 2022. Design optimization of cross-counter flow compact heat exchanger for energy recovery ventilator. Int. J. Air-Cond. Refrig. 20101333. 30 (1). Available from: https://www.springer.com/journal/44189. doi:10.1007/s44189-022-00016-2.

Lee, J., Lee, D.Y., 2013. Experimental study of a counter flow regenerative evaporative cooler with finned channels. Int. J. Heat Mass Transf. 00179310. 65, 173–179. doi:10.1016/j.ijheatmasstransfer.2013.05.069.

Li, W., Shi, W., Wang, J., Li, Y., Lu, J., 2021. Experimental study of a novel household exhaust air heat pump enhanced by indirect evaporative cooling. Energy Build. 03787788. 236, 110808. doi:10.1016/j.enbuild.2021.110808.

Mahmood, M.H., Sultan, M., Miyazaki, T., Koyama, S., Maisotsenko, V.S., 2016. Overview of the Maisotsenko cycle—a way towards dew point evaporative cooling. Renew. Sustain. Energy Rev. 18790690. 66, 537–555. Available from: https://www.journals.elsevier.com/renewable-and-sustainable-energy-reviews. doi:10.1016/j.rser.2016.08.022.

Mujahid Rafique, M., Gandhidasan, P., Rehman, S., Al-Hadhrami, L.M., 2015. A review on desiccant based evaporative cooling systems. Renew. Sustain. Energy Rev. 18790690. 45, 145–159. Available from: https://www.journals.elsevier.com/renewable-and-sustainable-energy-reviews. doi:10.1016/j.rser.2015.01.051.

Park, M.-H., Ryu, C.-M., Moon, S.J., Lee, D.-Y., 2025. Analysis of cooling performance of compact regenerative evaporative coolers for practical applications. Heat Mass Transf. 0947-7411. 61 (1). doi:10.1007/s00231-024-03535-3.

Patunkar, P., Dingare, S., 2022. Study of the experimental performance of dew point evaporative cooling arrangement for modified fluid flow passages. Front. Mech. Eng. 2297-3079. 8. doi:10.3389/fmech.2022.957898.

Ray, W.T., 1935. Conditioning liquids and air and other gases. U.S. Patent No. 1986529. Available from: https://patentimages.storage.googleapis.com/08/13/f9/e6ae534f44a8e2/US1986529.pdf.

Raza, H.M.U., Ashraf, H., Shahzad, K., Sultan, M., Miyazaki, T., Usman, M., Shamshiri, R.R., Zhou, Y., Ahmad, R., 2020. Investigating applicability of evaporative cooling systems for thermal comfort of poultry birds in Pakistan. Appl. Sci. 2076-3417. 10 (13), 4445. doi:10.3390/app10134445.

Raza, H.M.U., Sultan, M., Bahrami, M., Khan, A.A., 2021. Experimental investigation of evaporative cooling systems for agricultural storage and livestock air-conditioning in Pakistan. Build. Simul. 19968744. 14 (3), 617–631. Available from: http://www.springer.com/engineering/journal/12273?detailsPage=aboutThis. doi:10.1007/s12273-020-0678-2.

Sajjad, U., Abbas, N., Hamid, K., Abbas, S., Hussain, I., Ammar, S.M., Sultan, M., Ali, H.M., Hussain, M., Rehman, T.U., Wang, C.C., 2021. A review of recent advances in indirect evaporative cooling technology. Int. Commun. Heat Mass Transf. 07351933. 122. Available from: https://www.journals.elsevier.com/international-communications-in-heat-and-mass-transfer. doi:10.1016/j.icheatmasstransfer.2021.105140.

Shahzad, M.W., Lin, J., Xu, B.B., Dala, L., Chen, Q., Burhan, M., Sultan, M., Worek, W., Ng, K.C., 2021. A spatiotemporal indirect evaporative cooler enabled by transiently interceding water mist. Energy. 03605442. 217. Available from: https://www.journals.elsevier.com/energy. doi:10.1016/j.energy.2020.119352.

Sultan, M., El-Sharkawy, I.I., Miyazaki, T., Saha, B.B., Koyama, S., 2015. An overview of solid desiccant dehumidification and air conditioning systems. Renew. Sustain. Energy Rev. 18790690. 46, 16–29. Available from: https://www.journals.elsevier.com/renewable-and-sustainable-energy-reviews. doi:10.1016/j.rser.2015.02.038.

Sultan, M., Miyazaki, T., Koyama, S., Khan, Z.M., 2018. Performance evaluation of hydrophilic organic polymer sorbents for desiccant air-conditioning applications. Adsorpt. Sci. Technol. 20484038. 36 (1-2), 311–326. Available from: http://adt.sagepub.com/. doi:10.1177/0263617417692338.

Sultan, M., Miyazaki, T., 2017. Energy-efficient air-conditioning systems for nonhuman applications. In: Refrigeration.: IntechOpen, pp. 97-117. doi:10.5772/intechopen.68865.

Xu, P., Ma, X., Zhao, X., Fancey, K.S., 2016. Experimental investigation on performance of fabrics for indirect evaporative cooling applications. Build. Environ. 03601323. 110, 104–114. Available from: http://www.elsevier.com/inca/publications/store/2/9/6/index.htt. doi:10.1016/j.buildenv.2016.10.003.

Xuan, Y.M., Xiao, F., Niu, X.F., Huang, X., Wang, S.W., 2012. Research and application of evaporative cooling in China: a review (i)—research. Renew. Sustain. Energy Rev. 13640321. 16 (5), 3535–3546. doi:10.1016/j.rser.2012.01.052.

Xue, T., Wan, Y., Huang, Z., Chen, P., Lin, J., Chen, W., Liu, H., 2023. A comprehensive review of the applications of hybrid evaporative cooling and solar energy source systems. Sustainability. 2071-1050. 15 (24), 16907. doi:10.3390/su152416907.

Youbi-Idrissi, M., Macchi-Tejeda, H., Fournaison, L., Guilpart, J., 2007. Numerical model of sprayed air cooled condenser coupled to refrigerating system. Energy Convers. Manag. 01968904. 48 (7), 1943–1951. doi:10.1016/j.enconman.2007.01.025.

Zhao, X., Li, J.M., Riffat, S.B., 2008. Numerical study of a novel counter-flow heat and mass exchanger for dew point evaporative cooling. Appl. Therm. Eng. 13594311. 28 (14–15), 1942–1951. doi:10.1016/j.applthermaleng.2007.12.006.

Zohuri, B., 2018. Physics of cryogenics. In: Heat Exchangers, twelfth edn. University of New, Mexico, Albuquerque, pp. 299–330.

CHAPTER 4

CFD simulation study of a novel indirect evaporative cooler

Nida Imtiaz[a], Mazlan Abdul Wahid[a], Muhammad Zeshan[b], Muhammad Hamid Mahmood[b], Muhammad Sultan[b], and Muhammad Wakil Shahzad[c]
[a]Faculty of Mechanical Engineering, Universiti Teknologi Malaysia, Johor Bahru, Malaysia
[b]Department of Agricultural Engineering, Bahauddin Zakariya University, Multan, Pakistan
[c]Department of Mechanical and Construction Engineering, Northumbria University, Newcastle Upon Tyne, United Kingdom

4.1 Introduction

The rising earth temperature due to climate change increases the air-conditioning (AC) demand for the provision of thermal comfort conditions/environments in all applications. Recently, mechanical vapor compression systems capture the 90% AC market (Wu et al., 2024) in both the human and nonhuman applications. Mechanical vapor compression AC systems use the environmentally harmful chemical refrigerants and also consume excessive energy. Fifty (50%) percent of total energy is consumed by these energy-intensive systems in the building (Wu et al., 2024). In this regard, the trend of global AC stock, corresponding energy utilization, and emissions are shown in Fig. 4.1 (IEA, 2018; Jamil et al., 2023). The global stock of AC units in the next 20 years (2025–2045) is projected to be about 5.0 billion, accompanying energy utilization of 4.0 TWh and emissions of 120 Gt (IEA, 2018; Jamil et al., 2023). Moreover, the forecasted trends remain escalating in the next years (Fig. 4.1).

It is crucial now to explore other environmentally friendly cooling options for a sustainable greener future having net zero emissions. Evaporative coolers have reemerged as low cost, energy-efficient, and green cooling solutions for various applications. Indirect evaporative coolers (IECs) are gaining importance because they provide sensible cooling as compared to adiabatic cooling by direct evaporative coolers. Some of the promising applications of IEC are presented in Fig. 4.2 (Sajjad et al., 2021). The indirect evaporative cooling has been studied

Sustainable Hybrid Cooling and Dehumidification Strategies for Controlled Environments
https://doi.org/10.1016/B978-0-443-48555-8.00001-3

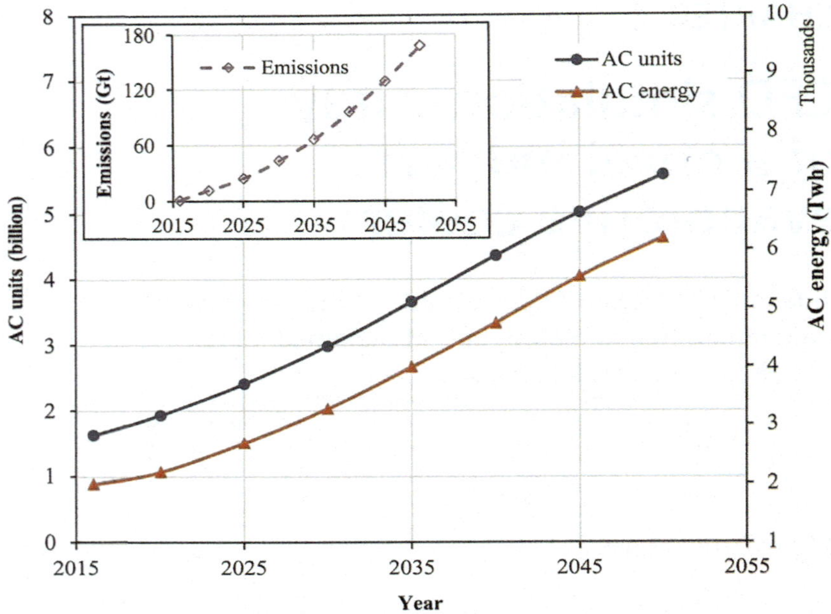

Fig. 4.1 Trend of global air-conditioning stock alongside emissions and energy utilization (IEA, 2018; Jamil et al., 2023). *(From IEA, 2018. The future of cooling opportunities for energy- efficient air conditioning. Available from: https://www.iea.org/data-and-statistics/charts/global-air-conditioner-stock-1990-2050 (Accessed 18 November 2025); Jamil, M.A., Ashraf, W.M., Imtiaz, N., Xu, B.B., Zubair, S.M., Yaqoob, H., Imran, M., Shahzad, M.W., 2024. Machine learning-based process design of a novel sustainable cooling system. Energy Convers. Manag. 01968904. 319. Available from: https://www.sciencedirect.com/science/journal/01968904. doi:10.1016/j.enconman.2024.118941.)*

Fig. 4.2 Representation of some promising applications of the IEC (Sajjad et al., 2021). *(From Sajjad, U., Abbas, N., Hamid, K., Abbas, S., Hussain, I., Ammar, S.M., Sultan, M., Ali, H.M., Hussain, M., Tauseef-ur-Rehman, Wang, C.C., 2021. A review of recent advances in indirect evaporative cooling technology. Int. Commun. Heat Mass Transf. 0735-1933. 122, 105140. doi:10.1016/j.icheatmasstransfer.2021.105140.)*

experimentally, analytically, and numerically by the researchers to analyze the thermodynamic performance (Alam et al., 2024; Sajjad et al., 2021). Its performance depends upon the ambient conditions, operating principle, flow configuration, geometric size, and development (etc.). The generalized performance parameters such as wet-bulb effectiveness, cooling capacity, and coefficient of performance (COP) are described in Eqs. (1–3) (Jamil et al., 2023; Xu et al., 2016). The optimized IEC design and associated performance compel to perform the simulation by considering different parameters instead of spending time on costly experimental prototypes. The computational fluid dynamics (CFD) tools such as ANSYS FLUENT and COMSOL are being widely used for simulation of IEC (Alam et al., 2024; Güzelel et al., 2024; Ma et al., 2023). Therefore, the CFD simulation of a humidifier-assisted novel indirect evaporative cooler (NIEC) is carried out in this study. The continuous phase CFD simulation of the experimental NIEC system (Jamil et al., 2023) is not reported in the reputed literature as per the authors' understanding.

$$\text{Wet bulb effectiveness} = \frac{T_{AA} - T_{SA}}{T_{AA} - T_{AA,wb}} \tag{1}$$

$$\text{Cooling performance} = m_{AA} c_p (T_{AA} - T_{SA}) \tag{2}$$

$$\text{Coefficient of performance} = \frac{\text{Cooling Performance}}{Ш_{FN} - Ш_{PP}} \tag{3}$$

where T, c_p, m, and Ш represent the temperature, specific heat, mass flow rate, and power, respectively. The subscripts *AA*, *SA*, *wb*, *FN*, and *PP* denote the ambient air, supply air, wet bulb, fan, and pump, respectively.

The CFD simulation of NIEC is performed and validated against the experimental data. The mesh independency test on different grid sizes is conducted for precise and stable simulation results. The effects of varying ambient air temperature, velocity, and channel gap on supply air temperature, pressure drop, and performance are investigated.

4.2 Methodology

4.2.1 Novel indirect evaporative cooler

A conventional IEC consists mainly of a heat exchanger core, fans, a water pump, nozzles, and a storage tank, as shown in Fig. 4.3A. The heat exchanger core contains a series of dry and wet channels. The dry and

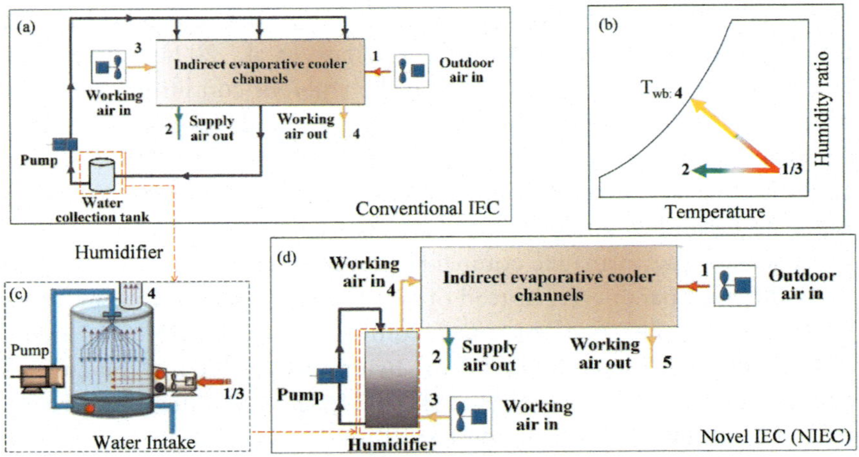

Fig. 4.3 Schematics of (A, B) conventional IEC and its psychrometric working, (C) Retrofitting of the humidifier, (D) Humidifier assisted Novel IEC, edited from (Jamil et al., 2023, 2024). *(From Jamil, M.A., Imtiaz, N., Ng, K.C., Xu, B.B., Yaqoob, H., Sultan, M., Shahzad, M.W., 2023. Experimental and parametric sensitivity analysis of a novel indirect evaporative cooler for greener cooling. Therm. Sci. Eng. Prog. 2451-9049. 42. Available from: https://www.journals.elsevier.com/thermal-science-and-engineering-progress. doi:10.1016/j.tsep.2023.101887.)*

wet channels are made/coated with hydrophilic and hydrophobic materials, respectively. The hydrophobic materials avoid the entry of water vapors in the dry channel, while the hydrophilic materials enhance the wicking of the wet channels. The water is accordingly supplied to wet channels from the storage tank by pump, and it remains recirculating for continuous wetting. Referring to the conventional IEC and corresponding psychrometric representation (Fig. 4.3A and B), the ambient air (1/3) simultaneously passes through the dry and wet channels. The product air in the dry channel after rejecting heat to the adjacent wet becomes sensibly cooled supply air (Güzelel et al., 2024). However, working air in the wet channel exits (Jamil et al., 2024) after accommodating both the sensible and latent heat loads.

The NIEC works on the conception of directly supplying the isenthalpic cooled working air to the wet channel. The NIEC was developed by Jamil et al. (2023) with the retrofitting of an external humidifier to the aluminum plate heat exchanger. The heat exchanger core of NIEC contains product (dry) and working (wet) channels. The working channels are not made or coated with any hydrophilic material as compared to wet channels of conventional IEC. Moreover, these working channels (NIEC) perform well without any issues related to microbial fouling, material degradation, and coating disintegration (etc.) as mostly encountered in the conventional IEC. The schematic describing the working of the humidifier and the NIEC is shown in Fig. 4.3C and D (Jamil et al., 2023, 2024). The ambient air (Alam et al., 2024) directly enters the product (dry) channel while the same outdoor air (IEA, 2018) passes through the humidifier in contrast to the conventional IEC. The humidifier externally sprays the water on the working air (IEA, 2018) and supplies the adiabatically cooled working air (Jamil et al., 2024) to the wet-bulb temperature of the ambient air (Alam et al., 2024). The cooled air (Jamil et al., 2024) then enters the working channel and counterflow to product channel air (Alam et al., 2024). The simultaneous heat and mass transfer in the working channel provides the sensible cooled supply air (Güzelel et al., 2024) at the exit of the product channel. The working air (Jamil et al., 2023) is ultimately out to the atmosphere, as shown in Fig. 4.3C (Jamil et al., 2023).

4.2.2 CFD simulation of NIEC

ANSYS FLUENT is used to simulate the performance of the NIEC heat exchanger to understand the heat & mass transfer in the product (dry) and working (wet) channels. FLUENT is a broadly used CFD tool for numerical simulations of thermal systems. It works on the concept of discretizing governing equations by applying the finite volume method. FLUENT offers steady-state and transient solvers to simulate the combined processes of fluid flow, species transport, evaporation, condensation, heat and mass transfer (etc.). Based on CFD, every conservation law governing the flow and transport phenomena containing mass, momentum, energy, and species attentiveness is able to show a unified mathematical structure known as the general transport equation. The general transport equation for steady-state flow used in the study is mentioned as Eq. (4) (Versteeg and Malalasekera, 2007). Accordingly, the governing equations (Eqs. 5–10) (Versteeg and Malalasekera, 2007; Riaz et al., 2023) demonstrate the basic mass, momentum, and energy transfer phenomena in the channels of the NIEC heat exchanger. Energy and mass conservation equations described the combined heat and moisture interchange in the channels. Moreover, the energy equation for incompressible flow includes both conduction and convection relevant for the temperature difference through the domain.

$$\frac{\partial}{\partial x_j}\left(\rho \mathfrak{v}_j \sigma\right) = \frac{\partial}{\partial x_j}\left(\frac{\partial \sigma}{\partial x_j}\right) + S_t \quad (4)$$

$$\nabla.(\rho \mathfrak{v}) = 0 \quad (5)$$

$$\text{Mass conservation, cartesian cordinates}: \frac{\partial(\rho u)}{\partial x} + \frac{\partial(\rho v)}{\partial y} + = 0 \quad (6)$$

$$\nabla.(\rho \mathfrak{v}.\mathfrak{v}) = \nabla.(\lambda) - \nabla p + \rho g \quad (7)$$

$$\text{Momentum conservation}, X: \rho\left(u\frac{\partial(u)}{\partial x} + v\frac{\partial(u)}{\partial y}\right) = \frac{\partial(-p + \lambda_{xx})}{\partial x} + \frac{\partial \lambda_{yx}}{\partial y} + \rho g_x \quad (8)$$

$$\text{Momentum conservation}, Y: \rho\left(u\frac{\partial(v)}{\partial x} + v\frac{\partial(v)}{\partial y}\right) = \frac{\partial \lambda_{xx}}{\partial x} + \frac{\partial(-p + \lambda_{yx})}{\partial y} + \rho g_y \quad (9)$$

$$\text{Energy conservation}: \nabla((p + \rho\varepsilon)\mathfrak{v}) = \nabla\left[-\sum_i j_i h_i + \mathfrak{v}.\lambda_{ef} + \nabla T k_{ef}\right] \quad (10)$$

where, σ represents the dissipation function, $\mathfrak{v}/u/v$ denotes the velocities in vector form or in the respective direction. λ represents the stress and its components are symbolized as λ_{xx} and λ_{yx}. Also, ρ, S_t, p, g, λ_{ef}, ε, k_{ef}, h_i, j_i, and T are the fluid density, diffusion coefficient, source term, static pressure, gravitational forces, stress effectiveness, energy per unit mass, thermal conductivity, specific enthalpy, diffusion flux, and temperature, respectively (Riaz et al., 2023).

4.2.2.1 Model description of NIEC

The NIEC heat exchanger consists of three channels, as one product channel is sandwiched between two working channels. The arrangement of these channels is presented by the schematics shown in Fig. 4.4. All the channels are made of aluminum sheet without using any coating/hydrophobic material inside them. The length and gap of channels are taken as

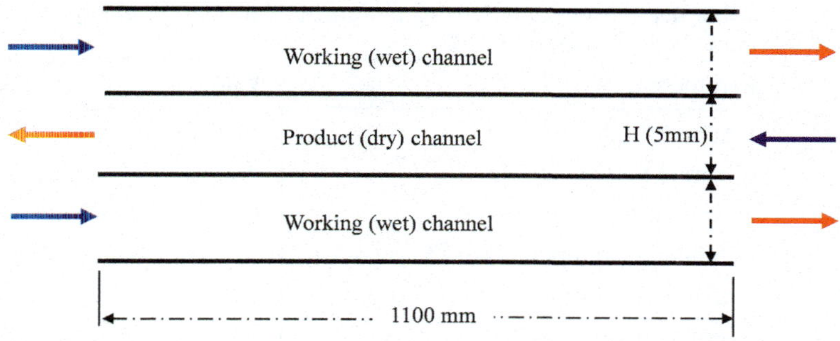

Fig. 4.4 Configuration of NIEC working and products channels for modeling in ANSYS.

1100 and 5 mm, respectively. The product and working channels are separated by the 0.025 mm thick wall of aluminum sheet. The middle and outside channels act as product and working channels, respectively. The outer walls of both working channels are considered insulated so there is no heat transfer from the outside/ambient conditions. The ambient air is directed to the product channel; however, the air flowing through the working channel is fully humid and cooled to the wet bulb temperature as it comes from the humidifier. The aluminum wall (0.025 mm thick) that separates the dry and wet channels allows the heat transfer between the channels. The counterflow of the product and working air enables the efficient heat and mass transfer in the channels. Therefore, the counterflow NIEC containing three aluminum channels of 1100 × 5 mm (length × gap/height) are developed in the design modeler of the ANSYS.

4.2.2.2 Simulation setup and validation

The boundary conditions are established for the developed geometry in order to simulate the NIEC heat exchanger. The product (dry) channel inlet is selected as velocity inlet, and thermal parameters like air temperature, velocity, and humidity are added. The working (wet) channel inlet is also designated as a velocity inlet for entering the air from the humidifier. The corresponding air temperature, velocity, and humidity are supplied at the inlet of the working channel. The air velocity is considered the same at the inlet of both the product and working channels. Heat transfer was made possible by the aluminum wall (0.025 mm thick) that separated the wet and dry channels. The aluminum wall is set up as a coupled wall with nonslippery conditions. On the air exit side, both channel outlets are configured as pressure outlets. The preset conditions for the initial

Table 4.1 Preset conditions to perform the simulation along with results validation of NIEC.

Parameters	Value
Channel length	1100 mm
Channel height/gap (H)	5 mm
Aluminum wall thickness	0.025 mm
Ambient air temperature (T_{AA})	37°C–43°C
Ambient air humidity ratio (X_{AA})	10–11 g/kg
Ambient air velocity (V_{in})	1.7 m/s

simulation and validation are presented in Table 4.1. Therefore, the simulations have been performed in steady-state conditions with the pressure-based solver setup in FLUENT. Energy and species transport capture the heat, mass, and moisture transfer between channels. The flow in channels is considered laminar, and the simple procedure was used for pressure and velocity coupling. The second-order upwind equation is used for the discretization of the momentum, energy, and species energy equations. Accordingly, some of the assumptions followed in the study to simplify the model are given below.

- The outer walls of the wet channels are considered insulated as there is no heat transfer from the outside/ambient air.
- The ambient air at certain temperatures and humidity enters the product channel while the humidifier exit air is supplied to the working channel.
- The air cooled to the wet-bulb temperature of the ambient air enters the working channel.
- The flow in both channels is considered steady-state and incompressible Newtonian fluid.

A mesh/grid independence test is normally performed to verify simulation accuracy. The grid size can be varied as fine, medium or large. Therefore, it can be selected based on the required simulation accuracy in the stipulated time frame. Fine mesh/grid size is generally considered for precise simulation results and validation. In this study, a mesh independence test is carried out by changing the element size from 2.0 to 0.25 mm with varying intervals. The quadrilateral structure with the same element size throughout the channels is considered for the development of a fine mesh as shown in Fig. 4.5A, and the geometry is represented by Fig. 4.5B. The multiple meshes are generated to precisely quantify the supply air

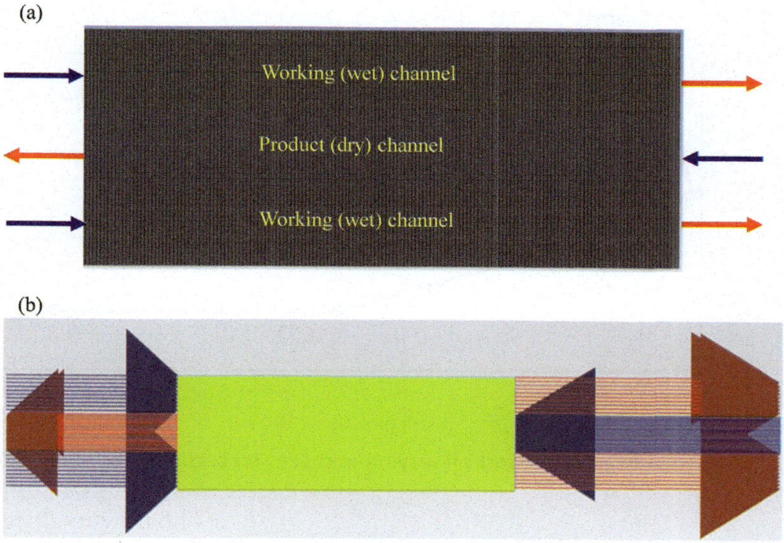

Fig. 4.5 Schematics of (A) mesh, and (B) geometry of NIEC developed in ANSYS.

temperature. Therefore, the test is completed as the simulated supply air temperature approaches the experimental one at the given mesh size.

The simulation model used in the study is validated against the experimental supply air temperature. The same experimental geometry and air conditions (Jamil et al., 2023) are used to simulate the model. The model is considered validated as the relative error between simulation and experimental result fall within ±5%. The same model is therefore used for further simulation by changing the ambient air temperature, velocity, and gap of the product channel.

4.3 Results and discussion

The effect of varying ambient air temperature, velocity, and channel gap on the supply air temperature and pressure drop is simulated for the counterflow NIEC. The grid independence test is performed to verify that the simulation results will not vary promisingly by changing the mesh numbers. The different numbers of meshes are generated at grid sizes 2.0–0.25 mm. A relatively fine mesh is generated throughout the channel at a grid size of 0.25 mm (Fig. 4.5). The supply air temperature (T_{SA}) is simulated as 25.20°C at a relatively larger mesh of 13,200–22,020 as shown in Fig. 4.6. It further increases to 25.40°C by increasing the mesh to 49,500. Therefore,

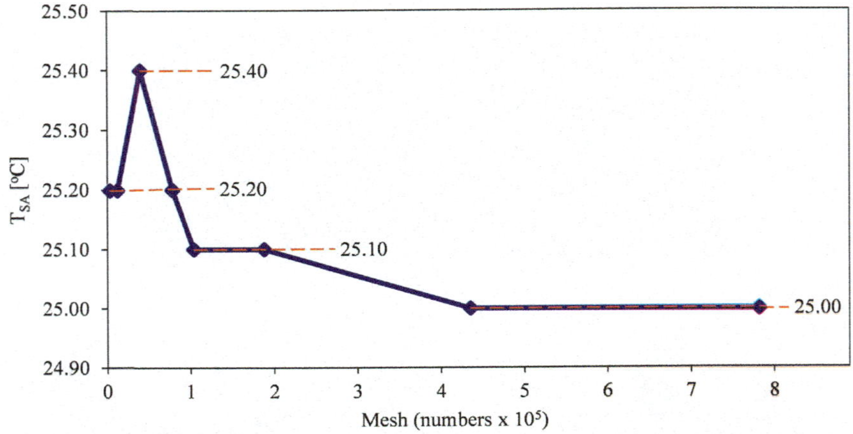

Fig. 4.6 Mesh independence test on varying grid sizes for accurate and stable simulation results.

grid size is reduced to produce higher numbers of mesh for stable simulated T_{SA}. However, it is interesting to note that the supply air temperature again falls to 25.20°C and remains at 25.10°C at the medium mesh numbers of 87,960 and up to 198,000, respectively. Finally, the fine mesh (445,500–792,000) created at a grid size of about 0.3–0.25 simulated the stable supply air temperature (25°C) corresponding to the experimental T_{SA}.

The developed simulation model of the NIEC is first validated against the experimental data before performing the further analyses. The validation of the simulation model is initially performed under the identical air conditions used in the experimental study (Jamil et al., 2023). In this regard, the simulation is performed at fine mesh numbers (445,500–792,000) under the ambient air temperature (T_{AA}) of 37.29°C, 40.23°C, 43°C and humidity ratio (X_{AA}) of 10–11 g/kg. The inlet air velocity (V_{in}) to product and working channels is taken to be the same as 1.7 m/s for the validation. The mesh size is chosen accordingly to keep the balance between model accuracy and simulation time. The supply air temperature is simulated to be about 1.15°C higher than the experimental T_{SA} at $T_{AA} \approx 37$°C as shown in Fig. 4.7. However, the difference between the simulated and experimental supply temperature decreases as the temperature increases from 37°C to 43°C, and it approaches less than ±0.2% at elevated T_{AA}. The maximum relative difference in both the temperatures falls within the ±5%; therefore, same model is further used to simulate the supply air temperature at varying T_{AA}, V_{in}, and product channel gap.

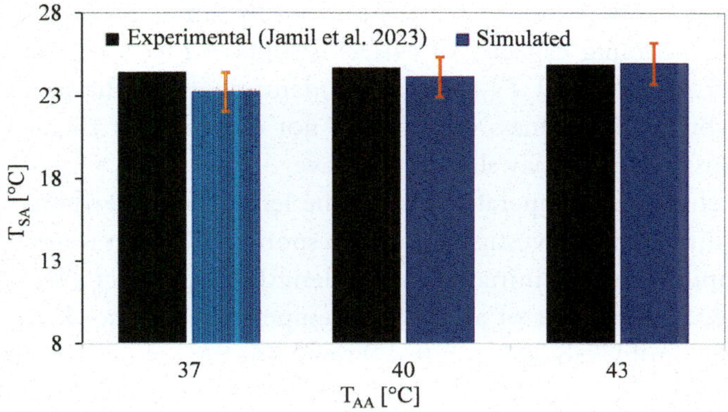

Fig. 4.7 Validation of the simulation results against the experimental data.

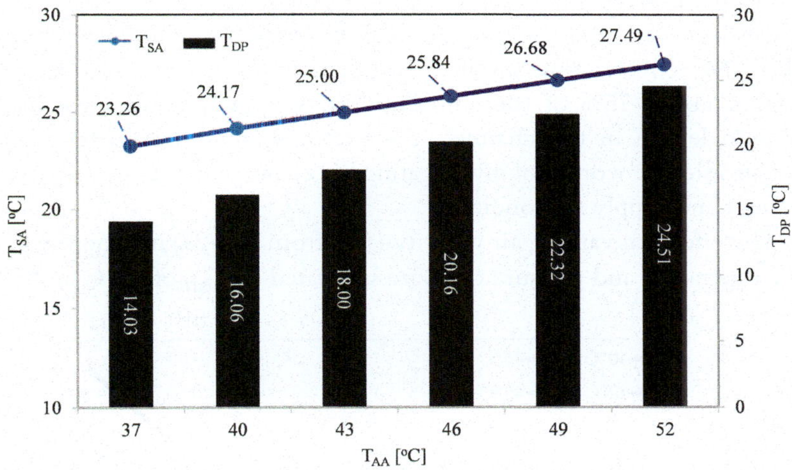

Fig. 4.8 Effects of ambient air temperature on the simulated supply air temperature and temperature drop.

The effects of elevated ambient air temperature at the same velocity and humidity ratio on the supply air temperature are simulated and compared with validated results. The temperature drop (T_{DP}) is also determined as the difference between T_{AA} and T_{SA}. In this regard, T_{AA} is varied as 46–49–52, but the humidity ratio (X_{AA}: 10–11 g/kg) and velocity (V_{in}: 1.7 m/s) remain the same as taken in the validation results. The simulated and experimental T_{DP} are found almost equal (18°C) at $T_{AA} \approx 43$°C as shown in Fig. 4.8. However, the NIEC supply air temperature increases to 25.84°C–27.49°C with a corresponding increase in T_{DP} of

20.16°C–24.51°C by varying the ambient air temperature from 46°C to 52°C. The cooling capacity of NIEC is supposed to be increased due to higher T_{DP} at elevated T_{AA} and other identical air conditions. Moreover, the wet-bulb effectiveness of NIEC is not decreased at the simulated air conditions; rather, it may slightly improve.

The change in temperature across the length of the product channel is further simulated to investigate the corresponding T_{DP}. It is noted that T_{AA} drops rapidly in the initial 200 mm length of product channel (T_{PC}); therefore, a decrement of about 5°C temperature is accordingly found at 52°C. The profiles shown in Fig. 4.9 represent the temperature drop in the center of the product channel gap (2.5 mm) across its length (1100 mm). The temperature drop profile at T_{AA} of 49°C is not shown in Fig. 4.9 because of its alike trend in between 46°C and 52°C. The temperature decline interval decreases at lower temperatures (40°C–49°C), but its trend remains the same due to counter air flow in the working channel of the NIEC. Moreover, all the simulated ambient air temperatures (40°C–52°C) drop to about 75% of their product air temperature after passing the 600 mm length in the channel. Therefore, such simulated temperature drop profiles are worthy of investigating the channel length under particular ambient and supply air conditions.

The effects of varying air velocity (V_{in}) from 0.8–2.0 m/s on the supply air temperature and pressure drop are simulated at T_{AA} of 43°C and 52°C.

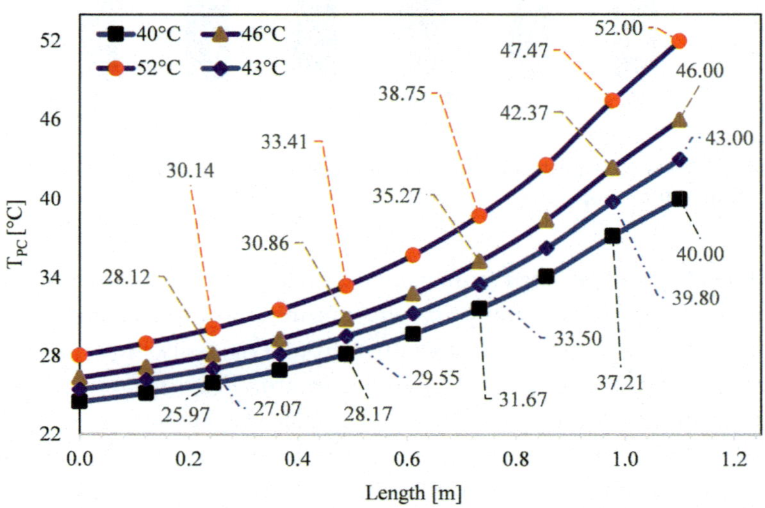

Fig. 4.9 Profiles represent the temperature drop in the center of the product channel across the length.

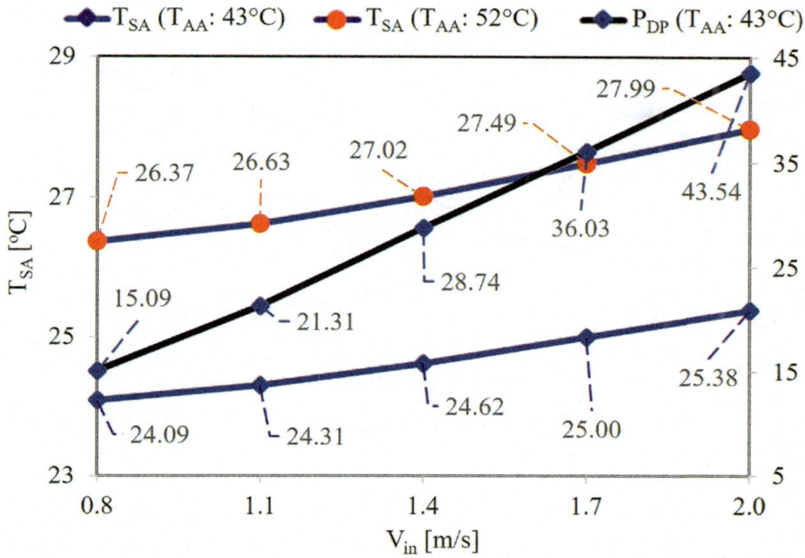

Fig. 4.10 Influence of varying air velocity on the simulated supply air temperature and pressure drop.

The supply air temperature lowers by decreasing the V_{in} from the reference 1.7 m/s and vice versa, as shown in Fig. 4.10. However, a temperature decrement of only about 1.0°C is found by lowering the V_{in} from 1.7 to 0.8 m/s at a T_{AA} of 43°C–52°C. The increased residing time results in higher evaporation and consequently decreases the T_{SA}. It is therefore anticipated that the corresponding wet-bulb effectiveness of NIEC will increase due to higher T_{DP}. However, the cooling capacity will be lowered due to a decrease in air mass flow rate. Moreover, higher cooling capacity can be achieved with increased V_{in}. On the other hand, pressure drop is obviously increased from 15 to 43 Pa by increasing the air velocity from 0.8 to 2.0 m/s.

The geometry of NIEC (length and height/gap) effects the heat and mass transfer between the channels. It is due to a change in the air flow state and/or flow rate by varying the channel length and/or channel height/gap. The height/gap of the product channel is varied in the study from 2.5 to 10.0 mm to simulate its influence on T_{SA} and P_{DP} as shown in Fig. 4.11. The supply air temperature increases, but the pressure decreases as the channel gap is increased from the reference 5.0 mm. The change is that both the T_{SA} and P_{DP} remain reasonable at increasing the channel gap. However, the pressure drop increases abruptly as the

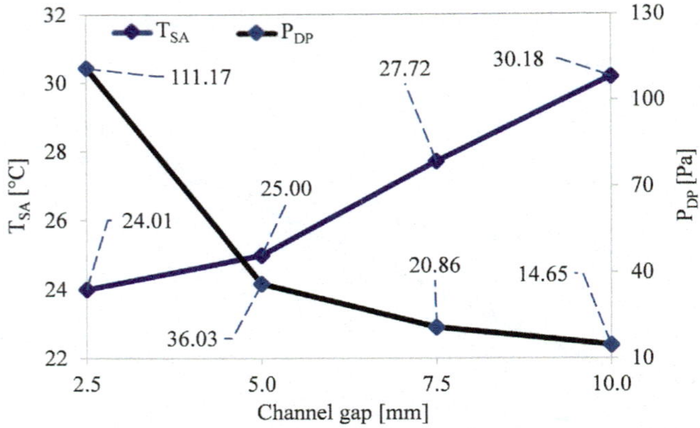

Fig. 4.11 Effects of varying height/gap of the product channel on the simulation results.

channel gap is reduced to 2.5 mm. The corresponding pressure drop is found to be three times higher, with only a 1.0°C temperature decrement, than 5.0 mm channel gap. The higher pressure drop increases the power consumption of the fan, which ultimately leads to the lower system COP. Therefore, the optimum cooling performance of NIEC can be achieved at a channel gap of 5.0 mm.

4.4 Conclusions

CFD simulation of a humidifier-assisted NIEC is carried out in this study. The effect of varying ambient air temperature, velocity, and channel gap on the supply air temperature and pressure drop is simulated for the counterflow NIEC. The grid independence test reveals that simulation results are not changing promisingly at a fine mesh of 0.25 mm grid size. The simulation model is validated at identical air conditions of the experimental data from the literature. The difference between the simulated and experimental supply decreases by increasing the temperature from 37°C to 43°C, and it approaches less than ±0.2% at 43°C. The change in temperature across the length of the product channel is also simulated to investigate the corresponding temperature drop. It is worthy to note that all the simulated ambient air temperatures (40°C–52°C) drop to about 75% of their product air temperature after passing the 600 mm length in the channel. The increasing air velocity and decreasing product channel gap simulated the higher pressure drop. The study highlights the importance of indirect evaporative cooling for the provision of thermal comfort conditions/environments in all applications. Therefore, the IEC design optimization and associated performance compel to perform the simulation by considering different parameters instead of spending time on costly experimental prototypes.

Acknowledgment

The United Kingdom Government funds this research (and/or) pilot activity through the Project, APP47457, titled "Super–efficient Sustainable Cooling Solution for All Applications (S2Cool)" under the Ayrton Challenge Programme of the UK Research and Innovation (UKRI); however, the views expressed do not necessarily reflect the UK Government's official policies. The Project, APP47457, is implemented/led by Northumbria University, UK, together with partners to develop sustainable cooling solution for several applications.

Funding statement

The United Kingdom Government funds this research (and/or) pilot activity through the Project, APP47457, titled "Super–efficient Sustainable Cooling Solution for All Applications (S2Cool)" under the Ayrton Challenge Programme of the UK Research and Innovation (UKRI); however, the views expressed do not necessarily reflect the UK Government's official policies. The Project, APP47457, is implemented/led by Northumbria University, UK, together with partners to develop sustainable cooling solution for several applications.

CRediT authorship contribution statement

Nida Imtiaz was involved in the conceptualization, methodology, software, formal analysis, investigation, writing of the original draft. **Mazlan Abdul Wahid, Muhammad Sultan, and Muhammad Wakil Shahzad** were involved in the conceptualization, methodology, validation, resources, visualization, supervision, project administration, writing (review and editing), funding acquisition. **Muhammad Zeshan and Muhammad Hamid Mahmood** were involved in the conceptualization, methodology, software, formal analysis, investigation, writing (review and editing).

Data availability statement

The datasets used and/or analyzed during the current study are available from the corresponding authors upon request.

Conflicts of interest

The authors declare no conflict of interest.

Declaration of generative AI and AI-assisted technologies in the writing process

No generative AI/AI-assisted technology is used in this paper.

References

Alam, M.S., Mohd Zubir, M.N.B., Muhamad, M.R.B., Kazi, S.N., Öztop, H.F., Abdullah, S., Shaikh, K., 2024. A technological review of dew point evaporative cooling: experimental, analytical, numerical and optimization perspectives. J. Build. Eng. 23527102. 91. Available from: https://www.sciencedirect.com/science/journal/23527102. doi:10.1016/j.jobe.2024.109544.

Güzelel, Y.E., Olmuş, U., Büyükalaca, O., 2024. Performance evaluation of different types of indirect evaporative coolers: a CFD-based comparative study. J. Build. Eng. 23527102. 96. Available from: https://www.sciencedirect.com/science/journal/23527102. doi:10.1016/j.jobe.2024.110399.

IEA, 2018. The future of cooling opportunities for energy- efficient air conditioning. Available from: https://www.iea.org/data-and-statistics/charts/global-air-conditioner-stock-1990-2050. (Accessed 18 November 2025).

Jamil, M.A., Ashraf, W.M., Imtiaz, N., Xu, B.B., Zubair, S.M., Yaqoob, H., Imran, M., Shahzad, M.W., 2024. Machine learning-based process design of a novel sustainable cooling system. Energy Convers. Manag. 01968904. 319. Available from: https://www.sciencedirect.com/science/journal/01968904. doi:10.1016/j.enconman.2024.118941.

Jamil, M.A., Imtiaz, N., Ng, K.C., Xu, B.B., Yaqoob, H., Sultan, M., Shahzad, M.W., 2023. Experimental and parametric sensitivity analysis of a novel indirect evaporative cooler for greener cooling. Therm. Sci. Eng. Prog. 24519049. 42. Available from: https://www.journals.elsevier.com/thermal-science-and-engineering-progress. doi:10.1016/j.tsep.2023.101887.

Ma, X., Shi, W., Yang, H., 2023. Improving the performance of indirect evaporative cooler for energy recovery from the perspective of nozzle configuration: a CFD model analysis. J. Build. Eng. 23527102. 76. Available from: http://www.journals.elsevier.com/journal-of-building-engineering/. doi:10.1016/j.jobe.2023.107195.

Riaz, M., Mahmood, M.H., Ashraf, M.N., Sultan, M., Sajjad, U., Hamid, K., Farooq, M., Wang, F., 2023. Experiments and CFD simulation of an air-conditioned tractor cabin for thermal comfort of tractor operators in Pakistan. Heliyon. 24058440. 9 (12). Available from: https://www.sciencedirect.com/science/journal/24058440. doi:10.1016/j.heliyon.2023.e23038.

Sajjad, U., Abbas, N., Hamid, K., Abbas, S., Hussain, I., Ammar, S.M., Sultan, M., Ali, H. M., Hussain, M., Tauseef-ur-Rehman, Wang, C.C., 2021. A review of recent advances in indirect evaporative cooling technology. Int. Commun. Heat Mass Transf. 07351933. 122, 105140. doi:10.1016/j.icheatmasstransfer.2021.105140.

Versteeg, H.K., Malalasekera, W., 2007. An Introduction to Computational Fluid Dynamics, Pearson Education Limited.

Wu, K., Wang, S., Lin, J., Shao, Y., Gao, F., Chua, K.J., 2024. The enhanced dew-point evaporative cooling with a macro-roughened structure. Int. J. Heat Mass Transf. 00179310. 219, 124898. doi:10.1016/j.ijheatmasstransfer.2023.124898.

Xu, P., Ma, X., Diallo, T.M.O., Zhao, X., Fancey, K., Li, D., Chen, H., 2016. Numerical investigation of the energy performance of a guideless irregular heat and mass exchanger with corrugated heat transfer surface for dew point cooling. Energy. 03605442. 109, 803–817. Available from: www.elsevier.com/inca/publications/store/4/8/3/. doi:10.1016/j.energy.2016.05.062.

CHAPTER 5

Investigation of heat-driven desiccant air conditioning systems for greenhouse farming in Pakistan

Aneela Sarwar[a], Muhammad Sultan[a], Muhammad Aleem[a], Takahiko Miyazaki[b], Md Shamim Ahamed[c], Muhammad Wakil Shahzad[d], Jie Liu[e], Zhaoli Zhang[f], and Hafiz Muhammad Ali[g]

[a]Department of Agricultural Engineering, Bahauddin Zakariya University, Multan, Pakistan
[b]Department of Advanced Environmental Science and Engineering, Faculty of Engineering Sciences, Kyushu University, Kasuga, Fukuoka, Japan
[c]Department of Biological and Agricultural Engineering, University of California, Davis, CA, United States
[d]Department of Mechanical and Construction Engineering, Northumbria University, Newcastle Upon Tyne, United Kingdom
[e]College of Engineering, Huazhong Agricultural University, Wuhan, China
[f]School of Mechanical Engineering, Southwest Jiaotong University, Chengdu, China
[g]Department of Mechanical Engineering, King Fahd University of Petroleum and Minerals, Dhahran, Saudi Arabia

5.1 Introduction

The second goal associated with the Sustainable Development Goals (SDGs) is to eradicate hunger. The FAO assessment of food production indicates that a 60% increase is necessary by 2050 to keep up with the exponential growth of the world population (Gitz et al., 2016). According to the Global Report on the Food Crisis, people in underdeveloped and developing countries are experiencing increasing levels of acute food insecurity. There is a total of 281.6 million people worldwide with severe food insecurity (GNAFC, 2024).

Economic shocks, war and instability, and extreme weather are the primary contributors to food scarcity. In Pakistan, the number of people experiencing food insecurity is also increasing over time, with 4.7 million, 8.6 million, and 11.8 million recorded in 2021–2023, respectively (GNAFC, 2024). The food security phase classification for Pakistan is illustrated in Fig. 5.1, which shows that the Punjab province has the least food insecurity among the provinces (Ali et al., 2017). To secure

Sustainable Hybrid Cooling and Dehumidification Strategies for Controlled Environments
https://doi.org/10.1016/B978-0-443-48555-8.00011-6

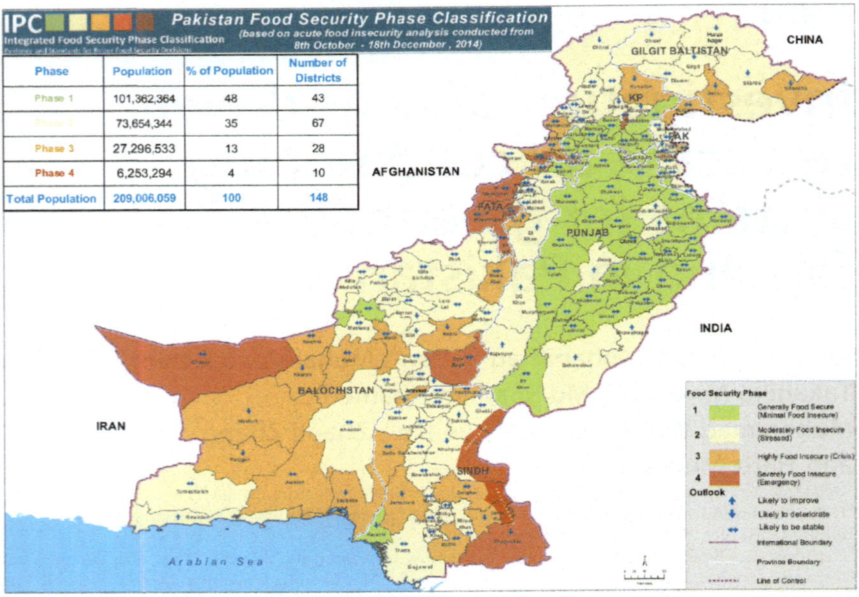

Fig. 5.1 Pakistan food security phase classification (Ali et al., 2017). *(From Ali, S., Liu, Y., Ishaq, M., Shah, T., Abdullah, Ilyas, A., Din, I., 2017. Climate change and its impact on the yield of major food crops: evidence from Pakistan. Foods. 2304-8158. 6 (6), 39. doi:10.3390/foods6060039.)*

the food supply of the rest of the country, now and in the future, it is necessary to take productive steps to address the impact of climate change. Greenhouse food production, or controlled environment agriculture (CEA), is among the most effective methods for meeting global food needs. Greenhouses can yield 15 times more per acre than traditional open field farming (Padmanabhan et al., 2016; Ahamed et al., 2023). Greenhouse farming is considered "protected" because it provides mechanical protection against external climatic conditions, including severe temperatures and humidity, while enabling greater control over CO_2 concentration. An experimental study reported a clear yield difference between capsicum production in open fields and in greenhouses Table 5.1.

Greenhouse agriculture is effective in areas with severe climate conditions, such as the Middle East, Gulf Cooperation Council, and North African countries (Maraveas et al., 2023). Arid environments are characterized by low precipitation and high evaporation rates, necessitating distinct humidity levels for optimal plant growth and productivity. The maximum average

Table 5.1 Comparison of capsicum yield in open fields versus greenhouses in the trans-Himalayan Ladakh region (Angmo et al., 2022).

Number of fruits and yield per plant	2017		2018	
	Open field	Greenhouse	Open field	Greenhouse
No. of marketable fruit	$1.9 \pm 0.9_{A}^{a}$	$5.9 \pm 1.3_{B}^{b}$	$1.9 \pm 0.4_{A}^{x}$	$6.8 \pm 0.5_{B}^{z}$
Marketable fruit weight (g)	$55 \pm 9_{A}^{a}$	$76.6 \pm 9.8_{B}^{b}$	$56.2 \pm 2.1_{A}^{x}$	$75 \pm 1.9_{B}^{y}$
Marketable yield (g)	$98.6 \pm 12.4_{A}^{a}$	$444.7 \pm 23.5_{D}^{c}$	$107.4 \pm 12.1_{A}^{x}$	$516.5 \pm 28.5_{E}^{z}$
No. of nonmarketable fruit	$0.8 \pm 0.2_{A}^{a}$	$0.7 \pm 0.2_{A}^{a}$	$0.9 \pm 0.1_{AB}^{y}$	$0.7 \pm 0_{A}^{x}$
Nonmarketable fruit weight (g)	$10.8 \pm 1.1_{A}^{a}$	$11.1 \pm 1.9_{A}^{a}$	$17.4 \pm 0.5_{A}^{z}$	$14 \pm 0.1_{A}^{y}$
Nonmarketable yield (g)	$8.9 \pm 2.2_{A}^{a}$	$7.6 \pm 3.3_{A}^{a}$	$15.5 \pm 2_{A}^{x}$	$9.9 \pm 0.3_{A}^{x}$

Note: Values expressed as mean ± SD In each row, distinct lowercase letters signify significant differences at $p \leq 0.05$, as determined by Tukey's test across the varying microclimate conditions. Angmo, P., Dolma, T., Phuntsog, N., Chaurasia, O.P., Stobdan, T., 2022. Effect of shading and high temperature amplitude on yield and phenolic contents of greenhouse capsicum (Capsicum annuum L.). Open Access Res. J. Biol. Pharm. 4, 30–39.

annual precipitation in these ecosystems varies from 150 to 450 mm, with periods of <50 mm persisting for 2 years (Angmo et al., 2022). Therefore, climate control strategies must be tailored to these conditions for optimal plant development and productivity.

Greenhouses, despite their high energy consumption, account for up to 25% of the operational costs of large vertical farms in the U.S.; improving crop yields in energy-efficient greenhouses has been recognized as a sustainable development goal for food production (Iddio et al., 2020). Adjusting temperature and humidity in the greenhouse is essential and can be achieved through ventilation or the application of covering materials (Kumar et al., 2009). Mechanical cooling systems regulate humidity; however, their efficacy is constrained by controlling optimal temperature and relative humidity (RH), which can adversely affect plant growth, induce fungal infestations, and cause dehydration (Sultan et al., 2016, 2021). Vapor pressure deficit (VPD) denotes the optimal temperature and RH for a plant, regardless of its developmental stage. It is the difference between the air's maximum capacity for water vapor retention and the current amount at a given temperature. Plants commence physiological degradation when they fail to satisfy the optimal VPD conditions (Sultan et al., 2016, 2021).

Elevated humidity levels in greenhouses are primarily due to plant transpiration, which is the principal source of moisture, despite the critical role of an appropriate growing environment in enhancing yield and quality (Wei et al., 2024). Over the last 50 years, the adoption of single- or double-layer polyethylene sheathing in greenhouses has enhanced moisture retention, leading to elevated relative humidity. This may result in plant loss owing to fungal infections, leaf necrosis, and calcium deficits, hence diminishing crop yield and quality (Campen et al., 2003; Körner and Challa, 2003). Dehumidification in greenhouses is crucial for optimal plant development. Multiple techniques are examined, encompassing conventional ventilation, cold-water condensation, hygroscopic dehumidification, and air-to-air heat exchange. Campen (Campen, 2009) posits that air-to-air heat exchangers represent the most competitive approach in cold regions. Nonetheless, little research has been conducted on the effectiveness of mechanical refrigeration dehumidifiers for moisture control in greenhouses, and no information is available on assessing dehumidification needs (Han, 2018). A moisture balance model is required to ascertain greenhouse dehumidification requirements. Plant evapotranspiration constitutes the principal source of moisture, whereas condensation on surfaces such as roofs and leaves, together with ventilation, infiltration, and dehumidification systems, facilitates moisture removal. Research on quantifying condensation rates in greenhouses and similar facilities is scarce. A cost-effective condensation detection system utilizing a leaf moisture sensor is proposed (Montross et al., 2006). A greenhouse experiment exhibited a system that accurately predicted the onset of condensation. However, no further investigations were conducted to determine the exact quantity of condensation. The observed condensation rate closely matched the computed rate, indicating that this model is appropriate for evaluating condensation on the inside surface of greenhouse glazing material (Han, 2018).

Energy-efficient dehumidification systems regulate humidity levels in greenhouses, with their capacity determined by the dehumidification load. Optimal capacity is crucial for cost-effective greenhouse production (Rahman et al., 2021). Several factors, including crop species, growth phase, duration, plant density, and greenhouse architecture, influence the efficacy of a dehumidifier. Projecting dehumidification needs in greenhouses requires the use of economic management systems that incorporate ventilation and renewable energy sources. Evaporative cooling (EC) and desiccant air-conditioning (DAC) provide economical and ecologically sustainable cooling solutions for many applications, including improving thermal comfort for humans and livestock, preserving agricultural goods, and supporting greenhouse operations. This chapter aims to provide a comprehensive overview of greenhouse dehumidification, including moisture generation and removal mechanisms, methods for estimating dehumidification load, and the potential of heat-driven desiccant cooling systems to maintain optimal temperature and humidity conditions and improve crop productivity.

5.2 Materials and methods

5.2.1 Description of a lab-scale greenhouse

A lab-scale greenhouse is designed and developed as shown in Fig. 5.2 at the Department of Agricultural Engineering, Bahauddin Zakariya University, Multan (Pakistan). The city is 215 m above sea level (Iqbal et al., 2019) and situated in a warm desert environment according to the Köppen-Geiger climate classification (Ashraf et al., 2021). The lab-scale greenhouse is an even-span structure with a total floor area of 4.46 m^2 and a volume of 9.18 m^3. The developed lab-scale unit has dimensions of 2.44 m (length),

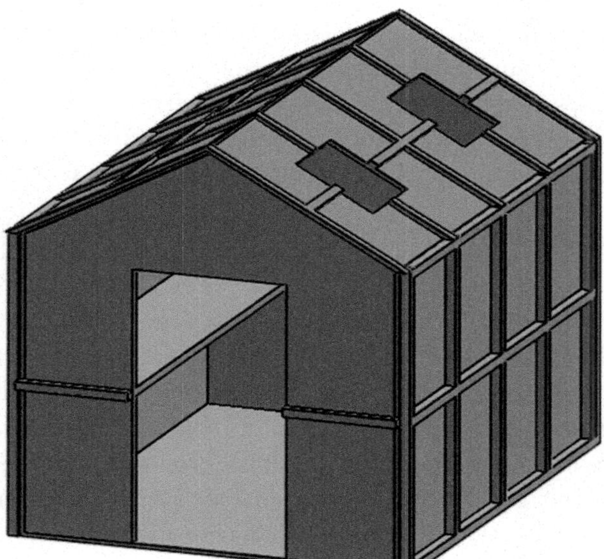

Fig. 5.2 Schematic diagram of lab-scale greenhouse located in Department of Agricultural Engineering, Bahauddin Zakariya University, Multan, Pakistan.

1.83 m (width), 1.83 m (height), and 0.46 m tapered height. Two exhaust fans with dash airflow rates are installed on the front and back walls.

5.2.2 Vapor pressure deficit

A vapor pressure deficit (VPD) is the difference between the maximum amount of water vapor that saturated air can hold at a given temperature and the actual water vapor content at that temperature. The VPD is a crucial environmental factor influencing photosynthesis, stomatal activity, and plant development in agriculture and horticulture. In the context of growing plants in greenhouses, AC systems and roof window opening and closing are used to regulate air temperature and RH (Inoue et al., 2021). Variations in VPD are a natural consequence of these processes. It is essential to clearly explain the specific effects of VPD variations on photosynthesis and plant development (Inoue et al., 2021). The greenhouse atmosphere increasingly accumulates water vapor as a result of these actions. The synthesis of carbohydrates in plants is primarily dependent on photosynthesis, a fact that remains significant despite the potential hazards posed by ongoing increase in humidity. These concerns encompass heightened vulnerability to pest invasions, fungal proliferation, insect infestations, and the potential for water vapor condensation and resulting leaks. Approximately 50% of the water vapor lost through evapotranspiration . Fig. 5.3 demonstrates the influence of VPD on the translocation of liquid from the earth's surface, through the plant and into the environment in vegetable crops. It also has an impact on plant behavior and leaf transpiration, as seen in Eq. (1) provided below (Yu et al., 2023). Accurate water management is a crucial advancement in greenhouse and indoor agriculture.

$$2nCO_2 + 4n(H_2O)\text{from roots} \rightarrow 2(CH_2O)_n + 2nO_2 + 2n(H_2O)\text{into air} \quad (1)$$

Apart from its impact on gas exchange, a deficient water environment also directly affects the efficiency of photosynthesis. This influence also affects the field of metabolic activities, particularly the Calvin-Benson cycle's enzymatic activity (Farquhar et al., 1989). As a result, the effects cascade into biomass output throughout the crop growth cycle. The stages of tomato plant growth are depicted in Fig. 5.4. The intricate interplay among photosynthetic efficiency, stomatal regulation, and water availability affects photosynthetic efficiency and plant growth. Plant-environment interactions are densely woven together by a network of interrelated elements that interact in this way.

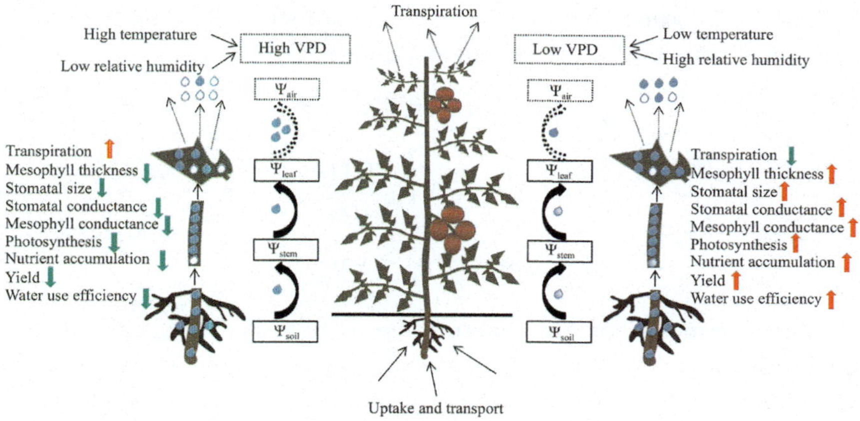

Fig. 5.3 Impact of higher and lower VPD on the plant. The passage of liquid water through the soil, stems, and leaves is indicated by solid black arrows, and the transport of gaseous water between the leaves and the air is indicated by dotted black arrows. In the text, Ψair, Ψleaf, Ψstem, and Ψsoil represent the water potential in the air, leaf, stem, and soil, respectively (Yu et al., 2023). *(From Yu, X., Zhang, Y., Zhao, X., Li, J., 2023. Systemic effects of the vapor pressure deficit on the physiology and productivity of protected vegetables. Veg. Res. 2769-0520. 3 (1). doi:10.48130/vr-2023-0020.)*

Fig. 5.4 Demonstration of the five growth stages of tomato, and the different levels of fruit ripeness (Shamshiri et al., 2018). *(From Shamshiri, R.R., Jones, J.W., Thorp, K.R., Ahmad, D., Man, H.C., Taheri, S., 2018. Review of optimum temperature, humidity, and vapour pressure deficit for microclimate evaluation and control in greenhouse cultivation of tomato: a review. Int. Agrophysics. 0236-8722. 32 (2), 287–302. Available from: http://www.degruyter.com/view/j/intag.2013.27.issue-1/issue-files/intag.2013.27.issue-1.xml. doi:10.1515/intag-2017-0005.)*

The VPD values for tomato, pea, cucumber, and strawberry were computed using Eqs. (2, 3) (Sultan et al., 2015, 2016).

$$\mathrm{VP_{sat}} = 610.7\ \exp^{\frac{7.5T}{237.3+T}} \tag{2}$$

$$\mathrm{VP_{air}} = \mathrm{VP_{sat}}\frac{\mathrm{RH}}{100} \tag{3}$$

$$\mathrm{VPD} = \mathrm{VP_{sat}} - \mathrm{VP_{air}} \tag{4}$$

The parameters of the equation are as follows: $\mathrm{VP_{sat}}$, or saturated environment vapor pressure (plant leaf), in kPa; vapor pressure of the greenhouse air is measured in kPa, C_g is a coefficient having value 610.7, and T is the leaf surface temperature recorded using an infrared gun. It was taken to be equivalent to the agricultural greenhouse air temperature for simplification; this value was taken to be equivalent to the agricultural greenhouse air temperature. The air relative humidity is given as RH, and the greenhouse's VPD is measured in kPa.

5.2.3 Greenhouse dehumidification requirement modeling

The principal source of moisture in greenhouses is plant transpiration. Supplementary moisture from evaporation of the growth medium and soil or floor surfaces; however, this is negligible compared with plant transpiration. The combination of these two sources is termed evapotranspiration. Moisture extraction occurs by ventilation, condensation, and dehumidification methods as given by Eq. (5) (Costantino et al., 2021). Ventilation may also bring moisture if the external air humidity exceeds that within the greenhouse. Condensation forms on the inside surface of the cover when the ambient temperature is low, extracting moisture from the air and releasing latent heat.

$$E_{\mathrm{dh_{req}}}(t) = E_{\mathrm{p}}(t) - E_{\mathrm{c}}(t) - E_{\mathrm{v}}(t) \tag{5}$$

where, $E_{\mathrm{dh_{req}}}(t)$ represent the dehumidification demand at time t in kg $\mathrm{h^{-1}\ m^{-2}}$, representing the amount of moisture that needs to be extracted from the dehumidifier; $E_{\mathrm{p}}(t)$, in kg $\mathrm{h^{-1}\ m^{-2}}$, indicates the moisture introduced to the greenhouse by evapotranspiration at time t; $E_{\mathrm{c}}(t)$, kg $\mathrm{h^{-1}\ m^{-2}}$, is the moisture added or removed to the air by the condensation process at time t, $E_{\mathrm{v}}(t)$, kg $\mathrm{h^{-1}\ m^{-2}}$, signifies the moisture removal or gain resulting from the exchange of air at time t; all variables are elaborated upon in detail as follows.

5.2.3.1 Evapotranspiration

Villarreal-Guerrero et al. (2012) examined greenhouse evapotranspiration rate prediction models, including those by Stanghellini (1987), Penman (1948), Monteith (1965), and Takakura et al. (2005). It was determined that Stanghellini's model exhibits superior overall performance and is utilized in the moisture balance model created by Han (2018). Moisture transpiration by plants constitutes majority of the water absorbed by the roots (Rosenberg et al., 1983). Evapotranspiration, the combined effect of transpiration and evaporation, can be determined using Eq. (6) (Stanghellini, 1987).

$$E_p = \frac{\Delta R_n + \left(2.\mathrm{LAI}.\rho_i.\frac{C_p}{r_a}\right) \times (e_s - e_i)}{\lambda.\left(\Delta + \gamma.\left(1 + \frac{r_c}{r_a}\right)\right)} \tag{6}$$

where, Δ is the slope of the water saturation vapor pressure curve in kPa K^{-1}; R_n is the net solar radiation inside the greenhouse in Wm^{-2}; when $R_n = 0$, E_p represents nighttime evapotranspiration; LAI is the plant leaf area index (dimensionless) i.e. the area of a single-sided green leaf relative to the space of the ground surface; ρi is the air density in kg m^{-3}; C_p is the specific heat of the water in J kg^{-1} K^{-1}; ra is the aerodynamic resistance, determined by the characteristic lengths of the leaves and the velocity of the inside air.in s m^{-1}; es is air water vapor pressure at saturation in kPa; ei is the indoor air water vapor pressure in kPa; λ is the latent heat of water vaporization in kJ kg^{-1}; γ is the psychrometric constant in kPa K^{-1}; and r_c is the internal canopy resistance to the water vapor diffusion from leaves in s m^{-1}.

The net solar irradiance inside the greenhouse (R_n) can be calculated using Eq. (7) (Villarreal-Guerrero et al., 2012).

$$R_n = 0.86 \times (1 - \exp(1 - 0.7 \times \mathrm{LAI})) \times I_s \tag{7}$$

where, I_s is the ambient solar irradiance (W m^{-2}). The internal canopy resistance (r_c) can be calculated by Eq. (8).

$$r_c = r_s\,\mathrm{LAI} \tag{8}$$

where, r_s is the stomatal resistance which is also solar irradiance dependent and computed by Eq. (9).

$$r_s = 200 \times \left(1 + \frac{1}{\exp^{(0.05 \times (I_s - 50))}}\right) \tag{9}$$

The aerodynamic resistance can be calculated by Eq. (10).

$$r_a = 220 \times \left(\frac{l_f^{0.2}}{u_i^{0.8}} \right) \tag{10}$$

where, l_f is the leaf characteristic length, in *m*; and u_i is the air speed (m s^{-1}). The leaf characteristic length indicates both the size and shape of the leaf calculated by Eq. (11).

$$l_f = \frac{2}{\left(\frac{1}{l}\right) + \left(\frac{1}{wd}\right)} \tag{11}$$

5.2.3.2 Condensation

The condensation rate E_c is directly correlated with the difference between the indoor air's water vapor pressure and the saturation water vapor pressure at the temperature of the interior cover surface. The rate of condensation is calculated by Eq. (12) (De Freitas and Schmekal, 2003).

$$E_c = 3600 \times \frac{0.671.h_{ci}.(e_i - e_{sc})}{C_p.P} = \frac{2.4 \times 10^3.h_{ci}.(e_i - e_{sc})}{C_p.P} \tag{12}$$

where, E_c denotes the condensation rate kg h^{-1} m^{-2}; h_{ci} is the convective heat transfer coefficient at the interior cover surface of the greenhouse W m^{-2} k^{-1}; e_{sc} signifies saturation air water vapor pressure at the inner cover surface temperature (kPa); and P is the atmospheric pressure in kPa.

Empirical data from tomato greenhouse research (Han, 2018) allows prediction of the condensation rate based on the temperature differential between the inside air and the inner cover surface, solar irradiation, and the airspeed adjacent to the cover. The empirical correlations for condensation rates during daylight and evening are given by Eqs. (13, 14) (Han, 2018).

$$E_{c-Day} = 0.74 + 0.43.\Delta T + 0.02 I_s \tag{13}$$

$$E_{c-Night} = -0.29 + 85.51.u_c + 0.45c\Delta T \tag{14}$$

where E_{c-Day} is the inner cover surface condensation on the day and $E_{c-Night}$ is on the night, in kg h^{-1} m^{-2}; ΔT is the difference between indoor and inner cover surface temperature, in *K*; and u_c represents the airspeed near the inner cover surface, in m s^{-1}.

5.2.3.3 Ventilation and moisture exchange

The transfer of moisture by air exchange, including ventilation and infiltration, depends on the disparity in humidity ratios or partial vapor pressures between external and internal air. The ventilation system, regulated by room temperature, enables moisture extraction calculated by Eq. (15) (Han, 2018).

$$E_v = \frac{0.62.\rho_i.q_v}{A_g.P} \times (e_i - e_o) \tag{15}$$

where, E_v denotes the moisture removal rate via ventilation in kg h^{-1} m^{-2}; ρ_i is the indoor air density in kg m^{-3}; q_v denotes the airflow rate resulting from ventilation or infiltration (m^{-3} h^{-1}); e_i and e_o are the indoor and outdoor air water vapor partial pressures (kPa); Ag is the greenhouse floor area (m^2); and P represents the atmospheric pressure (kPa).

5.2.3.4 Dehumidification requirement prediction

The dehumidification requirement is predicted by using Eqs. (16–20) (Han, 2018; Rahman et al., 2021).

$$E_{dh_{req}}(t) = \left(\frac{\Delta R_n + \left(2.LAI.\rho_i.\frac{C_p}{r_a}\right) \times (e_s - e_i)}{\lambda.\left(\Delta + \gamma.\left(1 + \frac{r_c}{r_a}\right)\right)} \right) - (0.74 + 0.43.\Delta T + 0.02 I_s) - \left(\frac{0.62.\rho_i.q_v}{A_g.P} \times (e_i - e_o) \right) \tag{16}$$

To simplify the calculation, the above equation is arranged as:

$$E_{dh_{req}}(t) = A + B.e_s - (B + F).e_i + F.e_o - E_c \tag{17}$$

$$A = \frac{3.6 \times R_n}{\lambda.\left(\Delta + \gamma.\left(1 + \frac{r_c}{r_a}\right)\right)} \tag{18}$$

$$B = \frac{3.6 \times \left(2.LAI.\rho_i.\frac{C_p}{r_a}\right)}{\left(2.LAI.\rho_i.\frac{C_p}{r_a}\right)} \tag{19}$$

$$F = \frac{0.62.\rho_i.q_v}{A_g.P} \tag{20}$$

If the inside temperature is considered constant throughout the day and night, the dehumidification load for any given external temperature and RH can be calculated by Eqs. (21–23) (Han, 2018; Rahman et al., 2021).

$$E_{dh_{req}}(t) = -Me_i + C \tag{21}$$

$$M = B + F \tag{22}$$

$$C = A + Be_s + Fe_o - E_c \tag{23}$$

5.2.4 Proposed DAC systems

Referring to Fig. 5.5A, proposed DAC is an open cooling cycle that includes a dehumidification unit, a sensible heat exchanger, a Maisotsenko cycle evaporative cooling (MEC), and a heating unit. The standalone DAC system comprises the components listed above, except the MEC unit. The DAC+MEC system (Fig. 5.5B) simultaneously handles latent and sensible loads and provides dehumidified, cooled air for greenhouse farming. An additional noteworthy aspect of this technology this technology is that the low-temperature heat (about 50-80 °C) necessary for regeneration can be achieved by combining it with a solar thermal facility or by harnessing waste heat from industrial or cogeneration activities. For performance evaluation of the proposed systems, steady-state models have been utilized, namely, the Beccali model (Beccali et al., 2003, 2004) for dehumidification aspects and the effectiveness model of the MEC unit (Sultan et al., 2018; Aleem et al., 2022; Ishaq et al., 2022; Hussain et al., 2022; Sultan et al., 2022). The Beccali model is given by Eqs. (24–27), while the effectiveness model is given by Eqs. (28, 29).

$$X_2 = \frac{e^{0.053T_2}\left(0.9428\text{RH}_{6/7} + 0.0572\text{RH}_1\right) - 1.7976}{18.671} \tag{24}$$

$$\frac{\left(\text{RH}_2 e^{0.053T_2} - 1.7976\right)}{18671.0} = \frac{\left(h_2 - 1.006T_2\right)}{2501 - 1.805T_2} \tag{25}$$

$$\text{RH}_2 = \left(0.9428\text{RH}_{6/7} + 0.0572\text{RH}_1\right) \tag{26}$$

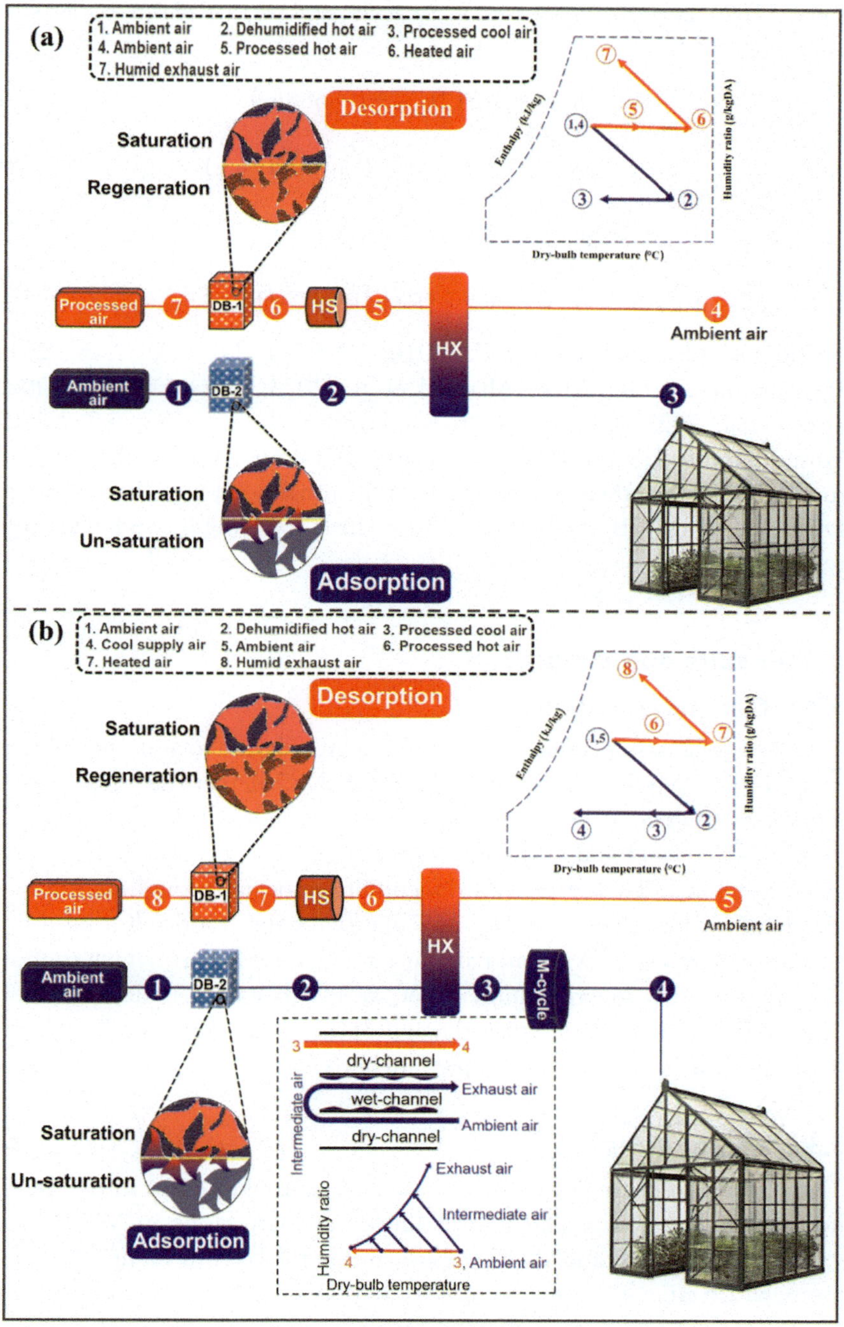

Fig. 5.5 Schematic representation of proposed DAC system (A) standalone DAC, and (B) Maisotsenko cycle-assisted DAC, reproduced from (Aleem et al., 2020). *(From Aleem, M., Hussain, G., Sultan, M., Miyazaki, T., Mahmood, M.H., Sabir, M.I., Nasir, A., Shabir, F., Khan, Z.M., 2020. Experimental investigation of desiccant dehumidification cooling system for climatic conditions of multan (Pakistan). Energies. 1996-1073. 13 (21). Available from: https://www.mdpi.com/1996-1073/13/21/5530. doi:10.3390/en13215530.)*

$$h_2 = \left(0.1312h_{6/7} + 0.8688h_1\right) \tag{27}$$

$$T_{DB,3} = T_{DB,2} - \varepsilon_{HX}\left(T_{DB,2} - T_{DB,1}\right) \tag{28}$$

$$T_{DB,4} = T_{DB,3} - \varepsilon_{wb}\left(T_{DB,3} - T_{WB,1}\right) \tag{29}$$

$$T_{DB,6} = T_{DB,5} - \varepsilon_{HX}\left(T_{DB,2} - T_{DB,5}\right) \tag{30}$$

where the subscripts (Gitz et al., 2016; GNAFC, 2024; Ali et al., 2017; Padmanabhan et al., 2016; Ahamed et al., 2023; Maraveas et al., 2023; Soussi et al., 2022) refer to air-state conditions as shown in Fig. 5.7. X is the humidity ratio (kg/kg), RH is the relative humidity (%), T is the temperature (°C), ε is the effectiveness, and h is the enthalpy (kJ/kg). The subscripts DB, HX, and WB refer to dry-bulb, heat exchanger, and wet-bulb, respectively.

5.3 Results and discussion

5.3.1 Climatic conditions

This study was conducted in an agricultural greenhouse to assess the climatic conditions in Multan, Pakistan. Figs. 5.6 and 5.7 depict the climatic conditions of Multan, Pakistan, as shown on a psychrometric chart. Fig. 5.6 illustrates that cold and arid conditions persist for over 90 h annually, but extremely hot and comparatively dry conditions occur throughout summer, albeit for a shorter duration. Fig. 5.7 illustrates the hourly distribution of climatic conditions in Multan, Pakistan, using an RH chart, which supports the assertion that severe summer (hot, arid) and winter (cold, arid) conditions prevail there. Consequently, appropriate management of the greenhouse must be prioritized to achieve higher yields.

5.3.2 VPD analysis

The evaluation of the VPD for tomato and cucumber crops in the greenhouse is being done. The figures illustrate fluctuations in VPD across various temperature and RH levels during both day and night. The ideal temperature range for tomato cultivation is 23°C–27°C during the day and 13°C–16°C at night, accompanied by an RH of 50%–80%. The analysis encompasses all temperature values and 5-point increments in RH. The VPD for the tomato crop is shown in Fig. 5.8. It can be seen that the trend of decreasing VPD values with rising RH holds irrespective of temperature

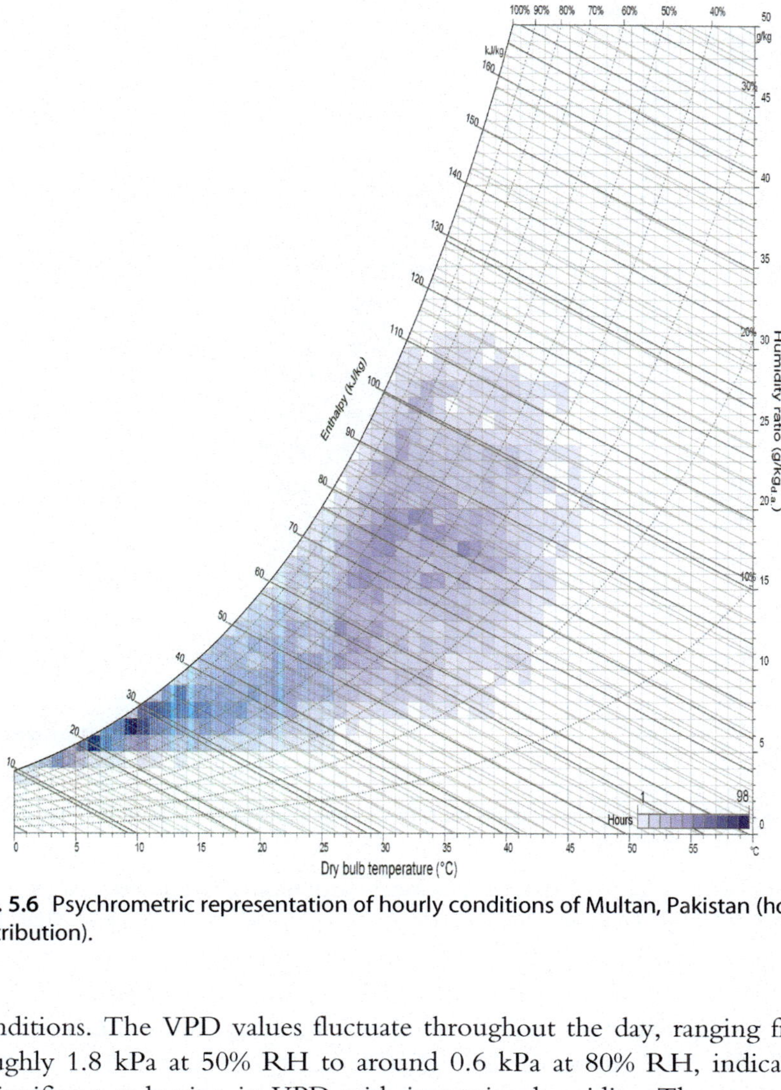

Fig. 5.6 Psychrometric representation of hourly conditions of Multan, Pakistan (hourly distribution).

conditions. The VPD values fluctuate throughout the day, ranging from roughly 1.8 kPa at 50% RH to around 0.6 kPa at 80% RH, indicating a significant reduction in VPD with increasing humidity. The nocturnal VPD measurements demonstrate a decline, ranging from around 0.8 kilopascals at 50% RH to about 0.4 kilopascals at 80% RH. The diverse temperature settings (up to 27°C during the day and 16°C at night) influence VPD measurements, with higher temperatures often resulting in higher VPD. The ideal temperature range for cucumber cultivation is 25°C–30°C during the daylight hours and 16-18°C at night, with a RH of 70%–90%. The analysis covers all temperature values and 5-point

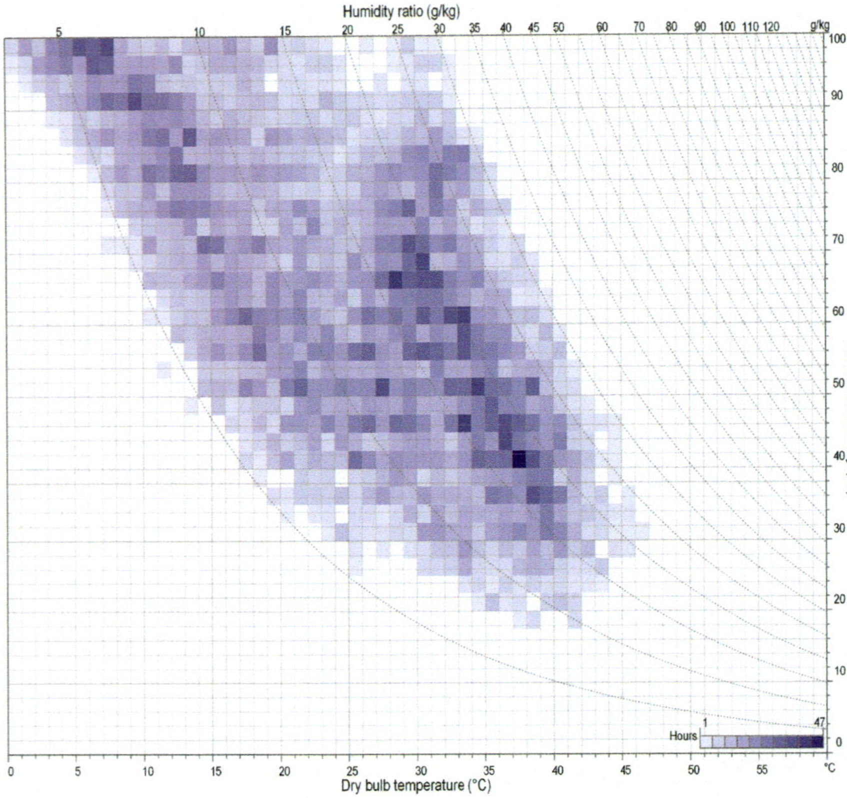

Fig. 5.7 RH chart of hourly distribution of climatic conditions of Multan, Pakistan.

intervals in RH. Similar observations are seen for the cucumber crop Fig. 5.9. The daytime VPD varies between around 1.2 kPa at 70% RH and around 0.3 kPa at 90% RH. The night-time VPD values decrease continuously, from approximately 0.6 kPa at 70% RH to around 0.2 kPa at 90% RH. The temperature range for cucumber cultivation is between 25°C and 29°C during the day and 16°C to 18°C at night. These temperature settings indicate that higher temperatures are associated with higher VPD.

The ideal temperature range for pea cultivation is 25-30°C during the day and 16-18°C at night, with a RH of 70%–80%. The analysis encompasses all temperature values and point intervals in RH. The VPD for the pea crop has a distinct pattern of diminishing VPD values as RH increases, regardless of temperature conditions, as shown in Fig. 5.10. Throughout the day, VPD varies from around 1.2 kPa at 70% RH to

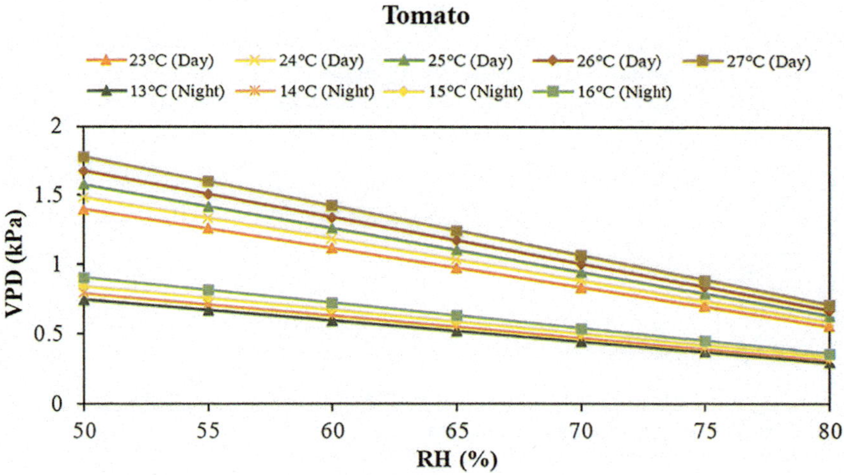

Fig. 5.8 VPD profiles for day and night for tomato.

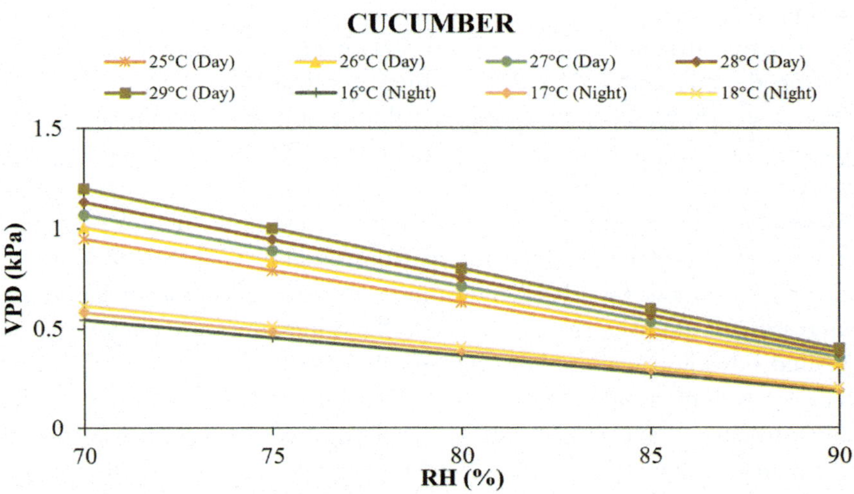

Fig. 5.9 VPD profiles for day and night for cucumber.

about 0.8 kPa at 80% RH, indicating a notable decrease in VPD as humidity increases. The VPD values are relatively low, ranging from about 0.6 kPa at 70% RH to 0.4 kPa at 80% RH. The varied temperature settings, spanning from 25°C to 30°C during the day and 16°C to 18°C at night, affect the VPD measurements. Generally, higher temperatures lead to an augmented VPD. The VPD decreases as RH increases and is affected by temperature

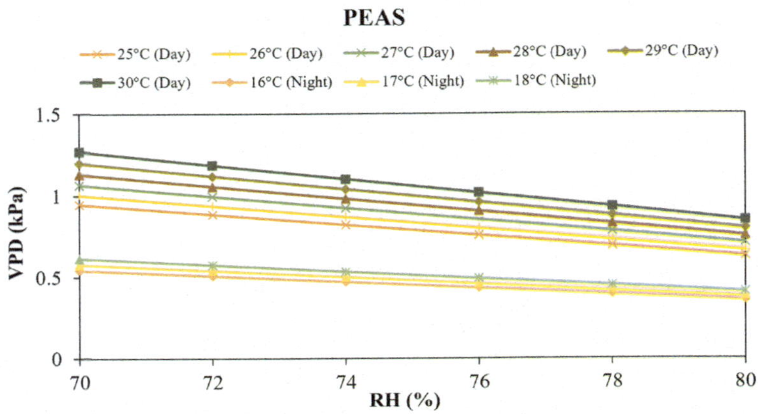

Fig. 5.10 VPD profiles for day and night for peas.

fluctuations. This persistent tendency underscores the essential need to maintain adequate VPD levels for various crops to ensure robust growth and productivity.

The optimal temperature range for strawberry production is 20°C–26°C during the day and 13°C–16°C at night, with a RH of 50%–65%. The analysis encompasses all temperature values and points of intervals in RH. Consistent with the preceding crop study, the VPD for the strawberry crop, shown in Fig. 5.11, rreveals a distinct tendency for VPD to decrease with increasing RH, irrespective of temperature conditions. The VPD values fluctuate throughout the day, ranging from roughly 1.7 kPa at 50% RH to around 0.8 kPa at 65% RH, indicating a notable reduction in VPD with increasing humidity. At night, the VPD values diminish, varying from around 0.9 kilopascals at 50% RH to about 0.5 kilopascals at 65% RH. This consistent trend underscores the critical need to sustain ideal VPD levels to ensure vigorous growth and yield of diverse crops. It is important to mentioned here that the ideal VPD levels to ensure optimal crop growth in the greenhouse varied with type of crops. This is because the optimal temperature and RH growth zones varied with crop type.

5.3.3 Greenhouse dehumidification requirements

Greenhouse dehumidicaiton requirements were evaluated for the summer months (June, July, and August). The Multan summer climatic data are presented in Fig. 5.12 which shows daily temperature (°C) and RH (%) for June, July, and August. Across the three months, the temperature remains

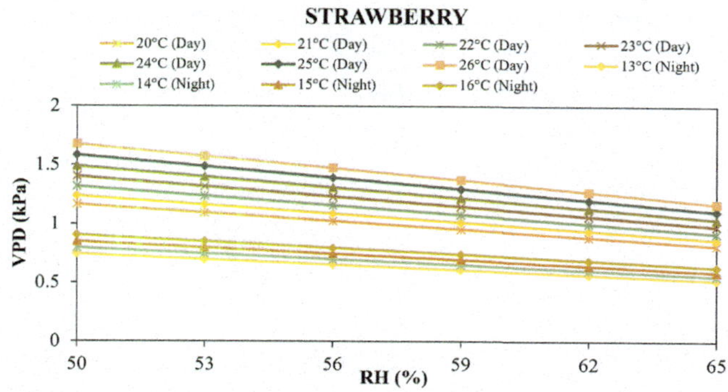

Fig. 5.11 VPD profiles for day and night for strawberry.

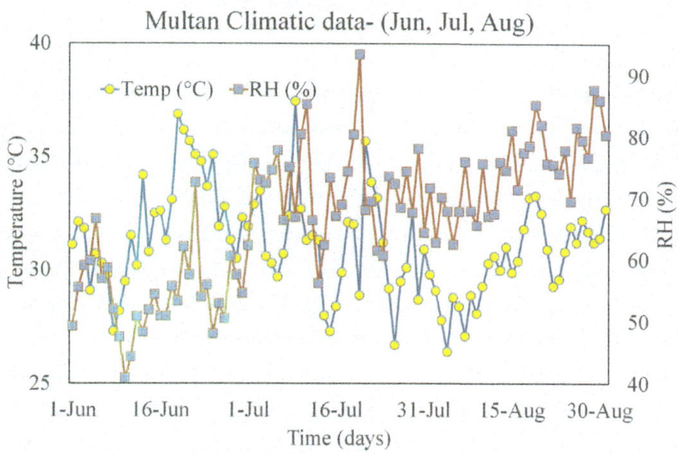

Fig. 5.12 Temperature and RH ranges in Multan during the summer. (June, July, and Aug).

relatively stable, hovering mostly between 30°C and 40°C. In contrast, the RH shows more variation, fluctuating between 40% and 90%, with notable peaks in July and August.

This data is crucial for modeling dehumidification requirements in greenhouses during the hot summer months. High temperatures and fluctuating humidity levels affect the internal greenhouse environment by increasing air moisture. The fluctuation in humidity, especially when coupled with elevated temperatures, underscores the need for dehumidification to sustain ideal conditions for plant development. The importance of this data lies in its ability to predict dehumidification load, as periods of high

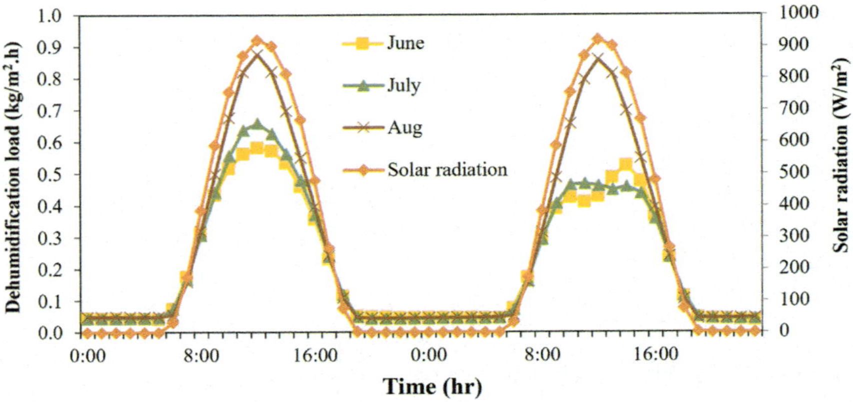

Fig. 5.13 Dehumidification requirements vs. solar radiation profiles in summer for two days of each month along with the solar radiation of Multan.

temperature and humidity require more intensive dehumidification to maintain the desired humidity and temperature levels in the greenhouse. The graph helps identify the critical days and periods when dehumidification systems will be under the most strain, especially during the humid peaks in July and August. Fig. 5.13 illustrates the relationship between dehumidification load and solar radiation over 24 hours on two days in June, July, and August. The x-axis denotes time of day; the left y-axis shows dehumidification load; and the right y-axis shows solar radiation. The solar radiation curve follows a consistent pattern across all three months, peaking around midday. The highest solar radiation intensity is observed in August, at approximately 950 W/m^2, while June and July show slightly lower values. The dehumidification load mirrors this pattern, increasing as solar radiation intensifies, but with a slight delay in its peak. During the day, as the greenhouse receives more sunlight, the need for dehumidification increases, particularly in the hotter months. August shows the highest dehumidification load, peaking at around 27 kg/h, reflecting the higher temperatures and solar radiation during this month. In contrast, June exhibits the lowest dehumidification load, consistent with its relatively lower solar radiation. Throughout the night, both solar radiation and dehumidification loads drop to near-zero levels, highlighting the reduced need for dehumidification when there is no solar heat input.

Fig. 5.14 shows the dehumidification requirements at different set points for each month. The dehumidification requirement follows a similar curve to solar irradiance but depends on the RH setpoints. When the RH setpoint is

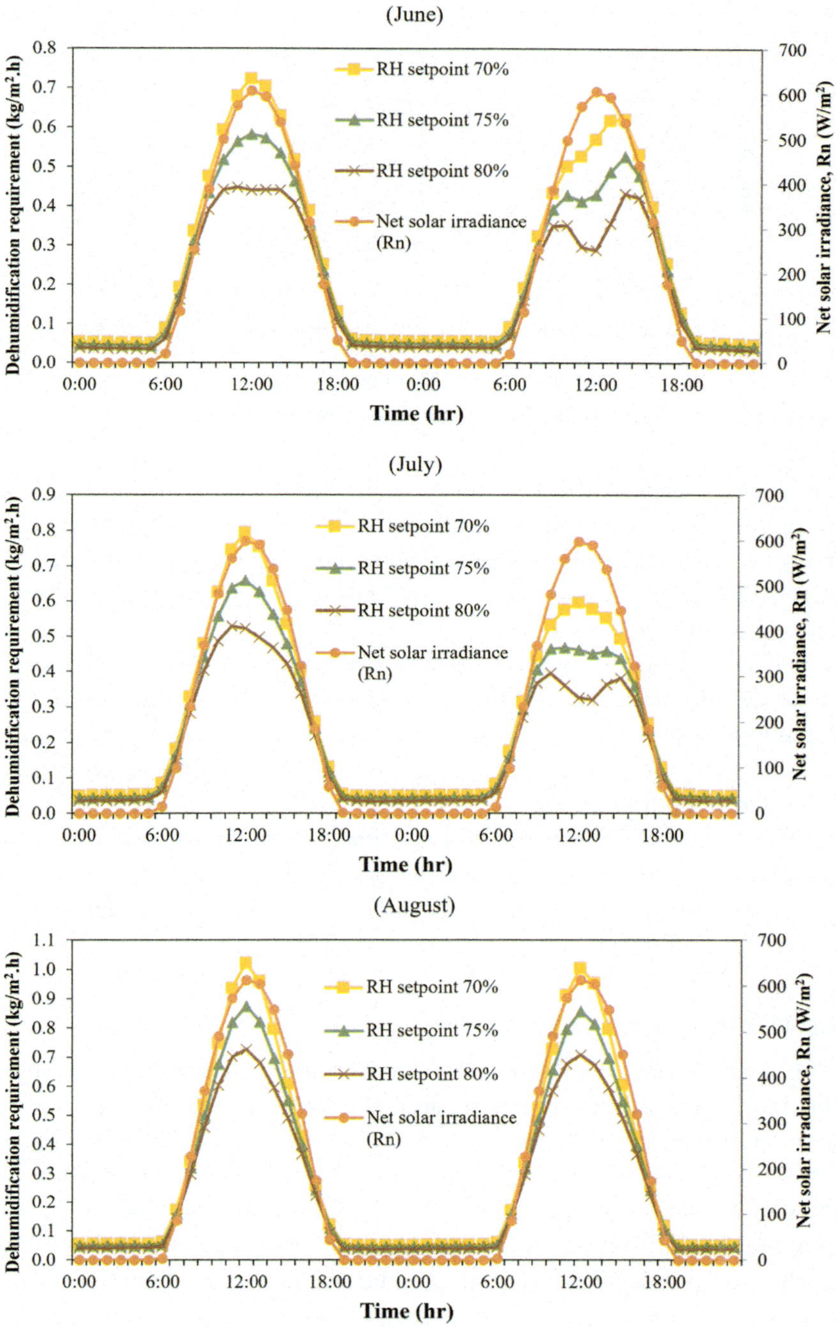

Fig. 5.14 Dehumidification load for different RH setpoints, by month, corresponding to net solar irradiance inside the greenhouse.

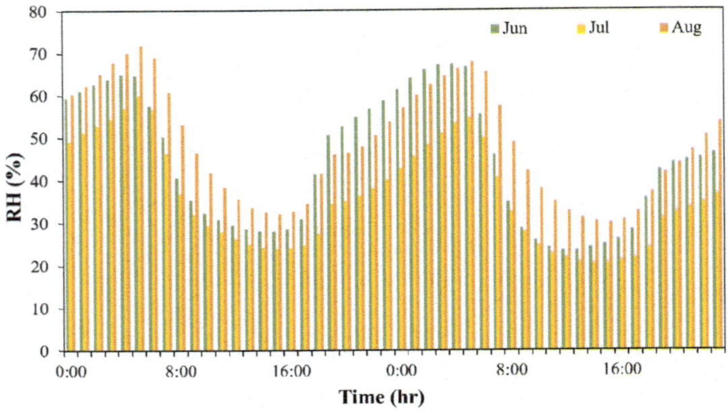

Fig. 5.15 Ambient RH profile of each studied summer month.

70% (orange squares), the dehumidification requirement is the highest, peaking at around 0.7 kg/m^2·h in the afternoon. For the 75% RH setpoint (gray triangles), the requirement is slightly lower, peaking around 0.6 kg/m^2·h. At 80% RH (yellow crosses), the dehumidification requirement is the lowest, peaking just above 0.4 kg/m^2·h.

This variation is expected because maintaining a lower RH requires more dehumidification, as more moisture must be removed from the greenhouse environment. The timing of the dehumidification peaks closely matches the peak solar irradiance, around midday, when the temperature and solar input are highest.

The dehumidification need in relation to solar irradiance within the greenhouse illustrates the influence of solar radiation on this requirement throughout the day. Furthermore, it demonstrates a similar dependence on additional input factors like RH as shown in Fig. 5.15.

5.3.4 DAC systems analyses

DAC is an effective method for regulating temperature and humidity in hot and humid environments. The analysis for DAC is conducted under the summer conditions in Multan, which define the required temperature and RH ranges. The following figure shows the temperature profile for the following conditions: 1) ambient, 2) air passes through desiccant air conditioning (DAC), 3) air passes through DAC and then from M-cycle EC. It is evident that, under these climatic conditions, a dehumidifier alone is not sufficient to achieve optimal growth conditions for tomato, cucumber, pea, and strawberry crops. The integration of DAC with MEC makes it feasible.

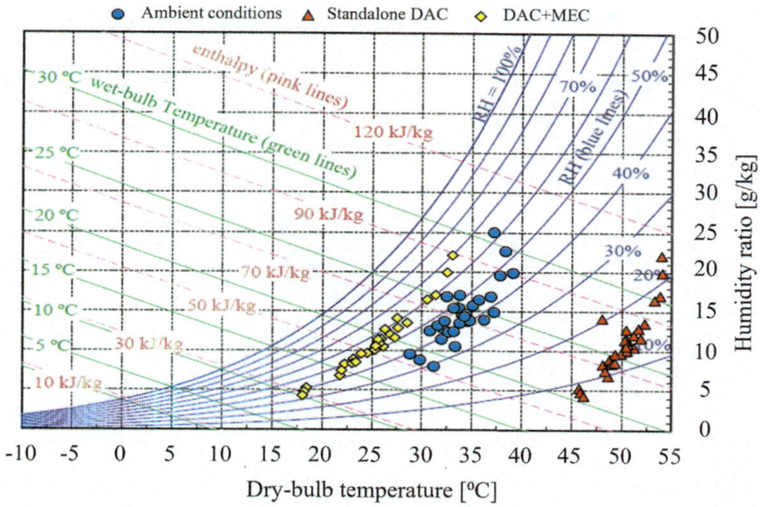

Fig. 5.16 Psychrometric presentation of standalone DAC and DAC+MEC systems for optimum greenhouse climate (June).

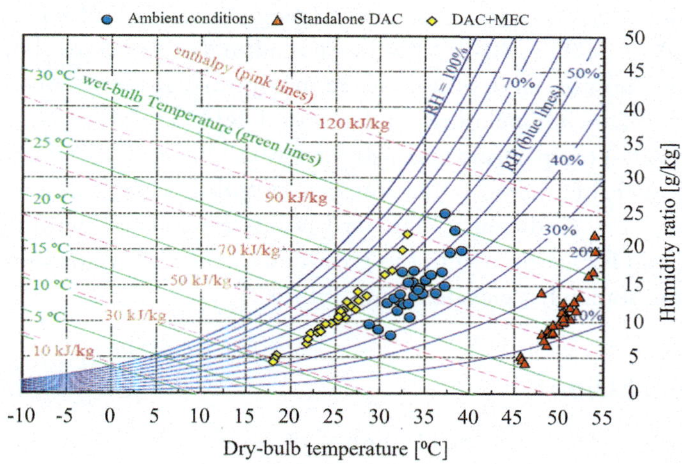

Fig. 5.17 Psychrometric presentation of standalone DAC and DAC+MEC systems for optimum greenhouse conditions (July).

Fig. 5.16 illustrates the distribution of 30 ambient temperature data points (for June) and their accompanying humidity ratios, which align with the optimum T and RH zone (i.e., T = 15-30°C, RH = 40-90%) of the agricultural greenhouse . Clearly, only a limited number of days fall within the optimal zone. This underscores the critical need to regulate temperature and RH in the agricultural greenhouse. Fig. 5.17 illustrates the distribution

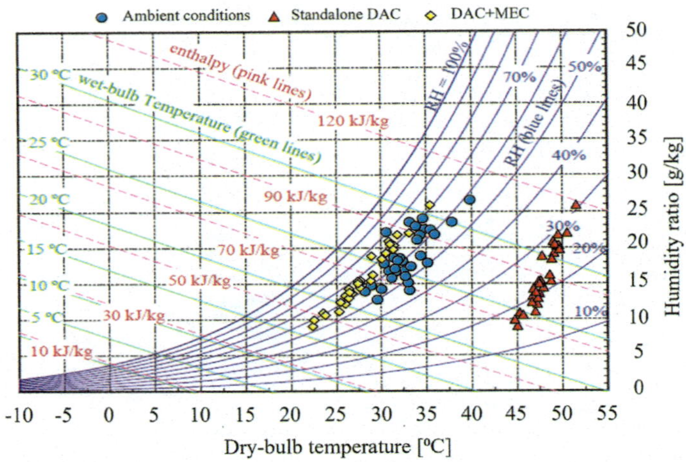

Fig. 5.18 Psychrometric presentation of standalone DAC and DAC+MEC systems for optimum greenhouse condition (August).

of 31 data points of July for the temperature and related humidity ratios of a standalone DAC system, aligned with the optimum T and RH zone (i.e., T = 15-30°C, RH = 40-90%) of the agricultural greenhouse. Throughout the whole months, all days are outside the optimal growing zone for plants due to elevated temperatures and reduced RH. Fig. 5.18 illustrates the distribution of temperature and associated humidity ratios for 31 data points for August from a DAC system integrated with a sensible heat exchanger operating at 90% efficiency. The data points are situated within the optimum T and RH zone within the greenhouse; yet, they remain inadequate due to the somewhat elevated temperature.

5.4 Conclusions

This study investigates the feasibility of a heat-driven DAC system for agricultural greenhouses in Pakistan, with the objective of enhancing environmental regulation for crop cultivation. The study examines the significant environmental challenges faced by greenhouses in arid areas such as Multan, Pakistan, where elevated temperatures and variable humidity adversely affect crops such as tomatoes, cucumbers, strawberries, and peas. The research employs a holistic methodology that integrates dehumidification, demand modeling, and thermodynamic analysis to enhance greenhouse climate control and ensure ideal conditions for crops. The primary goal is to simulate the dehumidification requirements of the chosen crops, accounting for their specific temperature and humidity requirements. The research indicates that temperature regulation alone is insufficient for effective cultivation in hot climates, as relative humidity is essential for plant transpiration, nutrient absorption, and overall physiological equilibrium. It is essential to maintain relative humidity between 60% and 80% for these crops, especially during the summer months when humidity levels in Multan vary considerably. The second aim is to examine an EC-assisted DAC system that incorporates desiccant materials to attain specified RH levels while minimizing thermal load in the greenhouse environment. The system design must utilize renewable or waste heat, in accordance with sustainable agriculture methods. Modeling desiccant dehumidification provides further insights into the system's capacity to maintain ideal conditions for crop cultivation. The desiccant system efficiently lowers relative humidity to the specified range, even during peak summer months, while maintaining a temperature favorable for crop development. The research emphasizes the advantages of integrating evaporative cooling with desiccant dehumidification in greenhouses. The EC is a prevalent method in warm areas; however, its effectiveness is constrained by ambient humidity. The use of desiccant dehumidification mitigates this constraint by lowering indoor relative humidity, thereby increasing the evaporative system's cooling efficiency. This hybrid method enhances temperature and humidity regulation, minimizes energy usage, and harnesses waste heat or solar thermal energy, rendering it a sustainable alternative for greenhouse climate management. The findings are substantial for Pakistan's agricultural economy, especially for greenhouse cultivation in arid areas. The heat-driven DAC system offers environmental and economic benefits, enabling farmers to maintain optimal growing conditions for high-value crops in challenging regions. The use of renewable energy sources aligns with global efforts to reduce greenhouse gas emissions and promote sustainable agricultural practices. The study provides substantial insights for future research and development in greenhouse air cooling, promoting more sustainable and effective agricultural practices in Pakistan and similar climatic regions.

Acknowledgment

This research work is completed with the financial support from the Higher Education Commission (HEC) of Pakistan through the research Project NRPU-17029 awarded to Bahauddin Zakariya University.

Funding statement

This research work is completed with the financial support from the Higher Education Commission (HEC) of Pakistan through the research Project NRPU-17029 awarded to Bahauddin Zakariya University.

CRediT authorship contribution statement

Aneela Sarwar was involved in the conceptualization, methodology, software, formal analysis, investigation, writing of the original draft. **Muhammad Sultan and Takahiko Miyazaki** were involved in the conceptualization, methodology, validation, resources, visualization, supervision, project administration, writing (review and editing), funding acquisition. **Muhammad Aleem, Md Shamim Ahamed, Muhammad Wakil Shahzad, Jie Liu, Zhaoli Zhang, and Hafiz Muhammad Ali** were involved in the conceptualization, methodology, software, formal analysis, investigation, writing (review and editing).

Data availability statement

The datasets used and/or analyzed during the current study are available from the corresponding authors upon request.

Conflicts of interest

The authors declare no conflict of interest.

Declaration of generative AI and AI-assisted technologies in the writing process

No generative AI/AI-assisted technology is used in this paper.

References

Ahamed, M.S., Sultan, M., Shamshiri, R.R., Rahman, M.M., Aleem, M., Balasundram, S.K., 2023. Present status and challenges of fodder production in controlled environments: a review. Smart Agric. Technol. 2772-3755. 3. Available from: www.sciencedirect.com/journal/smart-agricultural-technology/issues. doi:10.1016/j.atech.2022.100080.

Aleem, M., Hussain, G., Sultan, M., Miyazaki, T., Mahmood, M.H., Sabir, M.I., Nasir, A., Shabir, F., Khan, Z.M., 2020. Experimental investigation of desiccant dehumidification cooling system for climatic conditions of multan (Pakistan). Energies. 1996-1073. 13 (21). Available from: https://www.mdpi.com/1996-1073/13/21/5530. doi:10.3390/en13215530.

Aleem, M., Sultan, M., Mahmood, M.H., Miyazaki, T., 2022. Desiccant dehumidification cooling system for poultry houses in Multan (Pakistan) BT—energy-efficient systems for agricultural applications. In: Sultan, M., Miyazaki, T. (Eds.), Green Energy and

Technology. Springer International Publishing, Cham, pp. 19–42. doi:10.1007/978-3-030-86394-4_2.

Ali, S., Liu, Y., Ishaq, M., Shah, T., Abdullah, Ilyas, A., Din, I., 2017. Climate change and its impact on the yield of major food crops: evidence from Pakistan. Foods. 2304-8158. 6 (6), 39. doi:10.3390/foods6060039.

Angmo, P., Dolma, T., Phuntsog, N., Chaurasia, O.P., Stobdan, T., 2022. Effect of shading and high temperature amplitude on yield and phenolic contents of greenhouse capsicum (Capsicum annuum L.). Open Access Res. J. Biol. Pharm. 4, 30–39.

Ashraf, H., Sultan, M., Shamshiri, R.R., Abbas, F., Farooq, M., Sajjad, U., Md-Tahir, H., Mahmood, M.H., Ahmad, F., Taseer, Y.R., Shahzad, A., Niazi, B.M.K., 2021. Dynamic evaluation of desiccant dehumidification evaporative cooling options for greenhouse air-conditioning application in multan (Pakistan). Energies. 1996-1073. 14 (4). Available from: https://www.mdpi.com/1996-1073/14/4/1097/pdf. doi:10.3390/en14041097.

Beccali, M., Butera, F., Guanella, R., Adhikari, R.S., 2003. Simplified models for the performance evaluation of desiccant wheel dehumidification. Int. J. Energy Res. 0363-907X. 27 (1), 17–29. doi:10.1002/er.856.

Beccali, M., Adhikari, R.S., Butera, F., Franzitta, V., 2004. Update on desiccant wheel model. Int. J. Energy Res. 28, 1043–1049. doi:10.1002/er.1011.

Campen, J.B., Bot, G.P.A., de Zwart, H.F., 2003. Dehumidification of greenhouses at northern latitudes. Biosyst. Eng. 1537-5110. 86 (4), 487–493. doi:10.1016/j.biosystemseng.2003.08.008.

Campen, J., 2009. Dehumidification of Greenhouses. Wageningen University.

Costantino, A., Comba, L., Sicardi, G., Bariani, M., Fabrizio, E., 2021. Energy performance and climate control in mechanically ventilated greenhouses: a dynamic modelling-based assessment and investigation. Appl. Energy. 0306-2619. 288, 116583. doi:10.1016/j.apenergy.2021.116583.

De Freitas, C.R., Schmekal, A., 2003. Condensation as a microclimate process: measurement, numerical simulation and prediction in the Glowworm Cave, New Zealand. Int. J. Climatol. 23 (5), 557–575. doi:10.1002/joc.898.

Farquhar, G.D., Hubick, K.T., Condon, A.G., Richards, R.A., 1989. Carbon Isotope Fractionation and Plant Water-Use Efficiency. Stable Isotopes in Ecological Research21–40. doi:10.1007/978-1-4612-3498-2_2.

Gitz, V., Meybeck, A., Lipper, L., De, Y.C., Braatz, S., 2016. Climate change and food security: risks and responses. Food Agric. Organ. U. N. Rep. 110, 3–36.

Global Network Against Food Crises (GNAFC), 2024. Global report on food crises (GRFC). Available from: https://www.migrationdataportal.org/resource/global-report-food-crises-grfc-2024.

Han, J., 2018. Dehumidification technology evaluation and moisture balance modelling for greenhouse humidity control. University of Saskatchewan, Canada. Available from: https://harvest.usask.ca/items/6e3dccf7-d81f-4072-b0ec-3c960f5a4b68.

Hussain, G., Aleem, M., Sultan, M., Sajjad, U., Ibrahim, S.M., Shamshiri, R.R., Farooq, M., Khan, M.U., Bilal, M., 2022. Evaluating evaporative cooling assisted solid desiccant dehumidification system for agricultural storage application. Sustainability. 2071-1050. 14 (3), 1479. doi:10.3390/su14031479.

Iddio, E., Wang, L., Thomas, Y., McMorrow, G., Denzer, A., 2020. Energy efficient operation and modeling for greenhouses: a literature review. Renew. Sustain. Energy Rev. 1364-0321. 117, 109480. doi:10.1016/j.rser.2019.109480.

Inoue, T., Sunaga, M., Ito, M., Yuchen, Q., Matsushima, Y., Sakoda, K., Yamori, W., 2021. Minimizing VPD fluctuations maintains higher stomatal conductance and photosynthesis, resulting in improvement of plant growth in lettuce. Front. Plant Sci. 1664-462X. 12. doi:10.3389/fpls.2021.646144.

Iqbal, Z., Abbas, F., Ibrahim, M., Ayyaz, M.M., Ali, S., Mahmood, A., 2019. Surveillance of heavy metals in maize grown with wastewater and their impacts on animal health in peri-urban areas of multan, Pakistan. Pak. J. Agric. Sci. 2076-0906. 56 (2), 321–328. Available from: https://pakjas.com.pk/papers/2962.pdf. doi:10.21162/PAKJAS/19.8133.

Ishaq, M., Sultan, M., Aleem, M., Bilal, M., Ullah, H.S., Mahmood, M.H., et al., 2022. Desiccant dehumidification system for storage of fruits and vegetables BT—energy-efficient systems for agricultural applications. In: Sultan, M., Miyazaki, T. (Eds.), Green Energy and Technology. Springer International Publishing, Cham, pp. 65–83. doi:10.1007/978-3-030-86394-4_4.

Körner, O., Challa, H., 2003. Process-based humidity control regime for greenhouse crops. Comput. Electron. Agric. 0168-1699. 39 (3), 173–192. doi:10.1016/s0168-1699(03)00079-6.

Kumar, K.S., Tiwari, K.N., Jha, M.K., 2009. Design and technology for greenhouse cooling in tropical and subtropical regions: a review. Energy Build. 0378-7788. 41 (12), 1269–1275. doi:10.1016/j.enbuild.2009.08.003.

Maraveas, C., Karavas, C.-S., Loukatos, D., Bartzanas, T., Arvanitis, K.G., Symeonaki, E., 2023. Agricultural greenhouses: resource management technologies and perspectives for zero greenhouse gas emissions. Agriculture. 2077-0472. 13 (7), 1464. doi:10.3390/agriculture13071464.

Monteith, J.L., 1965. Evaporation and environment. Symp. Soc. Exp. Biol. 19, 205–234.

Montross, M.D., Duncan, G.A., Gates, R.S., 2006. Development and testing of a low-cost condensation detection system. Appl. Eng. Agric. 0883-8542. 22 (4), 603–608.

Padmanabhan, P., Cheema, A., Paliyath, G., 2016. Solanaceous fruits including tomato, eggplant, and peppers. In: Caballero, B., Finglas, P.M., Toldár, F. (Eds.), Encyclopedia of Food and Health.Elsevier BV, pp. 24–32. doi:10.1016/b978-0-12-384947-2.00696-6.

Penman, H.L., 1948. Natural evaporation from open water, bare soil and grass. Proc. R. Soc. Lond. A Math. Phys. Sci. 193, 120–145.

Rahman, M.S., Guo, H., Han, J., 2021. Dehumidification requirement modelling and control strategy for greenhouses in cold regions. Comput. Electron. Agric. 0168-1699. 187, 106264. doi:10.1016/j.compag.2021.106264.

Rosenberg, N.J., Blad, B.L., Verma, S.B., 1983. Microclimate: The Biological Environment, John Wiley & Sons.

Shamshiri, R.R., Jones, J.W., Thorp, K.R., Ahmad, D., Man, H.C., Taheri, S., 2018. Review of optimum temperature, humidity, and vapour pressure deficit for microclimate evaluation and control in greenhouse cultivation of tomato: a review. Int. Agrophysics. 0236-8722. 32 (2), 287–302. Available from: http://www.degruyter.com/view/j/intag.2013.27.issue-1/issue-files/intag.2013.27.issue-1.xml. doi:10.1515/intag-2017-0005.

Soussi, M., Chaibi, M.T., Buchholz, M., Saghrouni, Z., 2022. Comprehensive review on climate control and cooling systems in greenhouses under hot and arid conditions. Agronomy. 2073-4395. 12 (3). Available from: https://www.mdpi.com/2073-4395/12/3/626/pdf. doi:10.3390/agronomy12030626.

Stanghellini, C., 1987. Transpiration of Greenhouse Crops: An Aid to Climate Management, Wageningen University and Research.

Sultan, M., Miyazaki, T., Koyama, S., Khan, Z.M., 2018. Performance evaluation of hydrophilic organic polymer sorbents for desiccant air-conditioning applications. Adsorpt. Sci. Technol. 2048-4038. 36 (1–2), 311–326. Available from: http://adt.sagepub.com/. doi:10.1177/0263617417692338.

Sultan, M., Miyazaki, T., Saha, B.B., Koyama, S., Maisotsenko, V.S., 2015. Steady-state analysis on thermally driven adsorption air-conditioning system for agricultural greenhouses. Procedia Eng. 1877-7058. 118, 185–192. http://www.sciencedirect.com/science/journal/18777058. doi:10.1016/j.proeng.2015.08.417.

Sultan, M., Miyazaki, T., Saha, B.B., Koyama, S., 2016. Steady-state investigation of water vapor adsorption for thermally driven adsorption based greenhouse air-conditioning system. Renew. Energy. 1879-0682. 86, 785–795. Available from: http://www.journals.elsevier.com/renewable-and-sustainable-energy-reviews/. doi:10.1016/j.renene.2015.09.015.

Sultan, M., Ashraf, H., Miyazaki, T., Shamshiri, R.R., Hameed, I.A., 2021. Temperature and humidity control for the next generation greenhouses: overview of desiccant and evaporative cooling systems. In: Next-Generation Greenhouses for Food Security. IntechOpen, pp. 42-46 . doi:10.5772/intechopen.97273.

Sultan, M., Aleem, M., Miyazaki, T., 2022. Energy-efficient humidity pump system for poultry houses BT—synergy development in renewables assisted multi-carrier systems. In: Amidpour, M., Ebadollahi, M., Jabari, F., Kolahi, M.-R., Ghaebi, H. (Eds.), Green Energy and Technology. Springer International Publishing, Cham, pp. 431–457. doi:10.1007/978-3-030-90720-4_16.

Takakura, T., Takayama, K., Nishina, H., Tamura, K., Muta, S., 2005. Evapotranspiration estimate by heat balance equation. 2005 ASAE Annu. Meet., American Society of Agricultural and Biological Engineers. p. 1.

Villarreal-Guerrero, F., Kacira, M., Fitz-Rodríguez, E., Kubota, C., Giacomelli, G.A., Linker, R., Arbel, A., 2012. Comparison of three evapotranspiration models for a greenhouse cooling strategy with natural ventilation and variable high pressure fogging. Sci. Hortic. 0304-4238. 134, 210–221. doi:10.1016/j.scienta.2011.10.016.

Wei, X., Li, B., Lu, H., Guo, J., Dong, Z., Yang, F., Lü, E., Liu, Y., 2024. Distribution characteristics and prediction of temperature and relative humidity in a South China Greenhouse. Agronomy. 2073-4395. 14 (7), 1580. doi:10.3390/agronomy14071580.

Yu, X., Zhang, Y., Zhao, X., Li, J., 2023. Systemic effects of the vapor pressure deficit on the physiology and productivity of protected vegetables. Veg. Res. 2769-0520. 3 (1), 0. doi:10.48130/vr-2023-0020.

CHAPTER 6

Comparative thermal modeling of standalone desiccant and Maisotsenko-cycle assisted desiccant air conditioning systems for greenhouse climate control

Fakhar Abbas[a], Muhammad Sultan[a], Hafiz Muhammad Umar Raza[a], Muhammad Hamid Mahmood[a], Muhammad Ali Imran[a], Takahiko Miyazaki[b], Muhammad Wakil Shahzad[c], Jie Liu[d], and Zhaoli Zhang[e]

[a]Department of Agricultural Engineering, Bahauddin Zakariya University, Multan, Pakistan
[b]Department of Advanced Environmental Science and Engineering, Faculty of Engineering Sciences, Kyushu University, Kasuga, Fukuoka, Japan
[c]Department of Mechanical and Construction Engineering, Northumbria University, Newcastle Upon Tyne, United Kingdom
[d]College of Engineering, Huazhong Agricultural University, Wuhan, China
[e]School of Mechanical Engineering, Southwest Jiaotong University, Chengdu, China

6.1 Introduction

Growing plants in a greenhouse requires maintaining four essential microclimate parameters: temperature, relative humidity, light intensity, and carbon dioxide concentration (Badji et al., 2022). Traditional temperature/humidity control practices, such as shading practices; ventilation systems, such as natural and artificial ventilation; and fogging systems, are used, but these systems depend on outside climate conditions and can only partially control the temperature and humidity (Soussi et al., 2022). Temperature and humidity control are difficult in traditional greenhouse farming due to the constant interaction between external and internal environmental factors. Seasonal variations, regional location, and weather patterns are examples of external factors that might make precise control inside greenhouses challenging. Internally, soil moisture dynamics and plant transpiration make humidity control even more difficult (Agriculture et al., 2023). Due to inadequate temperature and humidity control inside the

Sustainable Hybrid Cooling and Dehumidification Strategies for Controlled Environments
https://doi.org/10.1016/B978-0-443-48555-8.00012-8

greenhouse, plants face issues such as heat stress, reduced growth rates of leaves and flowers, and overall significant yield losses (Sultan et al., 2014). Heat stress can negatively impact plant growth and productivity due to inadequate temperature control (Hatfield and Prueger, 2015).

This study focuses on the recent advancements in temperature and humidity control technologies for greenhouses, i.e., desiccant dehumidification and Maisotsenko cycle evaporative cooling technologies, respectively. The solid desiccant dehumidification systems are advanced systems that utilize hygroscopic materials, such as zeolite, silica gel, and metal–organic frameworks (MOFs), including CU-BTC, MIL-100, MOF-74, UiO-66 (Yang et al., 2024), CPO27(NI), MIL100(Fe), MIL-101(Cr), and aluminum fumarate (Aziz et al., 2022), to absorb moisture from the air and control the humidity level before entering the greenhouse. The solid desiccant wheel consists of a frame lined with a thin layer of desiccant material. The cross sectional area of the desiccant wheel is divided into two sections: one for dehumidifying the ambient air and another for regenerating the desiccant wheel. Desiccant-based dehumidifiers effectively reduce humidity and ensure plant health during high humidity periods (Sultan et al., 2016; Soussi et al., 2022). The ambient air is dehumidified due to adsorption through the desiccant materials as it passes through the solid desiccant. Due to the continuous dehumidification of the process air, the desiccant is fully saturated with moisture, and then it requires heat for regeneration (La et al., 2010). Waste heat, solar energy, and electric heaters can be used as thermal energy sources to regenerate the desiccant system (Lai et al., 2021). Fuel/electric energy consumption can be reduced using the solar system to regenerate the desiccant. Fuel consumption can be reduced by up to 20% using solar energy for desiccant regeneration (Yang et al., 2024). The average solar energy utilization rate can supply 25.14 kW of energy for desiccant regeneration each day (Kousar et al., 2021).

The Maisotsenko-cycle evaporative cooling (MEC) system is an advanced indirect evaporative cooling (IEC) system that can cool air down to the ambient wet-bulb and dew point temperatures (Buyadgie et al., 2015; Mahmood et al., 2016). The MEC system is a thermodynamic process that uses renewable energy derived from the latent heat of vaporization and redistributes the energy efficiently (Anisimov and Pandelidis, 2014; Mahmood et al., 2016). This system integrates heat transfer with evaporative cooling to push the temperature lower than direct evaporative cooling (DEC) and IEC systems. This system also produces saturated hot air as part of a heat

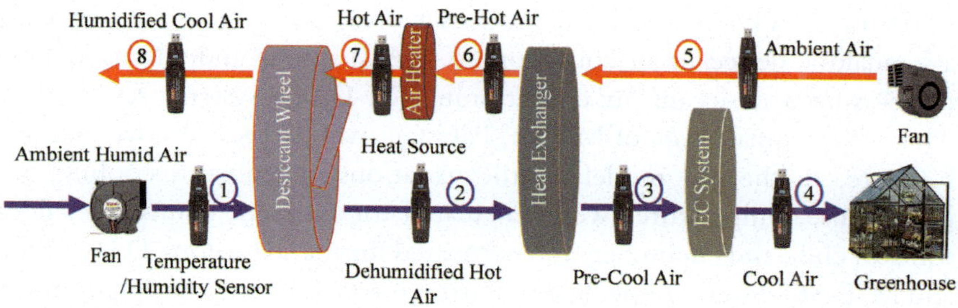

Fig. 6.1 Schematic diagram of an integrated desiccant-based evaporative cooling system for temperature and humidity of greenhouses.

recovery process, thereby improving system efficiency for various applications (Buyadgie et al., 2015).

Desiccant-assisted evaporative cooling systems represent a significant improvement in greenhouse temperature and humidity control. These systems efficiently control temperature and humidity by combining evaporative cooling and desiccant dehumidification (Sultan et al., 2015). Dehumidification removes moisture from the air, while evaporative cooling controls the temperature of the air. Zeolite, alumina, silica gel, hydratable salts, and their combinations are some adsorbent materials used in desiccants (Lai et al., 2021). These systems are suitable in hot and humid regions, where high ambient humidity may make traditional evaporative cooling less effective (Lai et al., 2021). These systems have no hazardous refrigerant and maintain a high air quality standard. A schematic diagram of an integrated desiccant and evaporative cooling system is shown in Fig. 6.1.

The working principle of desiccant-based evaporative cooling systems is that ambient air flows through a desiccant wheel, where latent heat causes the temperature to rise and adsorption reduces the air's moisture content. After that, the dehumidified air moves through a heat exchanger, where sensible cooling causes the temperature to drop while maintaining a consistent humidity level. After passing through evaporative cooling systems such as DEC, IEC, and MEC, the precooled air temperature is further dropped. After that, cooled air enters the greenhouses and provides the ideal temperature and humidity for plant growth. The existing air from the greenhouse flows through the heat exchanger, where its temperature rises due to sensible heating. The heated air then flows through the heat source, raising its temperature. After passing through the desiccant, the hot air's temperature drops, and its moisture content rises due to desorption. The exhaust air is released into the atmosphere (Lai et al., 2021).

In this study, the thermal modeling of a greenhouse was done for a standalone desiccant air conditioning (DAC) system and a Maisotsenko-cycle-assisted desiccant air conditioning (M-DAC) system, respectively. The yearly climate data of Multan, Pakistan, was gained from Meteonorm software. In thermal modeling, the variations in ambient, cooling, and greenhouse temperatures were calculated for different plants, i.e., flora, tomato, chili, and cucumber, and compared for the standalone DAC system and M-DAC system, respectively. Furthermore, the crop transpiration rate and sensible heat load of the greenhouse were calculated for different plants such as flora, tomato, chili, and cucumber, for both the standalone DAC system and the M-DAC system.

6.2 Materials and methods

6.2.1 Thermal modeling of the greenhouse

The following assumptions were considered by Banik and Ganguly (2017) at the stage of developing the thermal model for greenhouses. (1) The steady state condition was taken into consideration when developing the model. (2) It is assumed that the chemical dehumidification process is adiabatic. (3) It is assumed that the temperature of water at the heat exchanger's inlet is the same as the ambient air's wet-bulb temperature. (4) The mechanical parts of the proposed system do not absorb any energy. The energy balance equation (Badji et al., 2022) for the greenhouse considers the transpiration phenomenon, as given in the literature (Banik and Ganguly, 2017).

$$S_{gi} + U_g \Delta T = C - K_s \Delta T + \lambda E \tag{1}$$

where, S_{gi} shows the intensity of solar radiation (W/m^2), U_g shows the overall heat transfer coefficient (4.5 W/m^2), ΔT shows the temperature difference of greenhouse air and ambient air (K), K_s shows the sensible heat transfer coefficient of the greenhouse air (W/m^{2}K), λ shows the latent heat of vaporization of water (kJ/kg), and E shows the plant canopy transpiration rate (kg/m^{2}s).

The sensible heat transfer coefficient of the greenhouse can be calculated using the equation (Badji et al., 2022) as given in the literature (Bauri and Ganguly, 2013).

$$K_s = \frac{\rho_a C_{pa}(ACM)V_g}{(60A_{gf})} \tag{2}$$

where, ρ_a shows the density of air (kg/m^3), C_{pa} shows the specific heat of air (J/kg K), ACM shows the number of air exchanges per minute, Vg shows the volume of the greenhouse (m^3), and A_{gf} shows the greenhouse floor area (m^2).

The sensible heat load of a greenhouse and a constant (C) used to calculate the sensible heat load can be measured using equations (Agriculture et al., 2023; Sultan et al., 2014), respectively, as given in the literature (Banik and Ganguly, 2017).

$$H = C - K_s(T_a - T_g) \tag{3}$$

where, H shows the sensible heat load (W), T_g shows the greenhouse temperature (K), and C shows the constant used to calculate the sensible heat load (−).

$$C = K_s(T_a - T_4) \tag{4}$$

The intensity of solar radiation can be measured using equation (Hatfield and Prueger, 2015) as given in the literature (Banik and Ganguly, 2017).

$$S_{gi} = \frac{\Sigma S_i A_i}{\Sigma A_i}(1 - SF) \tag{5}$$

where, S_i shows the intensity of solar radiation (W/m^2) that is transmitted through the ith surface of the greenhouse, and SF shows the shading factor of shade used for the greenhouse.

The greenhouse crop transpiration rate can be calculated using equation (Dalai, 2020) as given in the literature (Boulard and Wang, 2000).

$$E = \frac{K_l(e_g - e_4)}{\lambda} \tag{6}$$

where, K_l shows the latent heat transfer coefficient (W/m^2Pa), e_{gh} shows the water vapor pressure inside the greenhouse air, and e_4 shows the water vapor pressure of the process air at the inlet of the greenhouse.

The latent heat transfer coefficient can be measured using the equation (Greenhouse et al., 2021) as given in the literature (Boulard and Wang, 2000; Mandal and Ganguly, 2022).

$$K_l = \frac{\lambda \zeta \rho_a (ACM) V_g}{(60A_{gf})} \tag{7}$$

where, λ shows the latent heat of vaporization of water (2260 kJ/kg), ζ shows the conversion factor (6.25×10^{-6} kg_w/kg_a Pa), ACM shows the number of air exchanges per minute, V_g shows the volume of the greenhouse (m^3), and A_{gf} shows the greenhouse floor area (m^2).

The water vapor pressure inside the greenhouse air and the process air at the inlet of the greenhouse can be measured using equations (Anon, 2020; Sultan et al., 2015), respectively, as given in the literature (Banik and Ganguly, 2017).

$$e_g = e^*(T_g)\left(\frac{RH_g}{100}\right) \tag{8}$$

$$e^*(T_g) = \exp\left(25.317 - \frac{5144}{T_g}\right) \tag{9}$$

$$e_4 = e^*(T_4)\left(\frac{RH_4}{100}\right) \tag{10}$$

$$e^*(T_4) = \exp\left(25.317 - \frac{5144}{T_4}\right) \tag{11}$$

where, $e^*(T_g)$ shows the inside greenhouse saturated vapor pressure of air (Pa), RH shows the relative humidity (%), and $e^*(T_4)$ shows the saturated vapor pressure of process air at the inlet of the greenhouse (Pa). The mean greenhouse temperature can be calculated using equation (Hua et al., 2024) as given in the literature (Banik and Ganguly, 2017).

$$T_g = T_a - \Delta T \tag{12}$$

$$\Delta T = \frac{[A - (BC)]}{[(DC) - E]} \tag{13}$$

where, A, B, C, D, and E are constants. The description of each constant is given below. The constant A can be calculated using equation (Sultan et al., 2016), and the constants B, C, D, and E can be calculated using equations (Mandal and Ganguly, 2022), respectively, as given in the literature (Banik and Ganguly, 2017).

$$A = \frac{\left[(\delta C) + \left(P_1 \frac{W_a}{r_a}\right)\right]}{P_2} \tag{14}$$

where, δ shows the slope of the water vapor saturation curve (Pa/K) at a temperature T(K), P_1 and P_2 show the constants, which can be calculated using equation (Yang et al., 2024; Bauri and Ganguly, 2013), respectively, as given in the literature (Banik and Ganguly, 2017); W_a shows the water vapor pressure of ambient air (Pa); and r_a shows the aerodynamic resistance (s/m) of greenhouse plants, respectively. The slope of the water vapor saturation curve can be measured using equation (La et al., 2010) as given in the literature (Bauri and Ganguly, 2013).

$$\delta = \left(\frac{5385}{T^2}\right)\left(2.229 * 10^{11}\right)\exp\left(\frac{-5385}{T}\right) \tag{15}$$

$$P_1 = 2I_{la}\rho_a C_{pa} \tag{16}$$

$$W_a = e^*(T_a) - e_a \tag{17}$$

$$e^*(T_a) = \exp\left(25.317 - \frac{5144}{T_a}\right) \tag{18}$$

$$e_a = e^*(T_a)\left(\frac{RH}{100}\right) \tag{19}$$

where, I_{la} = leaf area index (m^2/m^2), ρ_a shows the air density, and C_{pa} shows the specific heat of greenhouse air (Pa).

The aerodynamic resistance (s/m) of greenhouse plants can be calculated using equation (Buyadgie et al., 2015) as given in the literature (Boulard and Wang, 2000).

$$r_a = 220\left(\frac{L_c^{0.2}}{u_i^{0.8}}\right) \tag{20}$$

where, Lc shows the characteristic length of leaves (m), and u_i shows the average velocity of air (m/s), can be calculated using equation (Banik and Ganguly, 2017) as given in the literature (Banik and Ganguly, 2017).

$$u_i = \frac{ACM\left(\sqrt[3]{V_g}\right)}{60} \tag{21}$$

$$P_2 = \delta + \gamma\left(1 + \frac{r_s}{r_a}\right) \tag{22}$$

where, γ shows the psychrometric constant (Pa/K), and r_s shows the stomatal resistance of the target plantation can be calculated using equations (Boulard and Wang, 2000), as given in the literature (Boulard and Wang, 2000).

$$r_s = 220\left[1 + \frac{1}{\exp\left\{0.05\left(S_{gi} - 50\right)\right\}}\right] \tag{23}$$

$$B = \left(S_{gi} - C\right) \tag{24}$$

$$C = \left[1 - \left\{\frac{\left(\delta - \frac{P_1}{K_l r_a}\right)}{P_2}\right\}\right] \tag{25}$$

$$D = (U_g + K_s) \tag{26}$$

$$E = \frac{\left[\left(\frac{P_1\delta}{r_a}\right) - \delta K_s\right]}{P_2} \tag{27}$$

6.3 Results and discussion

This section presents the results of thermal modeling for the standalone DAC system and the M-DAC system. It presents the comparison results of variation in ambient temperature, DAC temperature, M-DAC temperature, and greenhouse temperature of the standalone DAC system and the M-DAC system, respectively. It presents the comparison results of crop transpiration rate and sensible heat load of the greenhouse for different plants such as flora, tomato, chili, and cucumber for both the standalone DAC system and the M-DAC system.

6.3.1 Temperature and relative humidity variations of the standalone DAC and the M-DAC systems

The variations in ambient temperature and relative humidity (RH); temperature and RH achieved by the standalone DAC system and the M-DAC system; and the greenhouse temperature for flora, tomato, chili, and cucumber, respectively, in May are shown in Fig. 6.2. The left side plot illustrates the variations in ambient temperature and RH, standalone DAC temperature and RH, and greenhouse inside temperature, respectively. The ranges of ambient temperature and RH are 25–41°C and 9%–35%; standalone DAC temperature and RH are 26°C–42°C and 11%–23%; and greenhouse temperature is 35°C–49.3°C, respectively, for the targeted plants. The right-side plot illustrates the variations in ambient temperature and RH, M-DAC temperature and RH, and inside greenhouse temperature, respectively. The ranges of the ambient temperature and RH are 25°C–40°C and 9%–35%; the M-DAC temperature and RH are 13°C–18°C and 34%–50%; and the greenhouse temperature is 23°C–28°C, respectively, for the targeted plants.

The variations in ambient temperature and RH; temperature and RH achieved by the standalone DAC system and the M-DAC system; and the greenhouse temperature for flora, tomato, chili, and cucumber, respectively, in June are shown in Fig. 6.3. The left side plot illustrates the variations in ambient temperature and RH, standalone DAC temperature and RH, and greenhouse inside temperature, respectively. The ranges of the ambient temperature and RH are 35°C–41.7°C and 15%–42%; standalone DAC temperature and RH are 36°C–42.3°C and 11%–16.1%; and

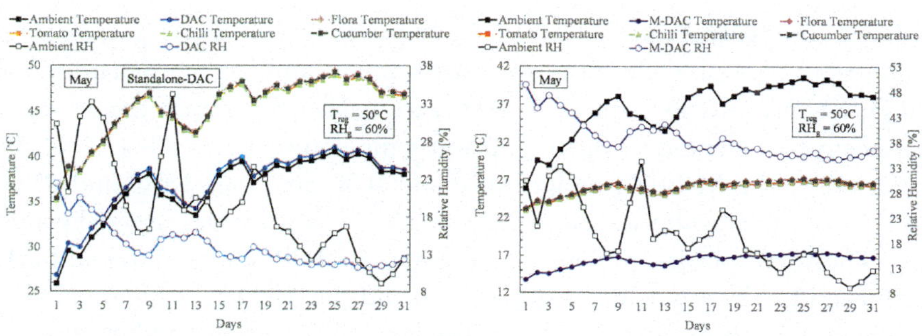

Fig. 6.2 Comparison of temperature and relative humidity variations of standalone DAC and M-DAC systems, respectively, at a constant regenerate temperature of 50°C and greenhouse inside relative humidity of 60% for flora, tomato, chilli, and cucumber, respectively, for May.

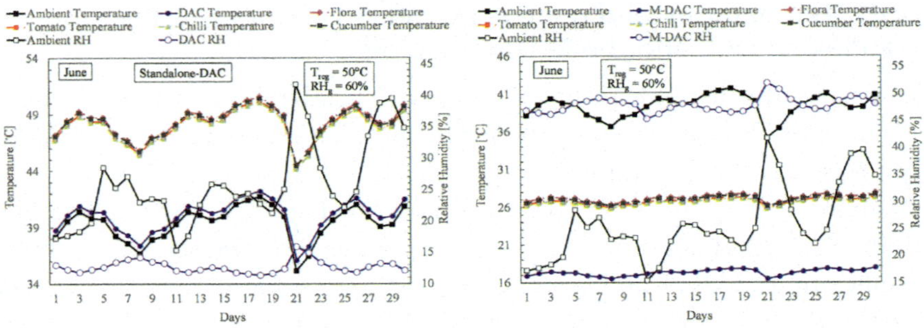

Fig. 6.3 Comparison of temperature and relative humidity variations of standalone DAC and M-DAC systems, respectively, at a constant regenerate temperature of 50°C and greenhouse inside relative humidity of 60% for flora, tomato, chilli, and cucumber, respectively, for June.

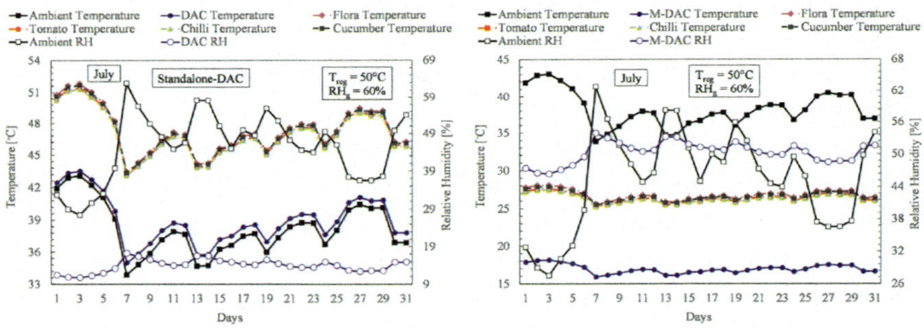

Fig. 6.4 Comparison of temperature and relative humidity variations of standalone DAC and M-DAC systems, respectively, at a constant regenerate temperature of 50°C and greenhouse inside relative humidity of 60% for flora, tomato, chilli, and cucumber, respectively, for July.

greenhouse temperature 44°C–51°C, respectively, for the targeted plants. The right-side plot illustrates the variations in ambient temperature and RH, M-DAC temperature and RH, and inside greenhouse temperature, respectively. The ranges of the ambient temperature are 35°C–41.7°C and 15%–42%; the M-DAC temperature and RH are 15°C–18°C and 45%–53%; and the greenhouse temperature is 26°C–28°C for the targeted plants.

The variations in ambient temperature and RH; temperature and RH achieved by the standalone DAC system and the M-DAC system; and the greenhouse temperature for flora, tomato, chili, and cucumber, respectively, in July are shown in Fig. 6.4. The left side plot illustrates the variations in ambient temperature and RH, standalone DAC temperature and RH, and greenhouse inside temperature, respectively. The ranges of the ambient temperature and RH are 33.8°C–43.1°C and 27.5%–63%; standalone DAC temperature and RH are 35°C–44°C and 10%–18%; and greenhouse temperature is 43°C–52°C, respectively, for the targeted plants. The right side plot illustrates the variations in ambient temperature and RH, the M-DAC temperature and RH, and the inside greenhouse temperature, respectively. The ranges of the ambient temperature and RH are 33.8°C–43.1°C and 27.5%–63%; the M-DAC temperature and RH are 13°C–18°C and 46.4%–54%; and the greenhouse temperature is 25.5°C–27.5°C, respectively, for the targeted plants.

The variations in ambient temperature and RH; temperature and RH achieved by the standalone DAC system and the M-DAC system; and the greenhouse temperature for flora, tomato, chili, and cucumber, respectively, in August are shown in Fig. 6.5. The left side plot illustrates the variations in ambient temperature and RH, standalone DAC temperature and RH, and greenhouse inside temperature, respectively. The ranges of the ambient temperature and RH are 29°C–36.4°C and 50%–86%; standalone DAC temperature and RH are 30°C–37°C and 15%–23%; and greenhouse temperature is 38°C–46°C, respectively, for the targeted plants. The righ side plots illustrate the variations in ambient temperature and RH, M-DAC temperature and RH, and inside greenhouse temperature, respectively. The ranges of the ambient temperature and RH are 29°C–36.4°C and 50%–86%; the M-DAC temperature and RH are 14.7°C–17°C and 51.7–58.3; and the greenhouse temperature is 24°C–26.5°C, respectively, for the targeted plants.

The variations in ambient temperature and RH; temperature and RH achieved by the standalone DAC system and the M-DAC system; and the

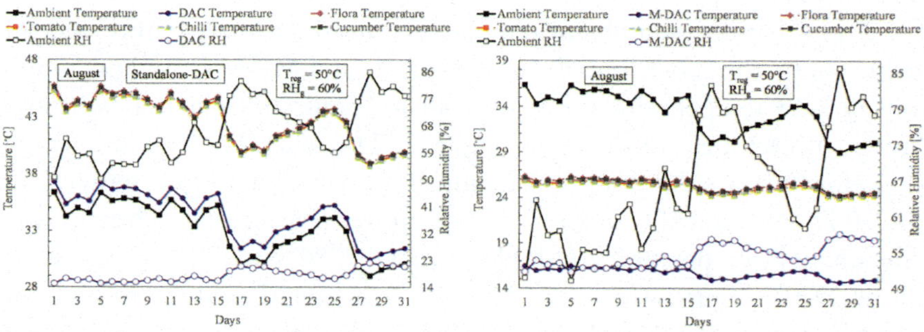

Fig. 6.5 Comparison of temperature and relative humidity variations of standalone DAC and M-DAC systems, respectively, at a constant regenerate temperature of 50°C and greenhouse inside relative humidity of 60% for flora, tomato, chilli, and cucumber, respectively, for August.

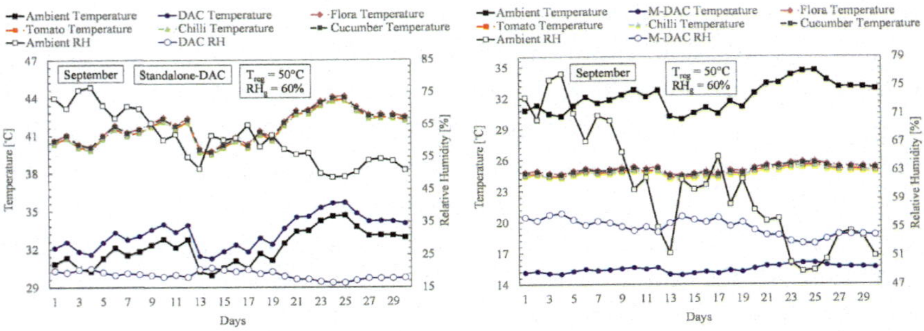

Fig. 6.6 Comparison of temperature and relative humidity variations of standalone DAC and M-DAC systems, respectively, at a constant regenerate temperature of 50°C and greenhouse inside relative humidity of 60% for flora, tomato, chilli, and cucumber, respectively, for September.

greenhouse temperature for flora, tomato, chili, and cucumber, respectively, in September, are shown in Fig. 6.6. The left-side plot illustrates the variations in ambient temperature and RH, standalone DAC temperature and RH, and greenhouse inside temperature, respectively. The ranges of the ambient temperature and RH are 29.9°C–34.7°C and 49%–77%; standalone DAC temperature and RH are 31°C–35.8°C and 16%–22%; and greenhouse temperature is 39.5°C–44.2°C, respectively, for the targeted plants. The right-side plot illustrates the variations in ambient temperature and RH, the M-DAC temperature and RH, and the inside greenhouse temperature, respectively. The ranges of the ambient temperature and RH are 29.9°C–34.7°C and 49%–77%; the M-DAC temperature and RH are 14.9°C–17°C and 52%–57%; and the greenhouse temperature is 23.4°C–25.3°C, respectively, for the targeted plants.

The variations in ambient temperature and RH; temperature and RH achieved by the standalone DAC system and the M-DAC system; and the greenhouse temperature for flora, tomato, chili, and cucumber, respectively, in October are shown in Fig. 6.7. The left-side plot illustrates the variations in ambient temperature and RH, standalone DAC temperature and RH, and greenhouse inside temperature, respectively. The ranges of the ambient temperature and RH are 27°C–33°C and 24%–47.1%; standalone DAC temperature and RH are 28°C–34°C and 17%–24%; and greenhouse temperature is 36.7°C–42.2°C, respectively, for the targeted plants. The right-side plot illustrates the variations in ambient temperature and RH, the M-DAC temperature and RH, and the inside greenhouse temperature, respectively. The ranges of the ambient temperature and RH

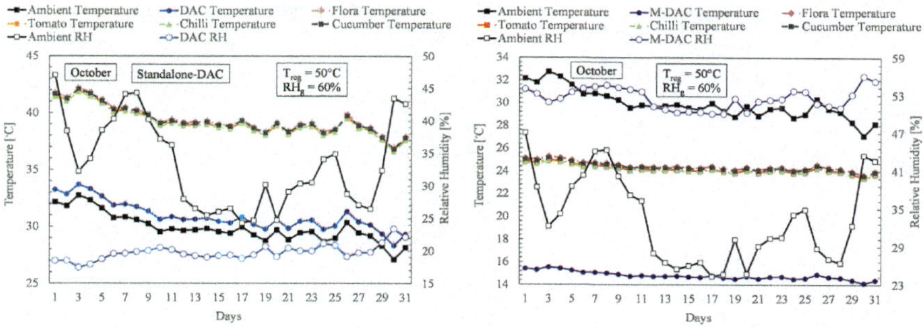

Fig. 6.7 Comparison of temperature and relative humidity variations of standalone DAC and M-DAC systems, respectively, at a constant regenerate temperature of 50°C and greenhouse inside relative humidity of 60% for flora, tomato, chilli, and cucumber, respectively, for October.

are 27°C–33°C and 24%–47.1%; the M-DAC temperature and RH are 14°C–16°C and 50%–56.2%; and the greenhouse temperature is 23.4°C–25.3°C, respectively, for the targeted plants.

6.3.2 Transpiration rate and sensible heat load of targeted plants

The variation in crop transpiration rate and sensible heat load of flora and tomato for standalone DAC and M-DAC systems, respectively, in May is shown in Fig. 6.8. The left side plots illustrate the variations in crop transpiration rate and sensible heat load of the flora and tomato for the standalone DAC system. The range of the crop transpiration rate is 3.2–7.8 kg/m^2/h, and the sensible heat load is 448–480 W/m^2 for both flora and tomato, respectively. The right side plots illustrate the variations in crop transpiration rate and sensible heat load of the flora and tomato for the M-DAC system. The range of the crop transpiration rate is 1.1–1.87 kg/m^2/h, and the sensible heat load is 520–556 W/m^2 for both flora and tomato, respectively.

The variation in crop transpiration rate and sensible heat load of chili and cucumber for standalone DAC and M-DAC systems, respectively, in May is shown in Fig. 6.9. The left side plots illustrate the variations in crop transpiration rate and sensible heat load of the chili and cucumber for the standalone DAC system. The range of the crop transpiration rate is 3.2–7.69 kg/m^2/h, and the sensible heat load is 474–484 W/m^2 for both chili and cucumber, respectively. The right side plots illustrate the variations in crop transpiration rate and sensible heat load of the chili and cucumber for the M-DAC system. The range of the crop transpiration rate is

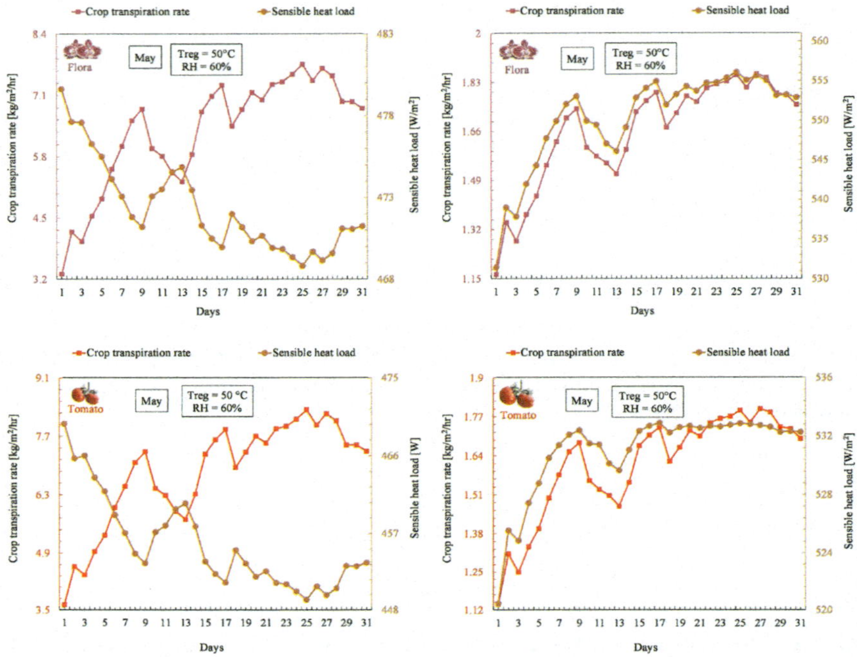

Fig. 6.8 Comparison of crop transpiration rate and sensible heat load variation of standalone DAC and M-DAC systems, respectively, at constant regenerate temperature of 50°C and greenhouse inside relative humidity of 60% for flora, and tomato, respectively, for May.

1.13–1.83 kg/m^2/h, and the sensible heat load is 515–545 W/m^2 for both chili and cucumber, respectively.

The variation in crop transpiration rate and sensible heat load of flora and tomato for standalone DAC and M-DAC systems, respectively, in June is shown in Fig. 6.10. The left side plots illustrate the variations in crop transpiration rate and sensible heat load of the flora and tomato for the standalone DAC system. The range of the crop transpiration rate is 5.6–8.24 kg/m^2/h, and the sensible heat load is 420–474 W/m^2 for both flora and tomato, respectively. The right side plots illustrate the variations in crop transpiration rate and sensible heat load of the flora and tomato for the M-DAC system. The range of the crop transpiration rate is 1.29–1.6 kg/m^2/h. and the sensible heat load is 505–558 W/m^2 for both flora and tomato, respectively.

The variation in crop transpiration rate and sensible heat load of chili and cucumber for standalone DAC and M-DAC systems, respectively, in June is

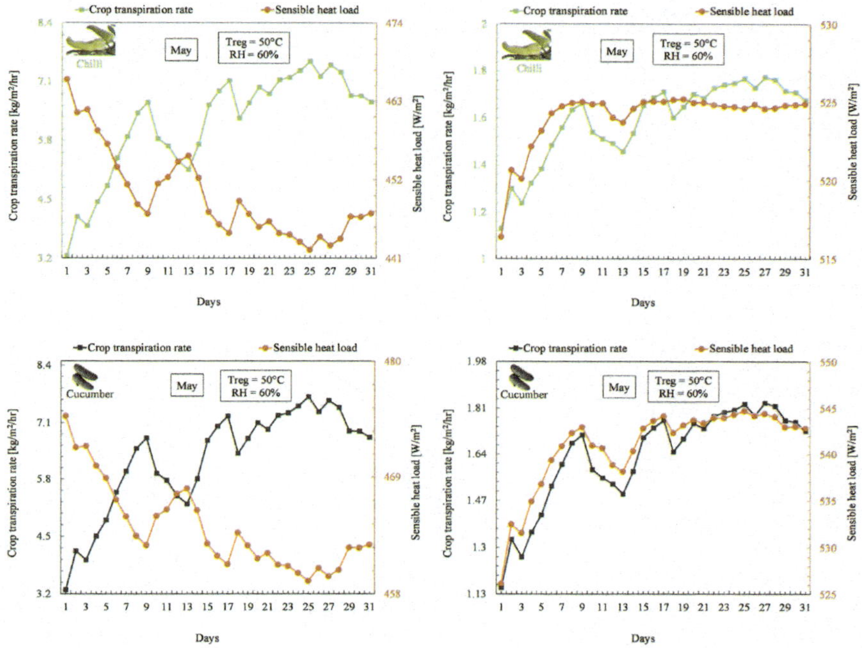

Fig. 6.9 Comparison of crop transpiration rate and sensible heat load variation of standalone DAC and M-DAC systems, respectively, at constant regenerate temperature of 50°C and greenhouse inside relative humidity of 60% for chilli and cucumber, respectively, for May.

shown in Fig. 6.11. The left side plots illustrate the variations in crop transpiration rate and sensible heat load of the chili and cucumber for the standalone DAC system. The range of the crop transpiration rate is 5.6–8.1 $kg/m^2/h$, and the sensible heat load is 419–475 W/m^2 for both chili and cucumber, respectively. The right side plots illustrate the variations in crop transpiration rate and sensible heat load of the chili and cucumber for the M-DAC system. The range of the crop transpiration rate is 1.15–1.53 $kg/m^2/h$, and the sensible heat load is 515–546 W/m^2 for both chili and cucumber, respectively.

The variation in crop transpiration rate and sensible heat load of flora and tomato for standalone DAC and M-DAC systems, respectively, in July is shown in Fig. 6.12. The left side plots illustrate the variations in crop transpiration rate and sensible heat load of the flora and tomato for the standalone DAC system. The range of the crop transpiration rate is 5.2–8.9 $kg/m^2/h$, and the sensible heat load is 444–475 W/m^2 for both flora and tomato, respectively. The right side plots illustrate the variations

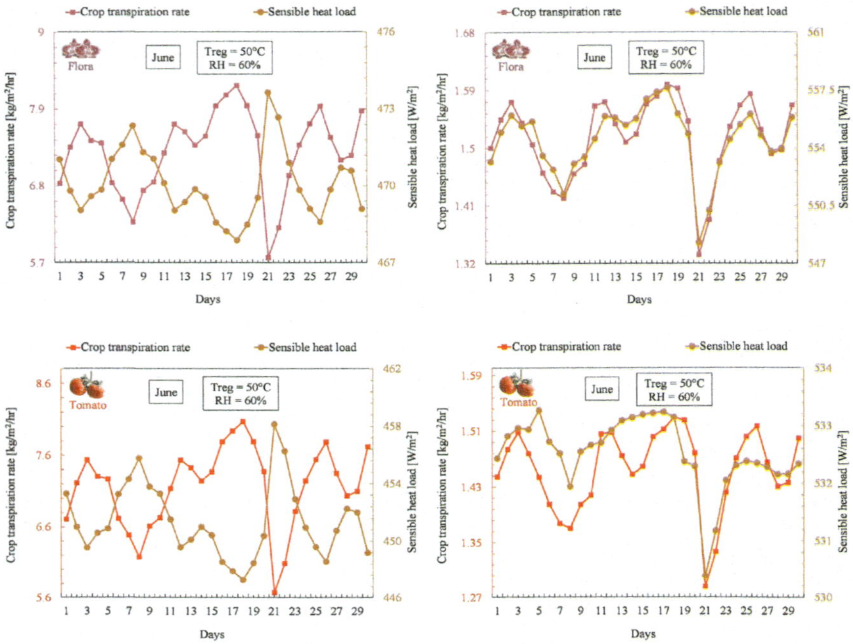

Fig. 6.10 Comparison of crop transpiration rate and sensible heat load variation of standalone DAC and M-DAC systems, respectively, at constant regenerate temperature of 50°C and greenhouse inside relative humidity of 60% for flora, and tomato, respectively, for June.

in crop transpiration rate and sensible heat load of the flora and tomato for the M-DAC system. The range of the crop transpiration rate is 1.2–1.64 kg/m^2/h, and the sensible heat load is 530–560 W/m^2 for both flora and tomato, respectively.

The variation in crop transpiration rate and sensible heat load of chili and cucumber for standalone DAC and M-DAC systems, respectively, in July is shown in Fig. 6.13 The left side plots illustrate the variations in crop transpiration rate and sensible heat load of the chili and cucumber for the standalone DAC system. The range of the crop transpiration rate is 5.2–8.74 kg/m^2/h, and the sensible heat load is 436–468 W/m^2 for both chili and cucumber, respectively. The right side plots illustrate the variations in crop transpiration rate and sensible heat load of the chili and cucumber for the M-DAC system. The range of the crop transpiration rate is 1.17–1.6 kg/m^2/h, and the sensible heat load is 523–547 W/m^2 for both chili and cucumber, respectively.

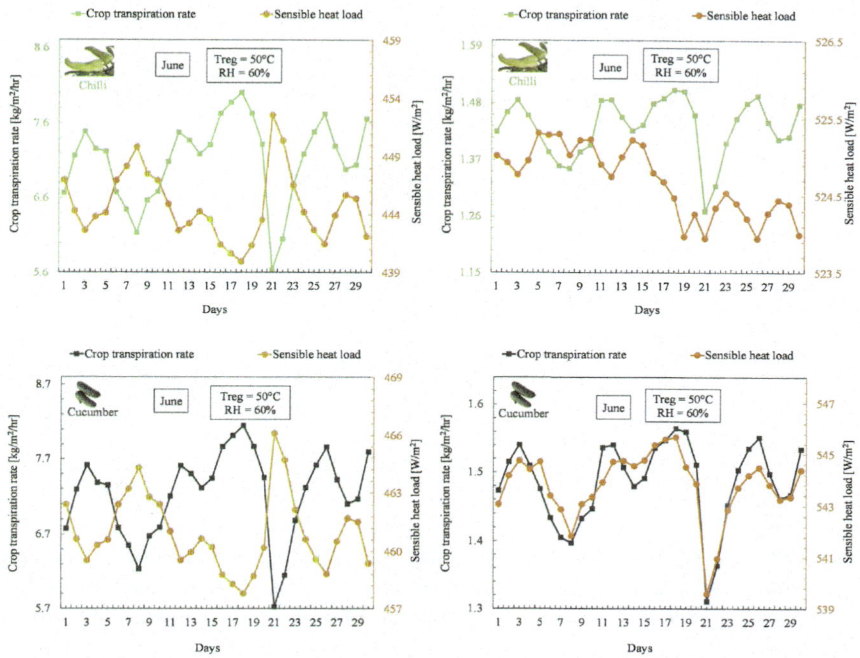

Fig. 6.11 Comparison of crop transpiration rate and sensible heat load variation of standalone DAC and M-DAC systems, respectively, at constant regenerate temperature of 50°C and greenhouse inside relative humidity of 60% for chilli and cucumber, respectively, for June.

The variation in crop transpiration rate and sensible heat load of flora and tomato for standalone DAC and M-DAC systems, respectively, in August is shown in Fig. 6.14. The left side plots illustrate the variations in crop transpiration rate and sensible heat load of the flora and tomato for the standalone DAC system. The range of the crop transpiration rate is 3.9–6.7 kg/m^2/h, and the sensible heat load is 456–478 W/m^2 for both flora and tomato, respectively. The right side plots illustrate the variations in crop transpiration rate and sensible heat load of the flora and tomato for the M-DAC system. The range of the crop transpiration rate is 1.03–1.35 kg/m^2/h, and the sensible heat load is 525–552 W/m^2 for both flora and tomato, respectively.

The variation in crop transpiration rate and sensible heat load of chili and cucumber for standalone DAC and M-DAC systems, respectively, in August is shown in Fig. 6.15. The left-side plots illustrate the variations in crop transpiration rate and sensible heat load of the chili and cucumber for

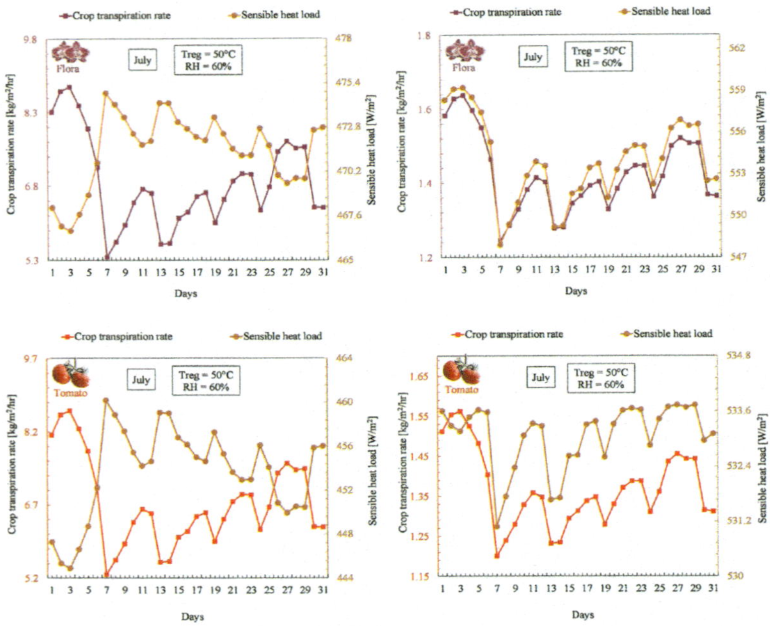

Fig. 6.12 Comparison of crop transpiration rate and sensible heat load variation of standalone DAC and M-DAC systems, respectively, at constant regenerate temperature of 50°C and greenhouse inside relative humidity of 60% for flora, and tomato, respectively, for July.

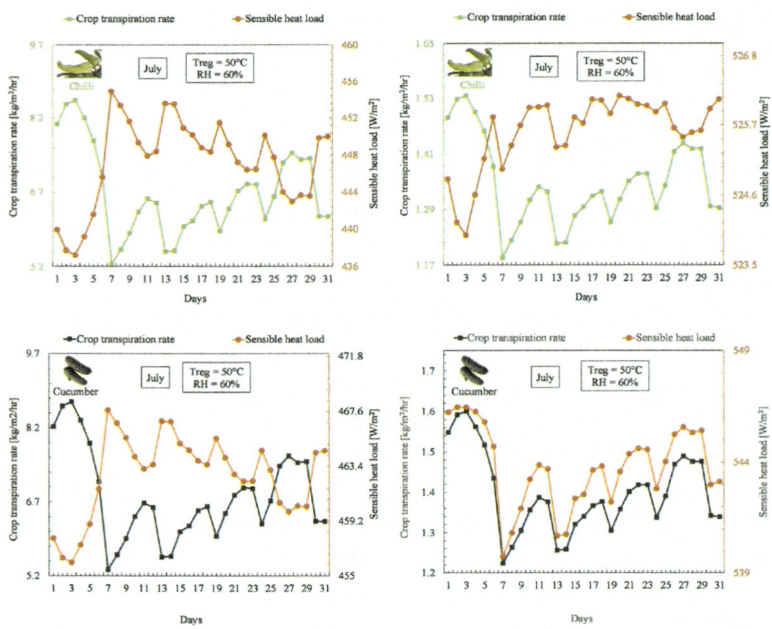

Fig. 6.13 Comparison of crop transpiration rate and sensible heat load variation of standalone DAC and M-DAC systems, respectively, at constant regenerate temperature of 50°C and greenhouse inside relative humidity of 60% for chilli, and cucumber, respectively, for July.

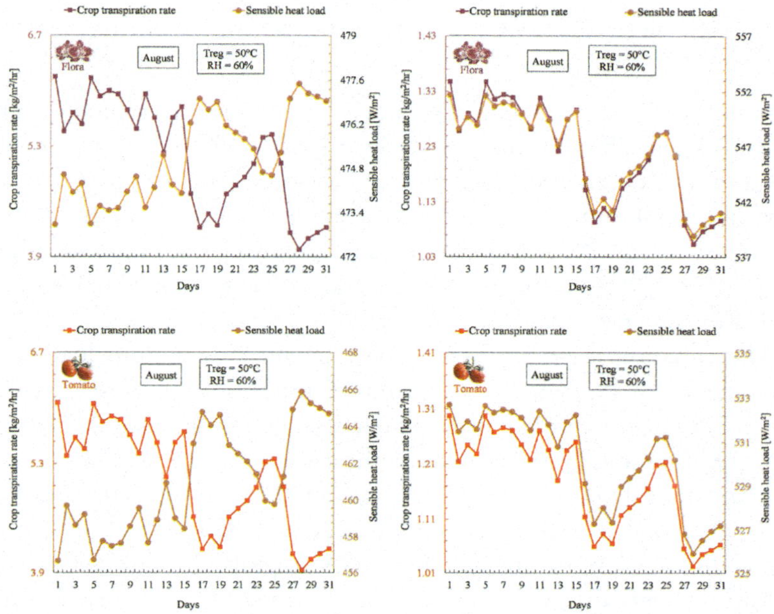

Fig. 6.14 Comparison of crop transpiration rate and sensible heat load variation of standalone DAC and M-DAC systems, respectively, at constant regenerate temperature of 50°C and greenhouse inside relative humidity of 60% for flora, and tomato, respectively, for August.

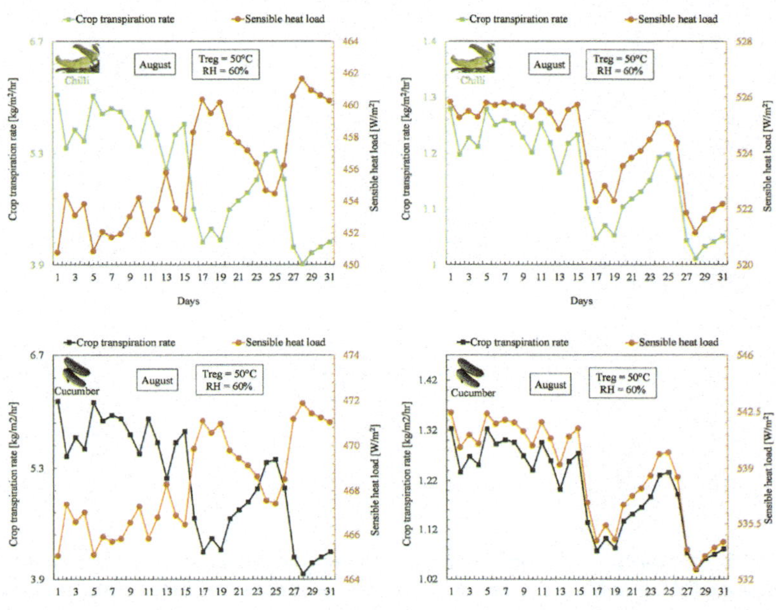

Fig. 6.15 Comparison of crop transpiration rate and sensible heat load variation of standalone DAC and M-DAC systems, respectively, at constant regenerate temperature of 50°C and greenhouse inside relative humidity of 60% for chilli, and cucumber, respectively, for August.

the standalone DAC system. The range of the crop transpiration rate is 3.9–6.13 kg/m^2/h, and the sensible heat load is 450–472 W/m^2 for both chili and cucumber, respectively. The right-side plots illustrate the variations in crop transpiration rate and sensible heat load of the chili and cucumber for the M-DAC system. The range of the crop transpiration rate is 1.0–1.33 kg/m^2/h, and the sensible heat load is 520–543 W/m^2 for both chili and cucumber, respectively.

The variation in crop transpiration rate and sensible heat load of flora and tomato for standalone DAC and M-DAC systems, respectively, in September is shown in Fig. 6.16. The left side plots illustrate the variations in crop transpiration rate and sensible heat load of the flora and tomato for the standalone DAC system. The range of the crop transpiration rate is 4.2–5.63 kg/m^2/h, and the sensible heat load is 458–477 W/m^2 for both flora and tomato, respectively. The right side plots illustrate the variations in crop transpiration rate and sensible heat load of the flora and tomato

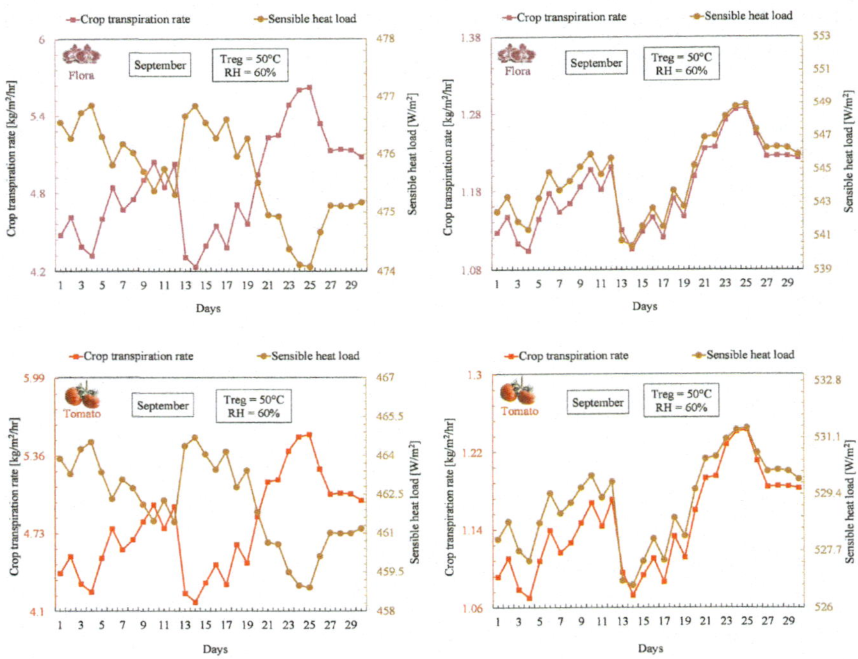

Fig. 6.16 Comparison of crop transpiration rate and sensible heat load variation of standalone DAC and M-DAC systems, respectively, at constant regenerate temperature of 50°C and greenhouse inside relative humidity of 60% for flora, and tomato, respectively, for September.

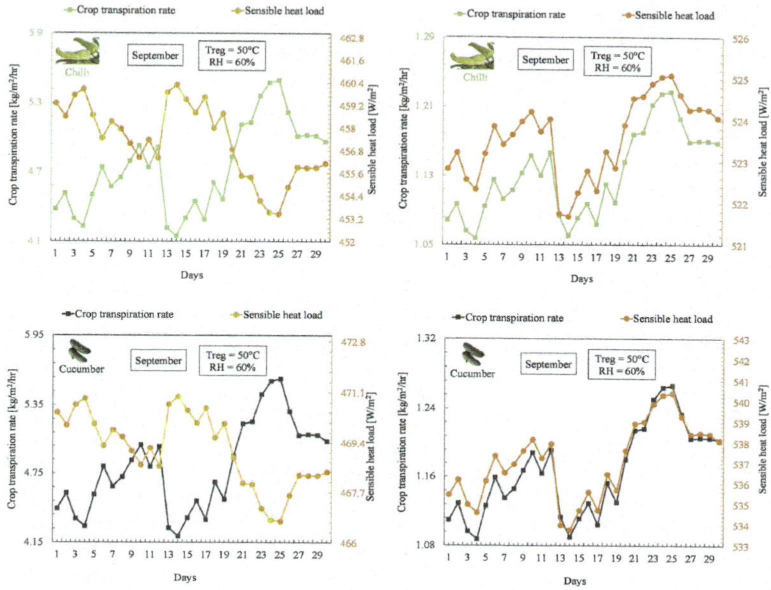

Fig. 6.17 Comparison of crop transpiration rate and sensible heat load variation of standalone DAC and M-DAC systems, respectively, at constant regenerate temperature of 50°C and greenhouse inside relative humidity of 60% for chilli and cucumber, respectively, for September.

for the M-DAC system. The range of the crop transpiration rate is 1.06–1.29 kg/m^2/h, and the sensible heat load is 526–549 W/m^2 for both flora and tomato, respectively.

The variation in crop transpiration rate and sensible heat load of chili and cucumber for standalone DAC and M-DAC systems, respectively, in September is shown in Fig. 6.17. The left side plots illustrate the variations in crop transpiration rate and sensible heat load of the chili and cucumber for the standalone DAC system. The range of the crop transpiration rate is 4.1–5.58 kg/m^2/h, and the sensible heat load is 452–471 W/m^2 for both chili and cucumber, respectively. The right side plots illustrate the variations in crop transpiration rate and sensible heat load of the chili and cucumber for the M-DAC system. The range of the crop transpiration rate is 1.05–1.27 kg/m^2/h, and the sensible heat load is 521–541 W/m^2 for both chili and cucumber, respectively.

The variation in crop transpiration rate and sensible heat load of flora and tomato for standalone DAC and M-DAC systems, respectively, in October is shown in Fig. 6.18. The left-side plots illustrate the variations

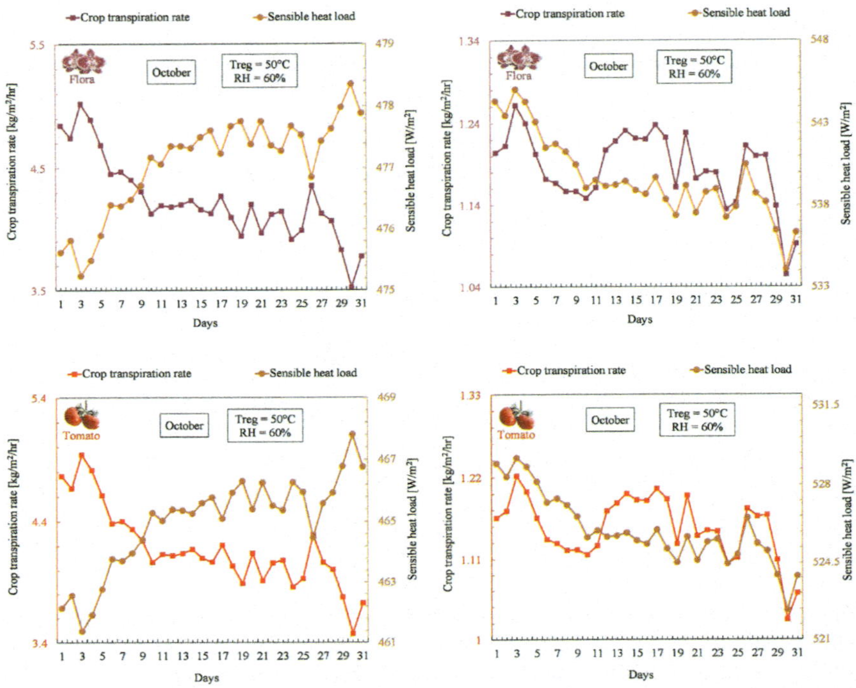

Fig. 6.18 Comparison of crop transpiration rate and sensible heat load variation of standalone DAC and M-DAC systems, respectively, at constant regenerate temperature of 50°C and greenhouse inside relative humidity of 60% for flora and tomato, respectively, for October.

in crop transpiration rate and sensible heat load of the flora and tomato for the standalone DAC system. The range of the crop transpiration rate is 3.4–5.1 kg/m^2/h, and the sensible heat load is 461–479 W/m^2 for both flora and tomato, respectively. The right-side plots illustrate the variations in crop transpiration rate and sensible heat load of the flora and tomato for the M-DAC system. The range of the crop transpiration rate is 1.0–1.26 kg/m^2/h, and the sensible heat load is 521–545 W/m^2 for both flora and tomato, respectively.

The variation in crop transpiration rate and sensible heat load of chili and cucumber for standalone DAC and M-DAC systems, respectively, in October is shown in Fig. 6.19. The left side plots illustrate the variations in crop transpiration rate and sensible heat load of the chili and cucumber for the standalone DAC system. The range of the crop transpiration rate is 3.4–5.0 kg/m^2/h, and the sensible heat load is 456–474 W/m^2 for both chili and cucumber, respectively. The right side plots illustrate the variations

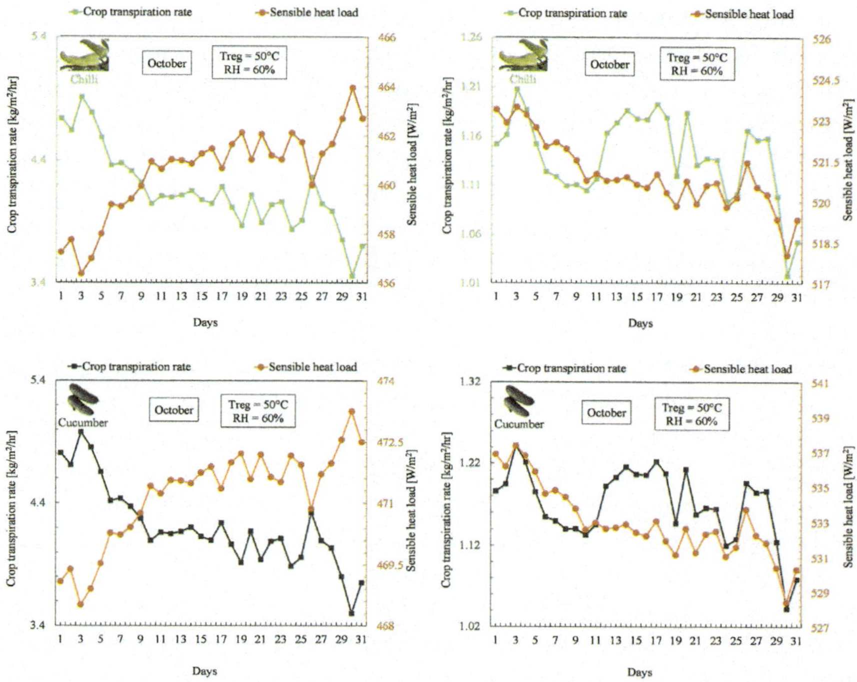

Fig. 6.19 Comparison of crop transpiration rate and sensible heat load variation of standalone DAC and M-DAC systems, respectively, at constant regenerate temperature of 50°C and greenhouse inside relative humidity of 60% for chilli and cucumber, respectively, for October.

in crop transpiration rate and sensible heat load of the chili and cucumber for the M-DAC system. The range of the crop transpiration rate is 1.0–1.21 kg/m^2/h, and the sensible heat load is 517–538 W/m^2 for both chili and cucumber, respectively.

6.4 Conclusions

In this study, the thermal modeling of a greenhouse was done for a standalone DAC system and an M-DAC system, respectively. The climate data of the summer season of Multan, Pakistan, was gained from Meteonorm software. In thermal modeling, the variations in ambient temperature, cooling temperature, and greenhouse temperature were calculated for different plants, i.e., flora, tomato, chili, and cucumber, and compared for the standalone DAC system and the M-DAC system, respectively. Furthermore, the crop transpiration rate and sensible heat load of the greenhouse were calculated for different plants such as flora, tomato, chili, and cucumber, for both the standalone DAC system and M-DAC system. The range of ambient temperature, temperature achieved by the standalone DAC system, and the greenhouse temperature for flora, tomato, chili, and cucumber were 30.16°C–38.19°C, 31.36°C–38.94°C, and 39.87°C–47.34°C, 39.65°C–47.02°C, 39.57°C–46.91°C, and 39.68°C–47.19°C, respectively. The range of crop transpiration rate and sensible heat load of flora, tomato, chili, and cucumber were 4.36–6.95 kg/m^2/h. and 470–477 W/m^2, 4.30–6.81 kg/m^2/h. and 453–468 W/m^2, 4.27–6.77 kg/m^2/h. and 446–460 W/m^2, 4.3–6.9 kg/m^2/h. and 457–471 W/m^2, respectively, for the standalone DAC system. The range of ambient temperature, temperature achieved by the M-DAC system, and greenhouse temperature for flora, tomato, chili, and cucumber were 22.85°C–31.53°C, 13°C–16°C, 22°C–-25°C, 21.9°C–24.4°C, 21.8°C–24.3°C, 21.9°C–24.5°C, respectively. The range of crop transpiration rate and sensible heat load of flora, tomato, chili, and cucumber were 1.1–1.5 kg/m^2/h. and 540–554 W/m^2, 1.12–1.45 kg/m^2/h. and 526–537 W/m^2, 1.12–1.43 kg/m^2/h. and 520–527 W/m^2, 1.13–1.48 kg/m^2/h. and 528–543 W/m^2, respectively, for the M-DAC system. Future research should focus on enhancing the design of advanced materials and systems to improve consistency and effectiveness. Dehumidification performance can be enhanced by utilizing innovative desiccant materials with higher moisture absorption capacities and lower regeneration energy requirements. Similarly, improving heat exchanger design, application of coatings, and water recycling features of evaporative cooling systems, efficient evaporative media (low-pressure drop and high saturation efficiency), optimized watering/spraying, shading, leakproof water tanks, and adjusting fan speed based on real-time conditions can minimize the loss of water associated with them and increase operational life.

Acknowledgment

This research work is completed with the financial support from the Higher Education Commission (HEC) of Pakistan through the research Project NRPU-17029 awarded to Bahauddin Zakariya University.

Funding statement

This research work is completed with the financial support from the Higher Education Commission (HEC) of Pakistan through the research Project NRPU-17029 awarded to Bahauddin Zakariya University.

CRediT authorship contribution statement

Fakhar Abbas was involved in the conceptualization, methodology, software, formal analysis, investigation, writing of the original draft. **Muhammad Sultan and Takahiko Miyazaki** were involved in the the conceptualization, methodology, validation, resources, visualization, supervision, project administration, writing (review, and editing), funding acquisition. **Hafiz Muhammad Umar Raza, Muhammad Hamid Mahmood, Muhammad Ali Imran, Muhammad Wakil Shahzad, Jie Liu, and Zhaoli Zhang** were involved in the conceptualization, methodology, software, formal analysis, investigation, writing (review, and editing).

Data availability statement

The datasets used and/or analyzed during the current study are available from the corresponding authors upon request.

Conflicts of interest

The authors declare no conflict of interest.

Declaration of generative AI and AI-assisted technologies in the writing process

No generative AI/AI-assisted technology is used in this paper.

References

Akpenpuun, T.D., Ogunlowo, Q.O., Na, W.H., Rabiu, A., Adesanya, M., Dutta, P., Zakir, E., Ogundele, O.M., Kim, H.T., Lee, H.W., 2023. Review of temperature management strategies and techniques in the greenhouse microenvironment. Adeleke Univ. J. Eng. Technol. 6, 126–147.

Anisimov, S., Pandelidis, D., 2014. Numerical study of the Maisotsenko cycle heat and mass exchanger. Int. J. Heat Mass Transf. 00179310. 75, 75–96. doi:10.1016/j.ijheatmasstransfer.2014.03.050.

Aziz, A.N., Mahmoud, S., Al-Dadah, R., Ismail, M.A., Mesfer, M.K.A., 2022. Numerical and experimental investigation of desiccant cooling system using metal organic framework materials. Appl. Therm. Eng. 13594311. 215. Available from: http://www.journals.elsevier.com/applied-thermal-engineering/. doi:10.1016/j.applthermaleng.2022.118940.

Badji, A., Benseddik, A., Bensaha, H., Boukhelifa, A., Hasrane, I., 2022. Design, technology, and management of greenhouse: a review. J. Clean. Prod. 09596526. 373, 133753. doi:10.1016/j.jclepro.2022.133753.

Banik, P., Ganguly, A., 2017. Performance and economic analysis of a floricultural greenhouse with distributed fan-pad evaporative cooling coupled with solar desiccation. Sol. Energy. 0038092X. 147, 439–447. doi:10.1016/j.solener.2017.03.057.

Bauri, B., Ganguly, A., 2013. Effect of crop transpiration on the microcilmate of a naturally ventilated greenhouse. Int. J. Emerg. Technol. Adv. Eng. 3, 337–343.

Boulard, T., Wang, S., 2000. Greenhouse crop transpiration simulation from external climate conditions. Agric. For. Meteorol. 01681923. 100 (1), 25–34. doi:10.1016/s0168-1923(99)00082-9.

Buyadgie, D., Buyadgie, O., Drakhnia, O., Brodetsky, P., Maisotsenko, V., 2015. Solar low-pressure turbo-ejector Maisotsenko cycle-based power system for electricity, heating, cooling and distillation. Int. J. Low Carbon Technol. 1748-1317. 10 (2), 157–164. doi:10.1093/ijlct/ctv012.

Dalai, S., 2020. Green-houses: Types and Structural Components. New Delhi Publishers. doi:10.30954/ndp-pcsa.2020.2.

Hatfield, J.L., Prueger, J.H., 2015. Temperature extremes: effect on plant growth and development. Weather Clim. Extrem. 22120947. 10, 4–10. Available from: http://www.journals.elsevier.com/weather-and-climate-extremes/. doi:10.1016/j.wace.2015.08.001.

Hua, Z., Cai, S., Hongyang, X., Yuan, W., Song, L., Zhengkai, T., 2024. Investigations of Silica/MOF composite coating and its dehumidification performance on a desiccant-coated heat exchanger. Energy. 03605442. 307, 132576. doi:10.1016/j.energy.2024.132576.

Kousar, R., Ali, M., Sheikh, N.A., Gilani, S.I.U.H., Khushnood, S., 2021. Holistic integration of multi-stage dew point counter flow indirect evaporative cooler with the solar-assisted desiccant cooling system: a techno-economic evaluation. Energy Sustain. Dev. 23524669. 62, 163–174. Available from: http://www.elsevier.com. doi:10.1016/j.esd.2021.04.005.

La, D., Dai, Y.J., Li, Y., Wang, R.Z., Ge, T.S., 2010. Technical development of rotary desiccant dehumidification and air conditioning: a review. Renew. Sustain. Energy Rev. 13640321. 14 (1), 130–147. doi:10.1016/j.rser.2009.07.016.

Lai, L., Wang, X., Kefayati, G., Eric, H., 2021. Evaporative cooling integrated with solid desiccant systems: a review. Energies. 1996-1073. 14 (18), 5982. doi:10.3390/en14185982.

Mahmood, M.H., Sultan, M., Miyazaki, T., Koyama, S., Maisotsenko, V.S., 2016. Overview of the Maisotsenko cycle – a way towards dew point evaporative cooling. Renew. Sustain. Energy Rev. 18790690. 66, 537–555. Available from: https://www.journals.elsevier.com/renewable-and-sustainable-energy-reviews. doi:10.1016/j.rser.2016.08.022.

Mandal, C., Ganguly, A., 2022. Thermo-economic analysis of two stages desiccant supported greenhouse cooling system for orchid cultivation in the tropical and sub-tropical region. Sci. Technol. Built Environ 28, 1237–1254. doi:10.1080/23744731.2022.2038491.

Masterpole, Z., 2020. Nonprofit use of deep winter greenhouses : minnesota case studies. Center for Urban and Regional Affairs: Minneapolis, MN, USA.

Mohammadi, S., Khalife, E., Kaveh, M., Sayyah, A. H.A., Nikbakht, A.M., Szymanek, M., Dziwulski, J., 2021. Comparison of optimized and conventional models of passive solar greenhouse—case study: The indoor air temperature, irradiation, and energy demand 14 (17), 5369. doi: 10.3390/en14175369.

Soussi, M., Chaibi, M.T., Buchholz, M., Saghrouni, Z., 2022. Comprehensive review on climate control and cooling systems in greenhouses under hot and arid conditions. Agronomy 20734395. 12 (3). Available from: https://www.mdpi.com/2073-4395/12/3/626/pdf. doi:10.3390/agronomy12030626.

Sultan, M., El-Sharkawy, I.I., Miyazaki, T., Saha, B.B., Koyama, S., 2015. An overview of solid desiccant dehumidification and air conditioning systems. Renew. Sustain. Energy Rev. 18790690. 46, 16–29. Available from: https://www.journals.elsevier.com/renewable-and-sustainable-energy-reviews. doi:10.1016/j.rser.2015.02.038.

Sultan, M., Miyazaki, T., Koyama, S., Saha, B.B., 2014. Utilization of desiccant air-conditioning system for improvement in greenhouse productivity: a neglected area of research in Pakistan. Int. J. Environ 4, 1–10.
Sultan, M., Miyazaki, T., Saha, B.B., Koyama, S., 2016. Steady-state investigation of water vapor adsorption for thermally driven adsorption based greenhouse air-conditioning system. Renew. Energy 18790682. 86, 785–795. Available from: http://www.journals.elsevier.com/renewable-and-sustainable-energy-reviews/. doi:10.1016/j.renene.2015.09.015.
Yang, B., Wang, C., Ji, X., Nie, J., Zhang, R., Yanmei, L., Chen, Q., 2024. A solar-assisted regenerative desiccant air conditioning with indirect evaporative cooling for humid climate region. Appl. Therm. Eng. 13594311. 243, 122598. doi:10.1016/j.applthermaleng.2024.122598.

CHAPTER 7

Advanced indirect evaporative cooling and desiccant dehumidification systems for greenhouse farming in arid and semi-arid climates

Hafiz Muhammad Umar Raza[a], Muhammad Sultan[a], Muhammad Aleem[a], Takahiko Miyazaki[b], Md Shamim Ahamed[c], Muhammad Wakil Shahzad[d], Zhaoli Zhang[e], and Hafiz Muhammad Ali[f]

[a]Department of Agricultural Engineering, Bahauddin Zakariya University, Multan, Pakistan
[b]Department of Advanced Environmental Science and Engineering, Faculty of Engineering Sciences, Kyushu University, Kasuga, Fukuoka, Japan
[c]Department of Biological and Agricultural Engineering, University of California, Davis, CA, United States
[d]Department of Mechanical and Construction Engineering, Northumbria University, Newcastle Upon Tyne, United Kingdom
[e]School of Mechanical Engineering, Southwest Jiaotong University, Chengdu, China
[f]Department of Mechanical Engineering, King Fahd University of Petroleum and Minerals, Dhahran, Saudi Arabia

7.1 Introduction

Photosynthesis and evapotranspiration are fundamental processes that contribute to plant growth by releasing water vapor into the atmosphere, thereby continuously raising the humidity levels within a greenhouse (Ashraf et al., 2021; Sultan et al., 2021; Achour et al., 2021). This elevated humidity can lead to potential issues such as insect infestations, fungal growth, and condensation. Photosynthesis is a crucial process in which plants convert carbon dioxide (CO_2) into carbohydrates using light energy from the sun (Sultan et al., 2016). As part of this process, plants expel a significant portion of the water they absorb through their roots into the air, as given by the following Eq. (1) (Sultan et al., 2021; Rawson et al., 1977; Sultan et al., 2014):

$$2nCO_2 + 4n(H_2O)\text{ from roots} \xrightarrow{\text{Photons}} 2(CH_2O)_n + 2nO_2 + 2n(H_2O)\text{ into air} \tag{1}$$

https://doi.org/10.1016/B978-0-443-48555-8.00014-1

Evapotranspiration encompasses both the evaporation occurring from the soil and the transpiration from plants (Sultan et al., 2016). For various plant species, the average rate of transpiration can vary between 150 and 450 g per day per square meter of leaf area, depending on specific conditions (Sultan et al., 2018a). While photosynthesis primarily occurs during daylight hours and under supplemental lighting, evapotranspiration is high under lighting conditions and low in the dark. Fig. 7.1 shows a general description of (A) evapotranspiration and (B) the photosynthesis process by a plant (Sultan et al., 2021). Adjusting the flow rate of CO_2-enriched outdoor air and the intensity of solar radiation can affect the efficiency of these processes, which in turn introduces variability in the sensible and latent cooling demands for air conditioning (Costantino et al., 2021).

From the previous discussion, it can be concluded that greenhouse farming is primarily depends on the optimal control of temperature and humidity, as

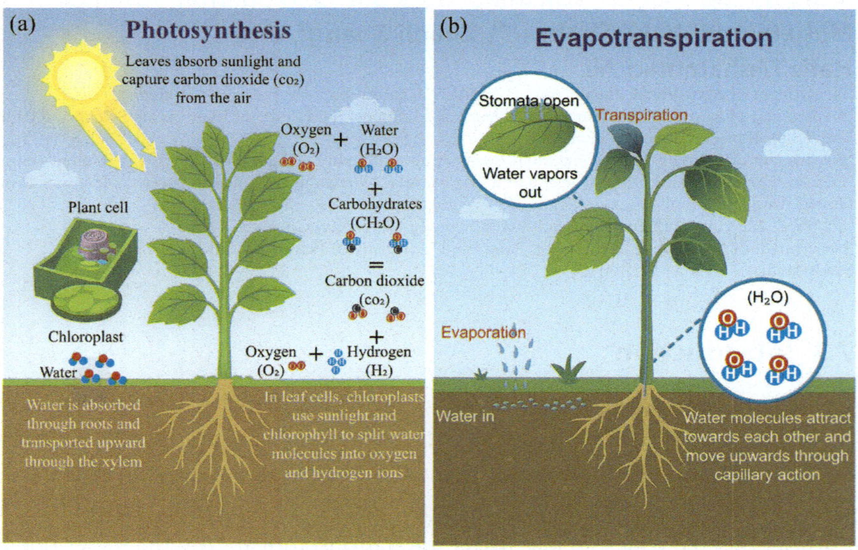

Fig. 7.1 General description of (A) evapotranspiration, and (B) the photosynthesis process by a plant. *(From Sultan, M., Ashraf, H., Miyazaki, T., Shamshiri, R.R., Hameed, I. A., 2021. Temperature and humidity control for the next generation greenhouses: overview of desiccant and evaporative cooling systems. IntechOpen. https://doi.org/10.5772/intechopen.97273; Ashraf, H., Sultan, M., Shamshiri, R.R., Abbas, F., Farooq, M., Sajjad, U., Md-Tahir, H., Mahmood, M.H., Ahmad, F., Taseer, Y.R., Shahzad, A., Niazi, B.M.K., 2021. Dynamic evaluation of desiccant dehumidification evaporative cooling options for greenhouse air-conditioning application in multan (Pakistan). Energies 19961073. 14 (4). Available from: https://www.mdpi.com/1996-1073/14/4/1097/pdf. doi:10.3390/en14041097)*

plants are highly susceptible to these conditions. Fig. 7.2A shows a psychrometric chart of optimal temperature and humidity zones for greenhouse farming. Hourly ambient climatic conditions of Multan are shown on the psychrometric chart, which highlights the critical need for a temperature and humidity control system for optimum greenhouse farming. The ideal temperature and humidity ranges are 15°C–30°C and 40%–90%, respectively (Ashraf et al., 2021). In terms of vapor pressure deficit (VPD), that is the difference between the saturated vapor pressure and the actual vapor pressure of air, the ideal range for the plants to grow inside the greenhouse is between 0.8 and 0.9 kPa; however, the normal greenhouse growth zones is between 0.4 and 1.2 kPa (Sultan et al., 2016; Amani et al., 2020; López-Morales et al., 2008). Fig. 7.2B shows the normal and ideal greenhouse growth zones and estimated vapor pressure deficit profiles based on the average climatic conditions of Multan.

The humidity level increases in the greenhouse owing to continuous transpiration by plants and condensation/dripping of water vapor through the greenhouse cover walls, which results in pest/fungus/disease attacks, thereby affecting overall yield (Amani et al., 2020; Ahamed et al., 2023). Fig. 7.3 shows possible impacts of low and high humidity levels on a plant inside the greenhouse. In view of the above scenario, an optimum temperature and humidity control unit is critical for developing greenhouse farming culture in Pakistan. Conventional technologies and practices used in Pakistan to control greenhouse temperature and humidity are either expensive or inefficient. In developed countries, vapor-compression-based technologies/systems are used; however, they consume a huge amount of electricity, which makes such systems economically unfeasible at least in the Pakistani situation (Sultan et al., 2018b; Mahmood et al., 2016b; Sultan et al., 2015; Bhat et al., 2024). In addition, these systems control humidity by cooling the air below the dew point temperature, thusdistinct control of temperature and humidity is not possible. The concept of evaporative cooling pad (swamp cooler) systems is not new in Pakistan. There are many companies and products available in Pakistan that control the air temperature via water vapor evaporation, and these devices are common for air conditioning in the summer season (Sultan et al., 2016; Mahmood et al., 2016b).

Advanced indirect evaporative cooling (Maisotsenko cycle, or simply M-Cycle) system and desiccant dehumidification are emerging technologies for temperature and humidity control that can be used in greenhouse farming. The advantage of the advanced indirect evaporative cooling concept is that it can control temperature without increasing absolute humidity, making it

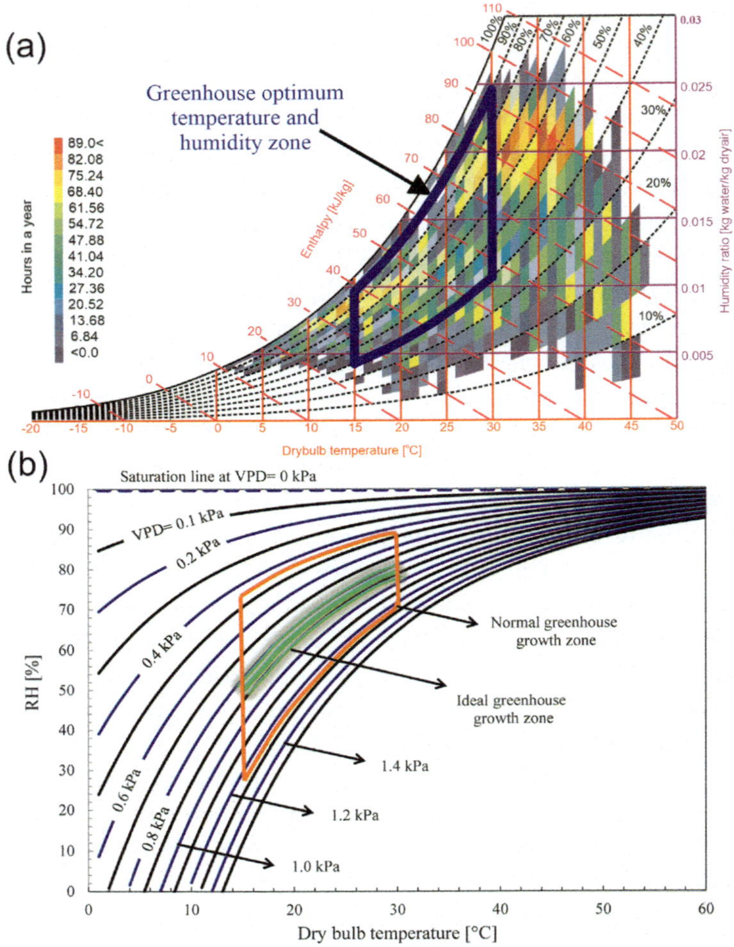

Fig. 7.2 (A) The psychrometric representation of the optimum temperature and humidity zone for greenhouse farming with hourly-based climatic conditions of Multan and (B) normal and ideal greenhouse growth zone based on vapor pressure deficit (VPD). *(From Sultan, M., Ashraf, H., Miyazaki, T., Shamshiri, R.R., Hameed, I.A., 2021. Temperature and humidity control for the next generation greenhouses: overview of desiccant and evaporative cooling ystems. IntechOpen. https://doi.org/10.5772/intechopen.97273; Ashraf, H., Sultan, M., Shamshiri, R.R., Abbas, F., Farooq, M., Sajjad, U., Md-Tahir, H., Mahmood, M. H., Ahmad, F., Taseer, Y.R., Shahzad, A., Niazi, B.M.K., 2021. Dynamic evaluation of desiccant dehumidification evaporative cooling options for greenhouse air-conditioning application in multan (Pakistan). Energies 19961073. 14 (4). Available from: https://www.mdpi.com/1996-1073/14/4/1097/pdf. doi:10.3390/en14041097; Sultan, M., Miyazaki, T., Saha, B. B., Koyama, S., 2016. Steady-state investigation of water vapor adsorption for thermally driven adsorption based greenhouse air-conditioning system. Renew. Energy 18790682. 86, 785–795. Available from: http://www.journals.elsevier.com/renewable-and-sustainable-energy-reviews/. doi:10.1016/j.renene.2015.09.015.)*

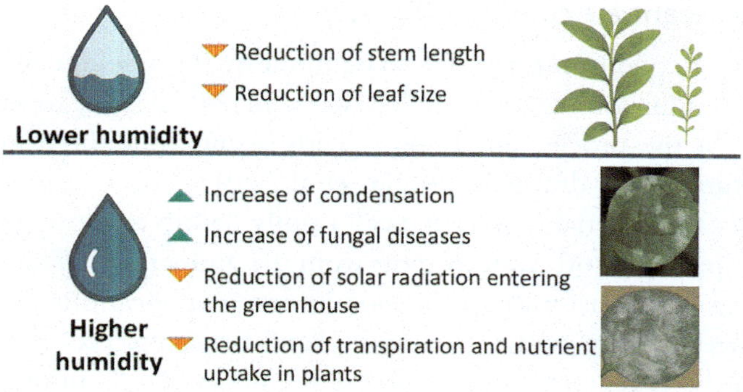

Fig. 7.3 Possible impacts of low and high humidity levels on a plant inside the greenhouse. *Data from: Amani, M., Foroushani, S., Sultan, M., Bahrami, M., 2020. Comprehensive review on dehumidification strategies for agricultural greenhouse applications. Appl. Therm. Eng. 181. doi:10.1016/j.applthermaleng.2020.115979.*

suitable for the monsoon (rainy) season as well. On the other hand, a desiccant dehumidification unit can control the humidity levels inside the greenhouse to prevent fungal/pest/disease attacks. These technologies have shown promising results in the literature (Sultan et al., 2016; Ashraf et al., 2021; Amani et al., 2020; Ahamed et al., 2023; Ahamed et al., 2023; Sultan et al., 2015) for optimal temperature and humidity control, energy consumption, and coefficient of performance in efficient greenhouse farming.

This study develops a lab-scale greenhouse by integrating a heat-driven desiccant dehumidification unit with an evaporative cooling unit. A heat-energy-driven heating unit has been integrated into the greenhouse to regenerate the desiccant dehumidification unit. The slope of the dehumidification line is calculated, and using this value, the current study investigates two types of air-conditioning systems: (i) a desiccant dehumidification air-conditioning (DAC) system, and (ii) M-Cycle-assisted desiccant air-conditioning (MDAC) system. The performance of the systems is evaluated across five cities in Pakistan for greenhouse farming applications.

7.2 Proposed air-conditioning systems

In this study, two types of air conditioning systems: (i) a desiccant air-conditioning (DAC) system, and (ii) M-Cycle-assisted desiccant air-conditioning (MDAC) system, are studied and examined for five cities in Pakistan for greenhouse applications.

7.2.1 Desiccant air-conditioning system

Desiccant air conditioning is used to dehumidify the ambient air. It consists of a dehumidification unit coupled with an AC system to control the humidity of the process air. It deals with latent (humidity) and sensible (temperature) loads distinctly (Sultan et al., 2015). In a dehumidification unit, desiccant material is used that can readily desorb at a low regeneration temperature and adsorb water vapors from the ambient air. The most often utilized desiccant materials for air conditioning applications include silica gel, synthetic zeolite, activated alumina, and bentonite clay (Aleem et al., 2020). In the regeneration cycle, the desiccant material can desorb at low regeneration temperature by using low-grade waste heat, biogas, and/or solar energy (Sultan et al., 2015; Sultan and Miyazaki, 2017; Sultan et al., 2020). The use of DAC is growing in areas with comparatively high humidity levels.

In principle, ambient air passes through a desiccant unit for dehumidification, where its temperature increases due to the heat of adsorption (air condition 1–2). In an ideal situation, this process will be isenthalpic (neglecting adsorption heat). In the second step, hot dehumidified air passes through the heat exchanger to reduce its temperature (air condition 2–3). On the other hand, the desiccant rotor/block will become saturated after some time. To regenerate the desiccant, ambient/process air passes through a heat exchanger to increase the temperature of the air (air condition 4). Further, this hot air passes through the heat source to reach the desired regeneration temperature (air condition 5-6). At the end, this hot regenerated air passes through the desiccant to remove the adsorbed water vapors for reuse (air condition 6-7). Fig. 7.4A shows the schematic and thermal process of DAC on a psychometric chart.

7.2.2 Maisotsenko cycle assisted desiccant air-conditioning system

Maisotsenko cycle assisted desiccant air-conditioning (MDAC) system combines the features of Maisotsenko cycle evaporative cooling (MEC) and DAC systems. The MDAC system is efficient in all types of climates. It can deal with the latent and sensible load of air-conditioning (AC) distinctly (Sultan et al., 2020). The latent load of the AC system is achieved by desiccant dehumidification, whereas the sensible load is achieved by MEC. In addition, the MDAC system can use the exhaust waste heat from the MEC wet-channel for regeneration of the desiccant rotor/block (Mahmood et al., 2016b, 2019). In principle, ambient air

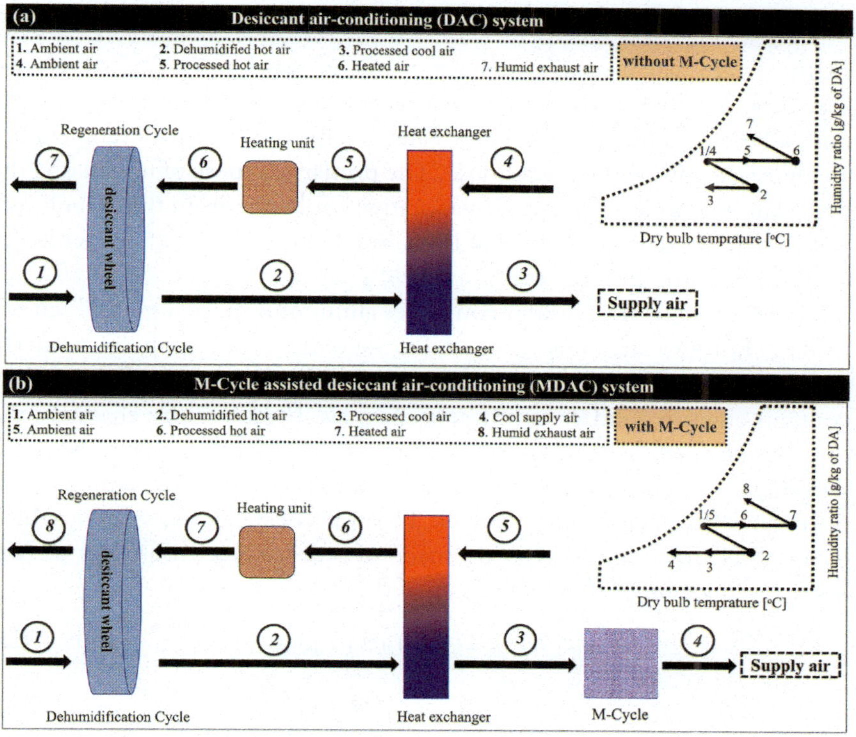

Fig. 7.4 Schematic and Psychrometric representation of the proposed systems: (A) desiccant air-conditioning (DAC) system, and (B) M-Cycle assisted desiccant air-conditioning (MDAC) system.

passes through a desiccant unit for dehumidification, where its temperature increases due to the heat of adsorption (air condition 1–2). In the second step, hot dehumidified air passes through the heat exchanger to reduce the temperature of the dehumidified air (air condition 2–3). In the third step, air passes through the MEC system to achieve the desired cooling conditions (air condition 3–4). On the other hand, the desiccant rotor/block will be saturated after some time. To regenerate the desiccant, ambient/process air passes through a heat exchanger to raise its temperature (air condition 5-6). Further, this hot air passes through the heat source to reach the desired regeneration temperature (air condition 6-7). At the end, this hot regenerated air passes through the desiccant to remove the adsorbed water vapor for reuse (air condition 7–8). Fig. 7.4B shows the schematic and thermal process of MDAC on a psychometric chart.

7.3 Materials and methods

In this study, a lab-scale greenhouse unit is developed. The developed greenhouse is integrated with a desiccant dehumidification unit, an MEC unit, a heating unit, and ventilation fans, to effectively control temperature and humidity inside the greenhouse. The primary purpose of integrating the desiccant dehumidification and evaporative cooling units in the greenhouse was to manage sensible and latent loads, respectively, inside the greenhouse to optimum crop production.

A developed MEC system comprises aluminum sheets upon which felt made of fibrous material is attached. In case of dry channels, holes are made in order to ease the air from the dry channels to the wet channel. Each sheet contains eight dry and 20 wet/water channels. In case of dry channels, the top five channels serve as product channels, while the last three serve as working channels. In each working channel, holes are made with a 4 mm diameter with a gap of one channel. The walls of the MEC system, which cover all channels and sheets, are made of stainless steel. Eqs. (2a) to (2d) show the inlet and outlet conditions of the MEC system.

$$(T_{in})_{dp} \leq (T_{out})_{db} \leq (T_{in})_{wb} \tag{2a}$$

$$RH_{out} > RH_{in} \tag{2b}$$

$$X_{out} = X_{in} \tag{2c}$$

$$h_{out} < h_{in} \tag{2d}$$

where, subscripts "in" and "out" show the inlet and outlet conditions, T_{db} is dry-bulb temperature, T_{wb} is wet-bulb temperature, T_{dp} is dew-point temperature, RH is relative humidity, X is humidity ratio, and h is enthalpy.

A developed desiccant dehumidification system contains a desiccant wheel/rotor, a motor, and a gear assembly to rotate the wheel and control temperature and humidity. The desiccant wheel undergoes two processes simultaneously, namely, dehumidification and regeneration. In dehumidification, the inlet/ambient/process air passes through a dedicated portion of the desiccant wheel, where it is dehumidified (its humidity ratio is reduced) due to the hydrophilic nature of the silica gel, while its temperature increases relative to the inlet air. On the other hand, the heated/regeneration air from the heating unit passes through the dedicated portion of the wheel to regenrate and desorb the adsorbed water vapors. The ambient or process air is directed towards the desiccant dehumidification unit, which supplies hot, dehumidified air after adsorbing water vapors. The hot and dehumidified air is directed toward the MEC unit, resulting in a decrease in temperature. The cooled air is supplied inside the greenhouse to control temperature and humidity. For regeneration of the desiccant dehumidification unit, ambient air is directed to the heating unit, thereby increasing its temperature. The heated air is directed towards the desiccant dehumidification unit to desorb the water vapors. Both dehumidification and regeneration processes occurred simultaneously in order to control temperature and humidity requirements inside the greenhouse. The experimental data of DAC system is further utilized to estimate output temperature from heat exchanger given by Eqs. (3a) and (3c). Likewise, the output or supply air conditions from M-Cycle EC are estimated as given by Eq. (3b).

$$(T_3)_{db} = (T_2)_{db} - \varepsilon_{HX}\left((T_2)_{db} - (T_1)_{db}\right) \tag{3a}$$

$$(T_4)_{db} = (T_3)_{db} - \varepsilon_{MEC}\left((T_3)_{db} - (T_1)_{wb}\right) \tag{3b}$$

$$(T_6)_{db} = (T_{1,5})_{db} + \varepsilon_{HX}\left((T_2)_{db} - (T_{1,5})_{wb}\right) \tag{3c}$$

where, 1–6 subscripts show the air states, T_{db} is dry-bulb temperature, T_{wb} is wet-bulb temperature, ε_{HX} shows the effectiveness of the heat exchanger, and ε_{MEC} shows the effectiveness of the M-Cycle air-conditioning system. Fig. 7.5 shows psychrometric representations of the air handling process of the temperature and humidity control system (desiccant dehumidification, evaporative cooling, heat exchanger, and heating units) inside the greenhouse. Additionally, a 3.2 kW heating was integrated to meet regeneration/heating requirements of the desiccant dehumidification unit and the greenhouse environment. The overall working strategy of the lab-scale greenhouse unit with a desiccant dehumidification cum evaporative cooling system to control air temperature and humidity inside the greenhouse is shown in Fig. 7.6.

During dehumidification, the slope of the dehumidification line on the psychrometric chart should ideally be isenthalpic. This indicates that when water vapor condenses into the liquid phase during the desiccant dehumidification process, the latent heat of condensation equals to the adsorption heat released. However, experimental observations indicate that the real adsorption heat is marginally higher than the ideal value due to the

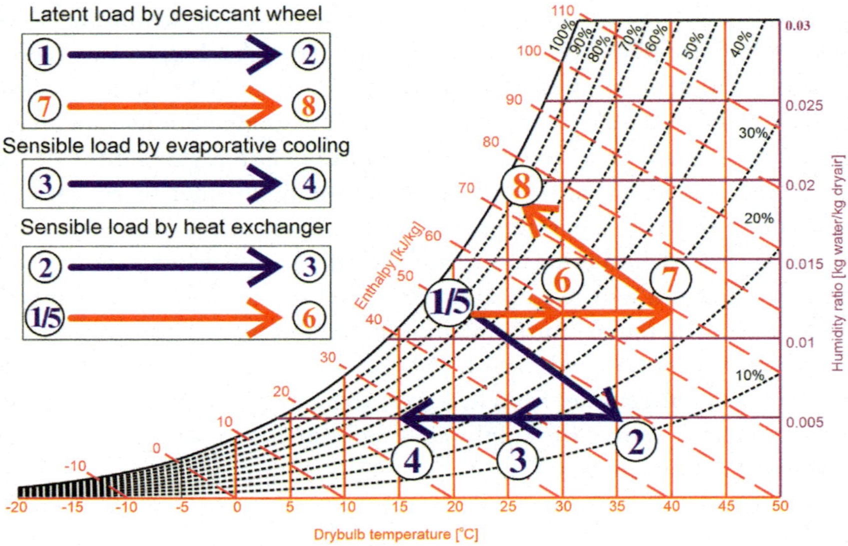

Fig. 7.5 The psychrometric representation of the air handling process for the developed lab-scale M-Cycle-assisted DAC system for the greenhouse unit.

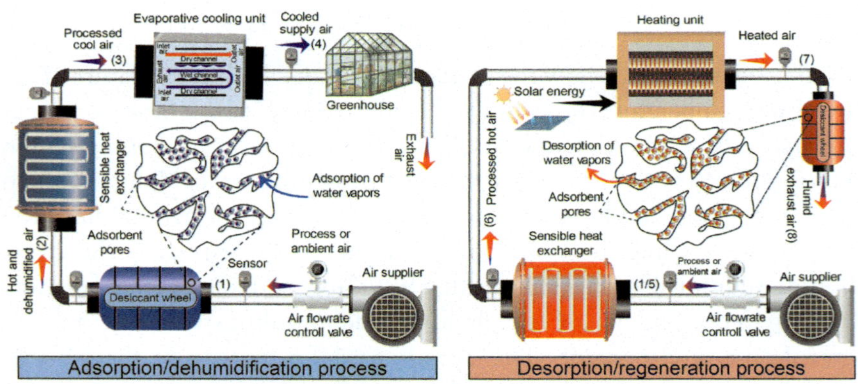

Fig. 7.6 The detailed schematics of the working strategy of the developed lab-scale greenhouse unit integrated with a desiccant dehumidification, and evaporative cooling units, to control air temperature and humidity inside the greenhouse.

contribution of isosteric heat of adsorption, and Eq. (4a) can be used to illustrates this behavior. The adsorption mechanism and the particular interactions between the desiccant and adsorbate are the primary causes of the difference between the actual and ideal dehumidification line slopes on the psychrometric chart. Therefore, it remains challenging to precisely predict the performance behavior of the desiccant unit. Consequently, numerous kinds of techniques have been developed to analyze the performance of the desiccant unit. The slope of the dehumidification line is experimentally measured, and several operational conditions are examined to develop a new correlation for the desiccant dehumidification process. Eq. (4b) provides the slope of the experimental or actual dehumidification line in terms of humidity ratio [g/kg DA].

$$\phi^* = \frac{\Delta}{Q_{st}}\phi_h \tag{4a}$$

$$\phi^* = \frac{X_{in} - X_{out}}{T_{db,\,out} - T_{db,\,in}} \tag{4b}$$

where, ϕ^* is the slope of dehumidification line [-], ϕ^h is the slope of the enthalpy line [-], Δ is the latent heat of vaporization [kJ/kg] and Q_{st} is the isosteric heat of adsorption [kJ/kg].

Fig. 7.7 shows the dehumidification line slope for the silica gel-based desiccant rotor. It is observed that the studied desiccant dehumidification system achieves a slope of approximate 0.27 for the dehumidification line at a regeneration temperature (T_{reg}) of 60°C. Furthermore, Eq. (5a) (Sultan et al., 2015) is used to determine the total amount of water adsorbed or desorbed (dehumidification potential), Eq. (5b) is used to calculate the

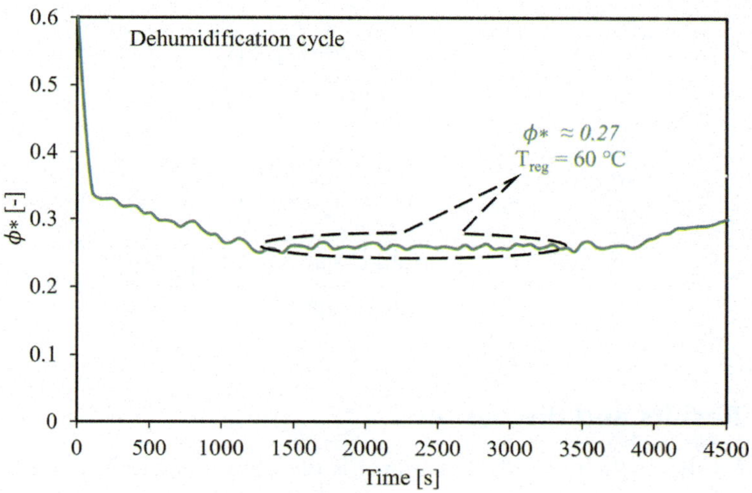

Fig. 7.7 Slope of the dehumidification line for silica gel-based desiccant rotor at a regeneration temperature (T_{reg}) of 60°C.

moisture removal rate, and Eqs. (6a) and (6b) is used to calculate the cooling capacity of DAC and MDAC systems.

$$\Delta X = X_{in} - X_{out} \tag{5a}$$

$$MRR = \dot{m}_{air}\left(X_1 - X_2\right) \tag{5b}$$

$$Q_{c,\,DAC} = h_1 - h_3 \tag{6a}$$

$$Q_{c,\,MDAC} = h_1 - h_4 \tag{6b}$$

VPD is an important parameter that shows the difference between the saturation vapor pressure, or the maximum amount of moisture that the air can hold at a specific temperature, and the actual vapor pressure, or the amount of moisture in the air at that particular time. It is a function of air temperature (T) and relative humidity. VPD can be calculated by using Eqs. (7a) to (7c) for the air T and relative humidity (RH) (Ashraf et al., 2021). The C_g is the coefficient equals to 610.7. The coefficient of performance (COP) of DAC and MDAC systems is calculated using Eqs. (8a) and (8b) (Mahmood et al., 2016a; Aleem et al., 2020):

$$\mathrm{VPD} = \mathrm{VP_{sat}} - \mathrm{VP_{act}} \tag{7a}$$

$$\mathrm{VP_{sat}} = \frac{C_g 10^{\frac{7.5T}{237.3+T}}}{1000} \tag{7b}$$

$$\mathrm{VP_{act}} = \frac{C_g 10^{\frac{7.5T}{237.3+T}}}{1000}\left(\frac{\mathrm{RH}}{100}\right) \tag{7c}$$

$$\mathrm{COP} = \frac{\dot{m}_{\mathrm{PA}}}{\dot{m}_{\mathrm{RA}}}\left(\frac{\text{Cooling capacity}}{\text{Heat input}}\right) = \frac{h_1 - h_3}{h_6 - h_5} \tag{8a}$$

$$\mathrm{COP} = \frac{\dot{m}_{\mathrm{PA}}}{\dot{m}_{\mathrm{RA}}}\left(\frac{\text{Cooling capacity}}{\text{Heat input}}\right) = \frac{h_1 - h_4}{h_7 - h_6} \tag{8b}$$

7.4 Results and discussion

Fig. 7.8 shows the monthly-based profile of dehumidification potential and moisture removal rate for five cities of Pakistan for the whole year. It is observed that both parameters depend on ambient climatic conditions and reach their highest values during the summer months, when humidity is high. The value of the dehumidification potential is achieved for Multan city, ranging from 4.93 g/kg to 2.02 g/kg; for Lahore city, it ranges from 5.48 g/kg to 2.50 g/kg; for Faisalabad city, it ranges from 5.24 g/kg to 2.16 g/kg; for Karachi city, it ranges from 13.8 g/kg to 3.73 g/kg; and for Peshawar city, it ranges from 5.3 g/kg to 1.4 g/kg. The figure also shows the behavior of the moisture removal rate for five cities for the whole year. The value of the moisture removal rate is achieved for Multan city, ranging from 1.15 kg/h to 0.6 kg/h; for Lahore city, it ranges from 1.67 kg/h to 0.76 kg/h; for Faisalabad city, it ranges from 1.6 kg/h to 0.6 kg/h; for

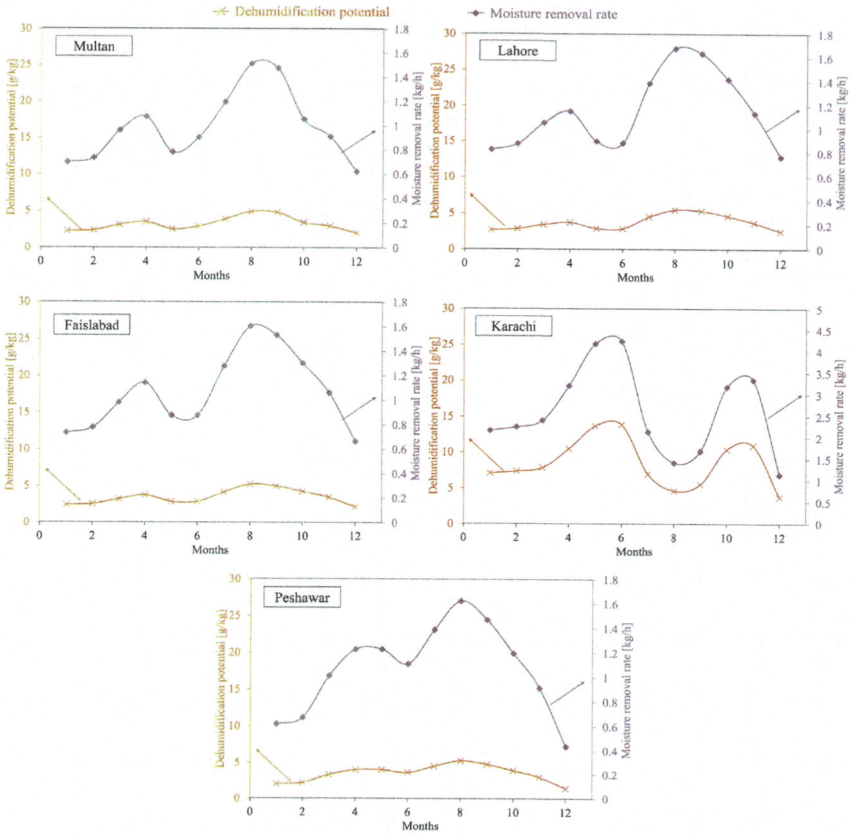

Fig. 7.8 Dehumidification potential and moisture removal rate for five cities of Pakistan throughout the year.

Karachi city, it ranges from 4.2 kg/h to 1.1 kg/h; and for Peshawar city, it ranges from 1.6 kg/h to 0.4 kg/h.

Fig. 7.9 shows the variation in temperature and relative humidity profiles of ambient, DAC, and MDAC air for five cities of Pakistan for the whole year. In Multan city, ambient air temperature ranges from 14.69°C to 39.38°C, and relative humidity ranges from 19.6% to 66.48%. Consequently, the product air T of the DAC system ranges from 40.4°C to 15.5°C, and RH ranges from 46.8% to 12.3%. In the case of the MDAC system, the product air T ranges from 28.16°C to 7.1 °C, and RH ranges from 68.9% to 21.5%. In Lahore, ambient air temperature ranges from 12.25°C to 38.44°C, and relative humidity ranges from 24.23% to

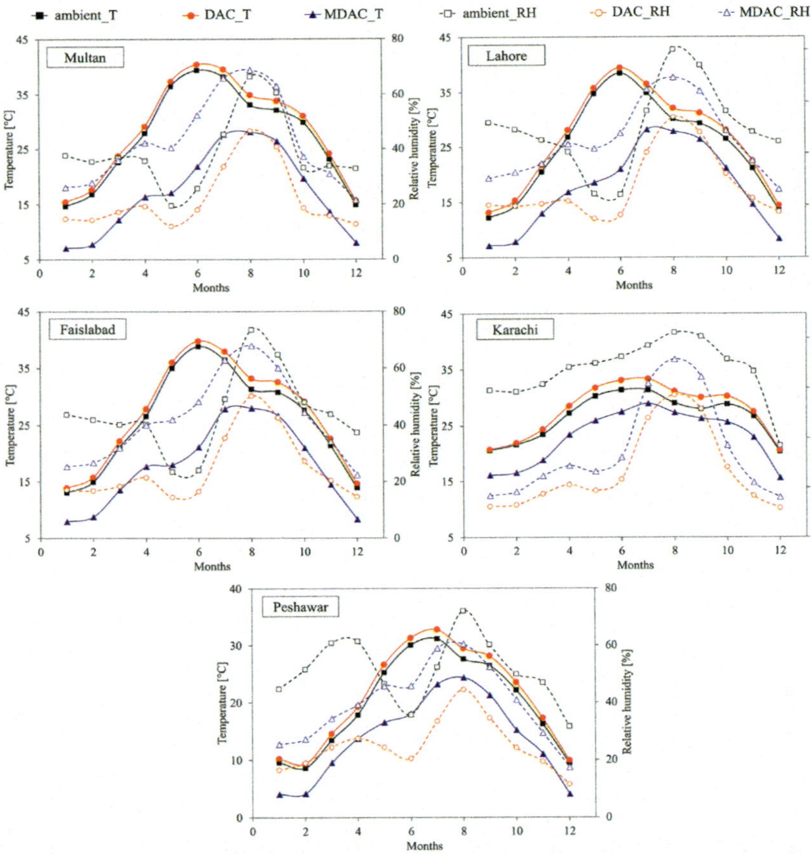

Fig. 7.9 Temperature and relative humidity profiles of ambient, DAC, and MDAC air for five cities of Pakistan throughout the whole year.

80.18%. Consequently, the product air T of the DAC system ranges from 39.51°C to 13.25°C, and RH ranges from 54.3% to 15.24%. In the case of the MDAC system, product air T values range from 28.23°C to 7.19°C, and RH values range from 69.55% to 26.25%. In Faisalabad city, ambient air temperature ranges from 38.81°C to 13.06°C, and relative humidity ranges from 73.57% to 23.5%. Consequently, the product air T of the DAC system ranges from 39.87°C to 13.95°C, and RH ranges from 50.5% to 14.6%. In the case of the MDAC system, the value of product air T ranges from 28.08°C to 8.02°C, and RH ranges from 68% to 22.3%. In Karachi, ambient air temperature ranges from 20.4°C to 31.39°C, and relative humidity ranges from 37.06% to 82.4%. Consequently, the product air T of the DAC system ranges from 33.42°C to 20.63°C, and RH ranges from 57.56% to 11.93%. In the case of the MDAC system, the value of product air T ranges from 29.02°C to 15.72°C, and RH ranges from 71.63% to 16.24%. In Peshawar city, ambient air temperature ranges from 31.17°C to 8.63°C, and relative humidity ranges from 72.08% to 31.55%. Consequently, the product air T of the DAC system ranges from 32.84°C to 9.4°C, and RH ranges from 44.6% to 11.6%. In the case of the MDAC system, the value of product air T ranges from 24.3°C to 4.07°C, and RH ranges from 60.5% to 17.4%. Fig. 7.10 shows the monthly based profile of cooling capacity for five cities of Pakistan for the whole year. For Multan city, the latent cooling capacity of the DAC system ranges from 10.7 kJ/kg to 4.35 kJ/kg, and for the MDAC system, it ranges from 26.17 kJ/kg to 12 kJ/kg. For Lahore city, the cooling capacity of the DAC system ranges from 11.8 kJ/kg to 5.38 kJ/kg, and for the MDAC system, it ranges from 25.19 kJ/kg to 11.43 kJ/kg. For Faisalabad city, the cooling capacity of the DAC system ranges from 11.37 kJ/kg to 4.65 kJ/kg, and for the MDAC system, it ranges from 25.33 kJ/kg to 11.03 kJ/kg. For Karachi city, the cooling capacity of the DAC system ranges from 33.6 kJ/kg to 9.23 kJ/kg, and for the MDAC system, it ranges from 39.35 kJ/kg to 13.5 kJ/kg. For Peshawar city, the cooling capacity of the DAC system ranges from 11.4 kJ/kg to 3.05 kJ/kg, and for the MDAC system, it ranges from 21.4 kJ/kg to 8.9 kJ/kg.

Fig. 7.11 shows the variation in Vapor pressure deficit (VPD) of ambient, DAC, and MDAC air for five cities of Pakistan for the whole year. In Multan city, the VPD values for ambient air range from 5.3 kPa to 1.04 kPa. Consequently, the VPD values for the DAC system range from 6.1 kPa to 1.4 kPa, and for the MDAC system, from 1.4 kPa to 0.7 kPa. In Lahore, the VPD values for ambient air range from 0.68 kPa to 5.14 kPa.

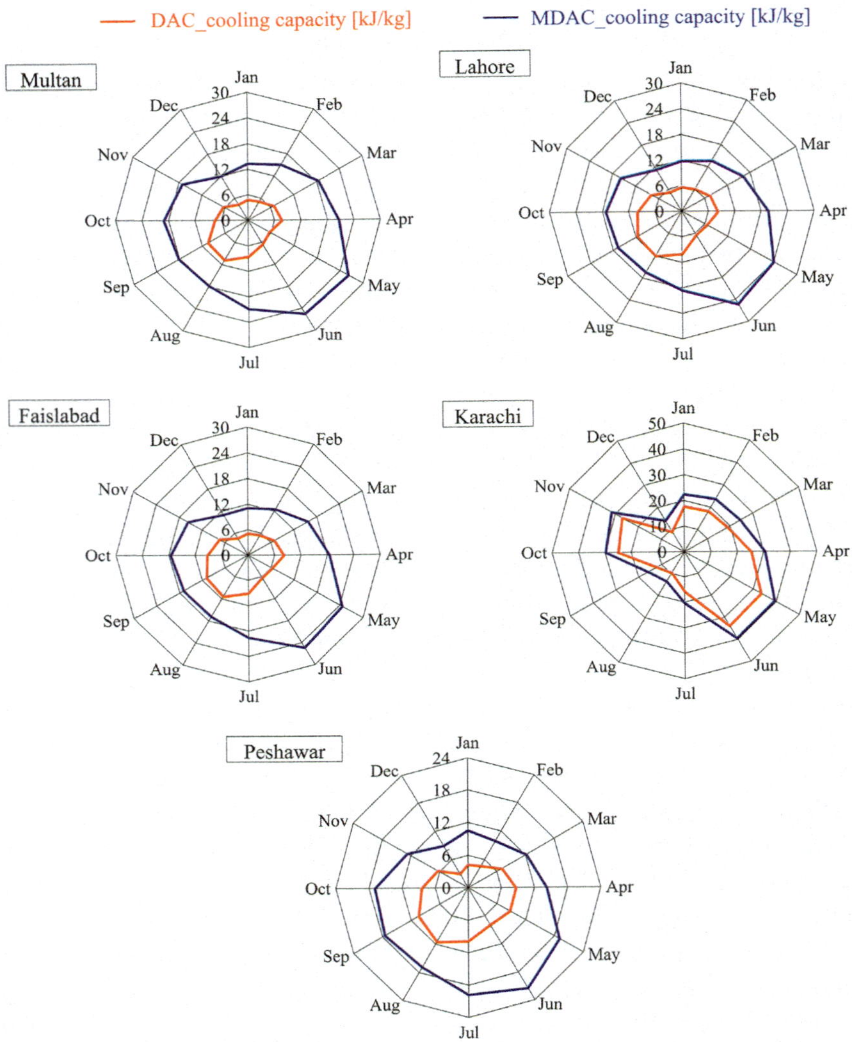

Fig. 7.10 Cooling capacity of DAC and MDAC systems for five major cities of Pakistan for throughout the year.

Consequently, the VPD values for the DAC system range from 5.9 kPa to 1.2 kPa, and for the MDAC system, from 1.3 kPa to 0.7 kPa. In Faisalabad city, the VPD values for ambient air range from 5.25 kPa to 0.84 kPa. Consequently, the VPD values for the DAC system range from 6.1 kPa to 1.3 kPa, and for the MDAC system, from 1.3 kPa to 0.8 kPa. In Karachi, the values of VPD in ambient air range from 1.5 kPa to 0.7 kPa. Consequently,

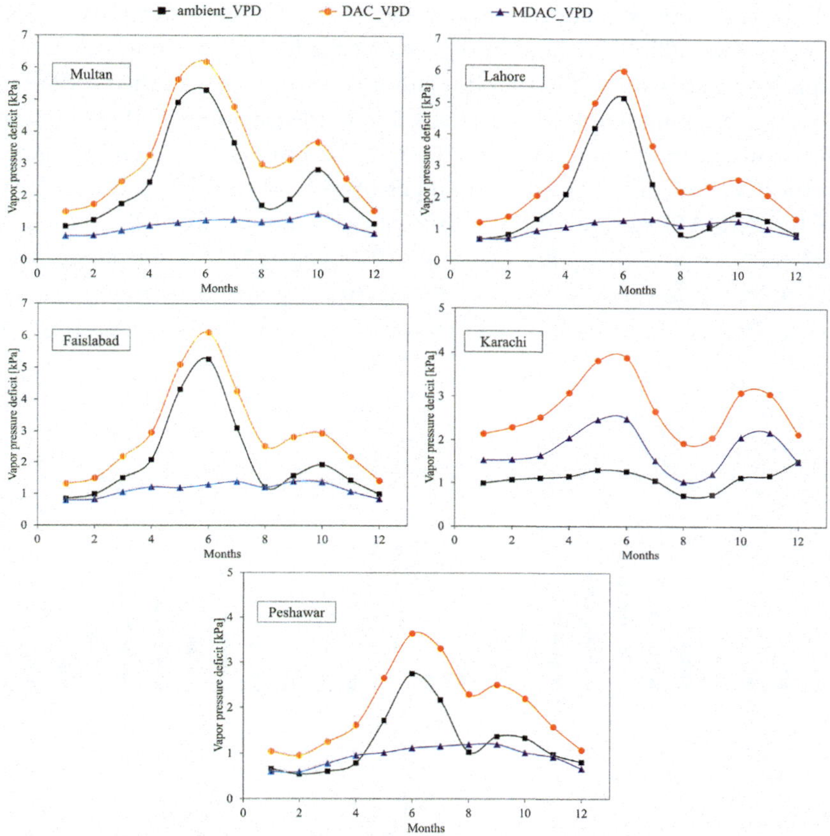

Fig. 7.11 Vapor pressure deficit of ambient, DAC, and MDAC air for five cities of Pakistan for the whole year.

the VPD values for the DAC system range from 3.8 kPa to 1.9 kPa, and for the MDAC system, from 2.4 kPa to 1.03 kPa. In Peshawar city, the VPD values for ambient air range from 0.5 kPa to 2.7 kPa. Consequently, the VPD values for the DAC system range from 3.6 kPa to 0.9 kPa, and for the MDAC system, from 1.2 kPa to 0.5 kPa. In addition, Appendix A1 shows vapor pressure deficit profiles of all five cities: (a) for the average range of climatic conditions, (b) for a standalone desiccant air conditioning system, and (c) for an M-Cycle-assisted desiccant air conditioning system. In addition, Fig. 7.12 shows VPD profiles of all five cities: (A) for the average range of climatic conditions, (B) for a standalone DAC system, and (C) for an MDAC system. Fig. 7.13 shows the coefficient of performance (COP) of the DAC and the MDAC system for studied five cities of Pakistan for the whole year. For Multan city, the value of the COP of the DAC system ranges from 0.98 to 0.1, and in the case of the MDAC system, the cooling capacity ranges from 2.3 to 0.3. For Lahore city, the value of the COP of the DAC system ranges from 0.98 to 0.13, and in the case of the MDAC system, the cooling capacity ranges from 2.06 to 0.29. For Faisalabad city, the value of the COP of the DAC system ranges from 0.97 to 0.11, and in the case of the MDAC system, the cooling capacity ranges from 2.12 to 0.28. For Karachi city, the value of the COP of the DAC system ranges from 2.5 to 0.24, and in the case of the MDAC system, the cooling capacity ranges from 2.97 to 0.37. For Peshawar city, the value of the COP of the

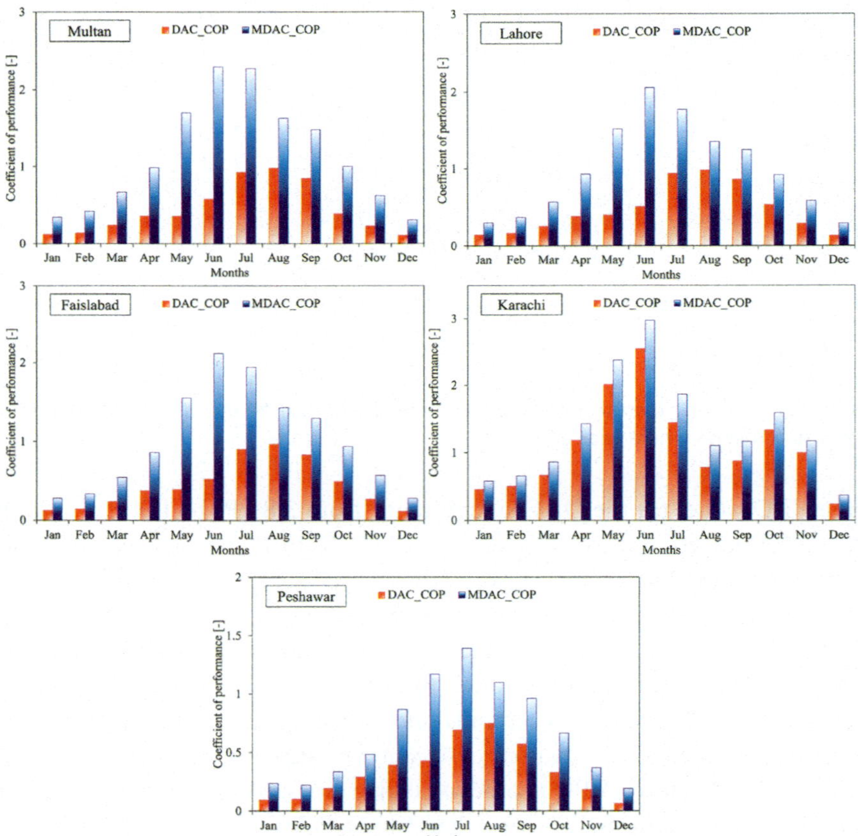

Fig. 7.12 Coefficient of performance of DAC and MDAC system for five big cities of Pakistan throughout the year.

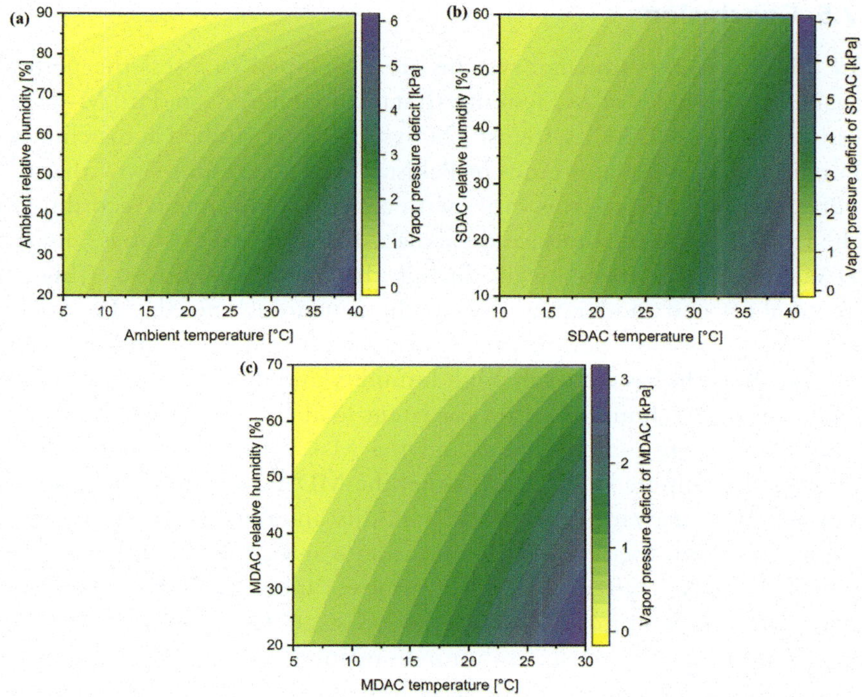

Fig. 7.13 Vapor pressure deficit profiles of studied five cities: (A) for the average range of climatic conditions, (B) for a standalone DAC system, and (C) for a MDAC system.

DAC system ranges from 0.75 to 0.06, and in the case of the MDAC system, the cooling capacity ranges from 1.39 to 0.19.

From the results, it is noticed that Karachi shows the highest values of dehumidification potential, moisture removal rate, cooling capacity, and COP and a low value of VPD for the whole year out of all cities due to its coastal location and high ambient humidity. Multan city shows moderate behavior due to its relatively hot and dry conditions. Lahore shows relatively higher values as compared to Multan due to its seasonal humidity variations and subtropical climate. Faisalabad exhibits a lower value of dehumidification potential and moisture removal rate due to its dry conditions and inland location. Peshawar shows a lower value of dehumidification potential and moisture removal rate due to its hot and semi-arid climate. These trends in both parameters indicate that the desiccant dehumidification system has the potential to perform better in hot and humid areas for managing latent load. On the other hand, the M-cycle-assisted air conditioning system has the potential to handle both sensible and latent loads.

7.5 Conclusion

In this study, experimental systems are developed to assess the performance of two types of AC systems: (i) standalone DAC , and (ii) M-Cycle assisted DAC (MDAC) systems. The systems were studied acrossclimatic conditions of five major cities of Pakistan, and the results show that the performance of the systems varies with climatic conditions. According to the results, Multan, Lahore, and Faisalabad are hot and dry cities, and the DAC system performs well during the monsoon season (July to September) and moderately during the remaining months. In case of the MDAC system, performance is very high under hot and dry conditions. Karachi has relatively high humidity and moderate temperatures and requiring dehumidification for latent load management. Therefore, the DAC system performs well and the M-Cycle has the potential to manage the sensible load. It shows that the MDAC system performs well in Karachi. Peshawar has moderate humidity and temperature in summer and shows the lowest moisture removal rate and cooling capacities. The standalone DAC system effectively removes moisture from the air and raises the temperature due to adsorption heat. However, the MDAC system is the most promising solution to maintain both sensible and latent loads. The performance of each system is based on key parameters, including dehumidification potential, moisture removal rate, temperature, relative humidity, VPD, and COP. In the case of desiccant dehumidification, the maximum variations in temperature range from 40.4°C to 9.4°C, relative humidity from 57.5% to 11.4%, moisture removal rate from 4.2 kg/h to 0.4 kg/h, VPD from 6.18 kPa to 0.19 kPa, and COP from 2.5 to 0.05 across the different climatic conditions of five cities in Pakistan. In case of MDAC system, the maximum variations in temperature range from 24.6°C to 2.2°C, relative humidity from 90.6% to 19.8%, VPD from 1.8 kPa to 0.2 kPa, and COP from 3.2 to 0.2 across the different climatic conditions of five cities of Pakistan. These results highlight the potential of hybrid air-conditioning systems for agricultural greenhouse crop production under controlled conditions in arid and semi-arid regions.

Funding statement

This research work is completed with the financial support from the Higher Education Commission (HEC) of Pakistan through the research Project NRPU-17029 awarded to Bahauddin Zakariya University.

Conflicts of interest

The authors declare no conflict of interest.

Declaration of generative AI and AI-assisted technologies in the writing process

No generative AI/AI-assisted technology is used in this paper.

CRediT authorship contribution statement

Hafiz Muhammad Umar Raza was involved in the conceptualization, methodology, software, formal analysis, investigation, writing of the original draft. **Muhammad Sultan and Takahiko Miyazaki** were involved in the conceptualization, methodology, validation, resources, visualization, supervision, project administration, writing (review and editing), funding acquisition. **Muhammad Aleem, Md Shamim Ahamed, Muhammad Wakil Shahzad, Zhaoli Zhang, and Hafiz Muhammad Ali** were involved in the conceptualization, methodology, software, formal analysis, investigation, writing (review and editing).

Acknowledgment

This work is part of the PhD. research of Engineer Hafiz M.U. Raza (first author) at the Department of Agricultural Engineering, Bahauddin Zakariya University, Multan, Pakistan. This research work is completed with the financial support from the Higher Education Commission (HEC) of Pakistan through the research Project NRPU-17029 awarded to Bahauddin Zakariya University.

References

Achour, Y., Ouammi, A., Zejli, D., 2021. Technological progresses in modern sustainable greenhouses cultivation as the path towards precision agriculture. Renew. Sustain. Energy Rev. 13640321. 147, 111251. doi:10.1016/j.rser.2021.111251.

Ahamed, M.S., Sultan, M., Shamshiri, R.R., Rahman, M.M., Aleem, M., Balasundram, S. K., 2023. Present status and challenges of fodder production in controlled environments: a review. Smart Agric. Technol. 27723755. 3. Available from: www.sciencedirect.com/journal/smart-agricultural-technology/issues. doi:10.1016/j.atech.2022.100080.

Aleem, M., Hussain, G., Sultan, M., Miyazaki, T., Mahmood, M.H., Sabir, M.I., Nasir, A., Shabir, F., Khan, Z.M., 2020. Experimental investigation of desiccant dehumidification cooling system for climatic conditions of multan (Pakistan). Energies. 1996-1073. 13 (21), 5530. doi:10.3390/en13215530.

Amani, M., Foroushani, S., Sultan, M., Bahrami, M., 2020. Comprehensive review on dehumidification strategies for agricultural greenhouse applications. Appl. Therm. Eng. 13594311. 181, 115979. doi:10.1016/j.applthermaleng.2020.115979.

Ashraf, H., Sultan, M., Shamshiri, R.R., Abbas, F., Farooq, M., Sajjad, U., Md-Tahir, H., Mahmood, M.H., Ahmad, F., Taseer, Y.R., Shahzad, A., Niazi, B.M.K., 2021. Dynamic evaluation of desiccant dehumidification evaporative cooling options for greenhouse air-conditioning application in multan (Pakistan). Energies 19961073.

14 (4).Available from: https://www.mdpi.com/1996-1073/14/4/1097/pdf. doi:10.3390/en14041097.

Bhat, S.A., Sajjad, U., Hussain, I., Yan, W.M., Raza, H.M.U., Ali, H.M., Sultan, M., Omar, H., Azam, M.W., Bozzoli, F., Huang, N.F., 2024. Experiments and modeling on thermal performance evaluation of standalone and M-cycle based desiccant air-conditioning systems. Energy Rep. 23524847. 11, 1445–1454. Available from: https://www.sciencedirect.com/science/journal/23524847. doi:10.1016/j.egyr.2024.01.019.

Costantino, A., Comba, L., Sicardi, G., Bariani, M., Fabrizio, E., 2021. Energy performance and climate control in mechanically ventilated greenhouses: a dynamic modelling-based assessment and investigation. Appl. Energy 03062619. 288. Available from: https://www.journals.elsevier.com/applied-energy. doi:10.1016/j.apenergy.2021.116583.

López-Morales, V., López-Ortega, O., Ramos-Fernández, J., Muñoz, L.B., 2008. JAPIEST: an integral intelligent system for the diagnosis and control of tomatoes diseases and pests in hydroponic greenhouses. Expert Syst. Appl. 09574174. 35 (4), 1506–1512. doi:10.1016/j.eswa.2007.08.098.

Sultan, M., Miyazaki, T., 2017. Energy-efficient air-conditioning systems for nonhuman applications. In: Ekren, O. (Ed.), Refrigeration, InTech. doi:10.5772/intechopen.68865.

Mahmood, M.H., Sultan, M., Miyazaki, T., Koyama, S., 2016a. Desiccant air-conditioning system for storage of fruits and vegetables: Pakistan preview. Evergreen 24325953. 3 (1), 12–17. Available from: http://www.tj.kyushu-u.ac.jp/leading/en/c_publication/EG2016-3-1content/Pages%2012-17.pdf. doi:10.5109/1657381.

Mahmood, M.H., Sultan, M., Miyazaki, T., Koyama, S., Maisotsenko, V.S., 2016b. Overview of the Maisotsenko cycle – a way towards dew point evaporative cooling. Renew. Sustain. Energy Rev. 18790690. 66, 537–555. Available from: https://www.journals.elsevier.com/renewable-and-sustainable-energy-reviews. doi:10.1016/j.rser.2016.08.022.

Mahmood, M.H., Sultan, M., Miyazaki, T., 2019. Significance of temperature and humidity control for agricultural products storage: overview of conventional and advanced options. Int. J. Food Eng. 15563758. 15 (10). Available from: https://www.degruyter.com/journal/key/ijfe/html. doi:10.1515/ijfe-2019-0063.

Rawson, H.M., Begg, J.E., Woodward, R.G., 1977. The effect of atmospheric humidity on photosynthesis, transpiration and water use efficiency of leaves of several plant species. Planta. 0032-0935. 134 (1), 5–10. doi:10.1007/bf00390086.

Sultan, M., El-Sharkawy, I.I., Miyazaki, T., Saha, B.B., Koyama, S., 2015. An overview of solid desiccant dehumidification and air conditioning systems. Renew. Sustain. Energy Rev. 18790690. 46, 16–29. Available from: https://www.journals.elsevier.com/renewable-and-sustainable-energy-reviews. doi:10.1016/j.rser.2015.02.038.

Sultan, M., El-Sharkawl, I.I., Miyazaki, T., Saha, B.B., Koyama, S., 2014. Experimental study on carbon based adsorbents for greenhouse dehumidification. Evergreen 24325953. 1 (2), 5–11. Available from: https://ncrs.cm.kyushuu.ac.jp/ncrs2/380.html. doi:10.5109/1495157.

Sultan, M., Miyazaki, T., Mahmood, M.H., Khan, Z.M., 2018a. Solar assisted evaporative cooling based passive air-conditioning system for agricultural and livestock applications. J. Eng. Sci. Technol. 18234690. 13 (3), 693–703. Available from: http://jestec.taylors.edu.my/index.htm.

Sultan, M., Miyazaki, T., Saha, B.B., Koyama, S., 2016. Steady-state investigation of water vapor adsorption for thermally driven adsorption based greenhouse air-conditioning system. Renew. Energy 18790682. 86, 785–795. Available from: http://www.journals.elsevier.com/renewable-and-sustainable-energy-reviews/. doi:10.1016/j.renene.2015.09.015.

Sultan, M., Miyazaki, T., Koyama, S., 2018b. Optimization of adsorption isotherm types for desiccant air-conditioning applications. Renew. Energy 18790682. 121, 441–450. Available from: http://www.journals.elsevier.com/renewable-and-sustainable-energy-reviews/. doi:10.1016/j.renene.2018.01.045.
Sultan, M., Ashraf, H., Miyazaki, T., Shamshiri, R.R., Hameed, I.A., 2021. Temperature and Humidity Control for the Next Generation Greenhouses: overview of Desiccant and Evaporative Cooling Systems. IntechOpen. doi:10.5772/intechopen.97273.
Sultan, M., Niaz, H., Miyazaki, T., 2020. Investigation of Desiccant and Evaporative Cooling Systems for Animal Air-Conditioning. IntechOpen. doi:10.5772/intechopen.88945.

CHAPTER 8

Recent developments in membrane energy recovery ventilators for optimum channel geometry, operating parameters, and membrane materials

Muhammad Asif[a], Muhammad Sultan[a], Zahid Mahmood Khan[a], Muhammad Wakil Shahzad[b], Faheem Hassan Akhtar[c], Hadeed Ashraf[a], Naveed Husnain[d], Uzair Sajjad[e], Muhammad Farooq[f], and Muhammad Yasar Javaid[g]

[a]Department of Agricultural Engineering, Bahauddin Zakariya University, Multan, Pakistan
[b]Department of Mechanical and Construction Engineering, Northumbria University, Newcastle Upon Tyne, United Kingdom
[c]Department of Chemistry and Chemical Engineering, Lahore University of Management Sciences (LUMS), Lahore, Pakistan
[d]Department of Mechanical Engineering, Bahauddin Zakariya University, Multan, Pakistan
[e]Interdisciplinary Research Center for Sustainable Energy Systems (IRC-SES), King Fahd University of Petroleum and Minerals, Dhahran, Saudi Arabia
[f]Department of Mechanical Engineering, University of Engineering and Technology, Lahore, Pakistan
[g]Department of Mechanical Engineering and Technology, Government College University Faisalabad, Faisalabad, Pakistan

8.1 Introduction

Globally, the energy consumption for domestic use and the residential/building sector has always been a serious concern, which consumes ~36% of the total energy (Global CCS, 2021). In the world, as per climatic and geological conditions, energy consumption pattern vary; the regions having severe climatic conditions are most affected by energy consumption (Hansen et al., 2006; Arrow, 2007). Simultaneously cooling and heating are required in different regions of the world. So, it complexes energy consumption across the world. Energy consumption in the residential, commercial/public services, and agricultural sectors is 43%, 22%, and 12%, respectively, which costs a lot of carbon emissions, which is ~3201 $MtCO_2$ (International Energy Agency, 2021). Fig. 8.1 shows energy

Sustainable Hybrid Cooling and Dehumidification Strategies for Controlled Environments
https://doi.org/10.1016/B978-0-443-48555-8.00002-5

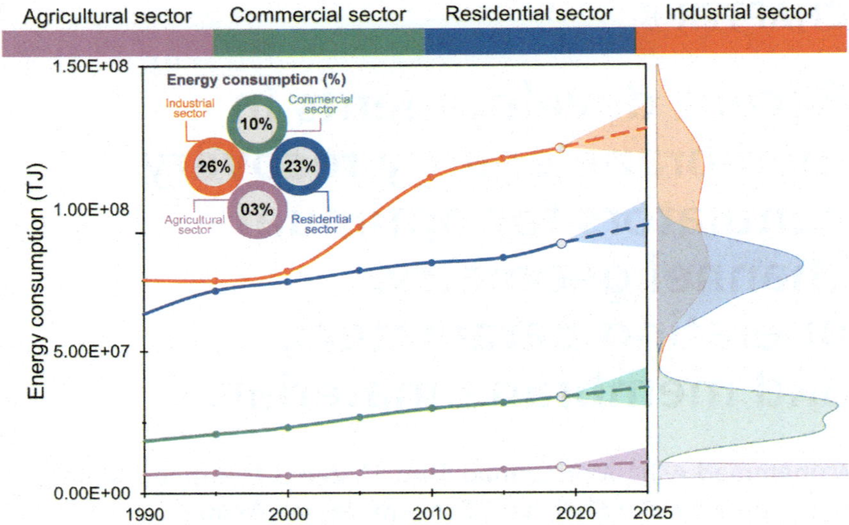

Fig. 8.1 Energy consumption in the world by the industrial, commercial, residential, and agricultural sectors. *(International Energy Agency, 2021. https://www.iea.org/data-and-statistics/data-tools/energy-statistics-data-browser.)*

consumption in the different sectors of the world. It elaborates on energy consumption by the different sectors, i.e., industrial, commercial, residential, and agricultural sectors.

Energy consumption in space heating/cooling, cooking, lighting, electrical appliances, water heating, and others, which are 55%, 8%, 5%, 10%, 9%, and 5%, respectively. Fig. 8.2 shows energy consumption in the residential sector and emphasizes the significance of energy consumption by the HVAC system. Therefore, several international firms are working efficiently to reduce energy consumption in buildings (Pérez-Lombard et al., 2008; Santamouris et al., 2001; Kingma and Lichtenbelt, 2015; European Commission, 2020). Fig. 8.3A shows how the indoor air gets deteriorated as per the effect of human activity. It also explains the hazardous particles that decorate the indoor air that could potentially harm humans or other living habitats. Fig. 8.3B shows the health risk due to indoor air quality. Any human lives 90% of his life indoors. So, it is also necessary to improve indoor air quality using advanced technologies. Energy from the conditioned space is wasted in the form of stale air that needs to be replaced with fresh air.

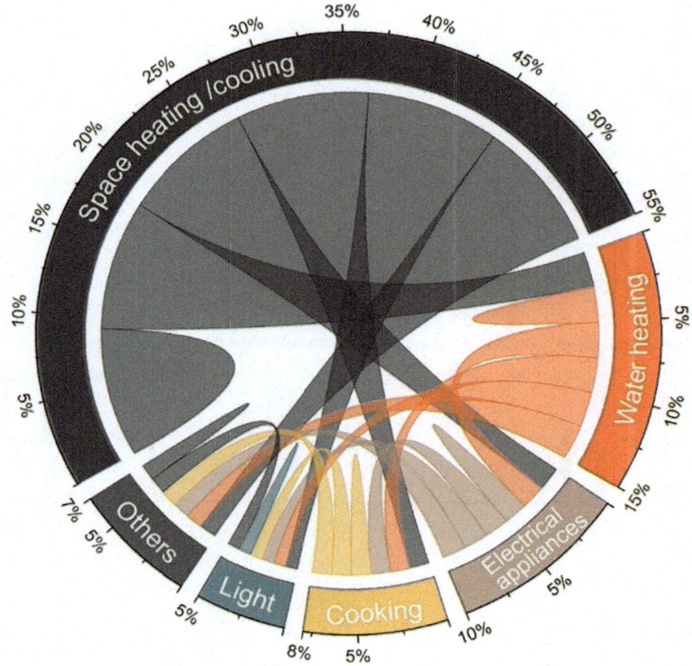

Fig. 8.2 Percentage of energy consumption in residential sector. *(International Energy Agency, 2021. https://www.iea.org/data-and-statistics/data-tools/energy-statistics-data-browser.)*

Generally, the conventional standalone vapor compression system is used for space heating and cooling, which is responsible for greenhouse gas emissions and has higher energy consumption; therefore, alternative green cooling technologies could be preferred, i.e., evaporative air conditioning and desiccant air conditioning technologies (Sultan et al., 2015; Mahmood et al., 2016; Noor et al., 2020). In all the above scenarios, membrane-based air conditioning could be a possible option to reduce energy consumption.

Energy recovery systems are classified as heat recovery ventilators (HRVs) and enthalpy exchangers. HRVs recover sensible heat, but enthalpy recovery exchangers recover both sensible and latent heat. Fig. 8.4 shows the classification of energy recovery systems. HRVs are further classified as plate and pipe heat exchangers. Enthalpy exchangers are further subclassed into energy recovery wheels, run-around coils, and membrane-based energy recovery ventilators (MERVs). An MERV could

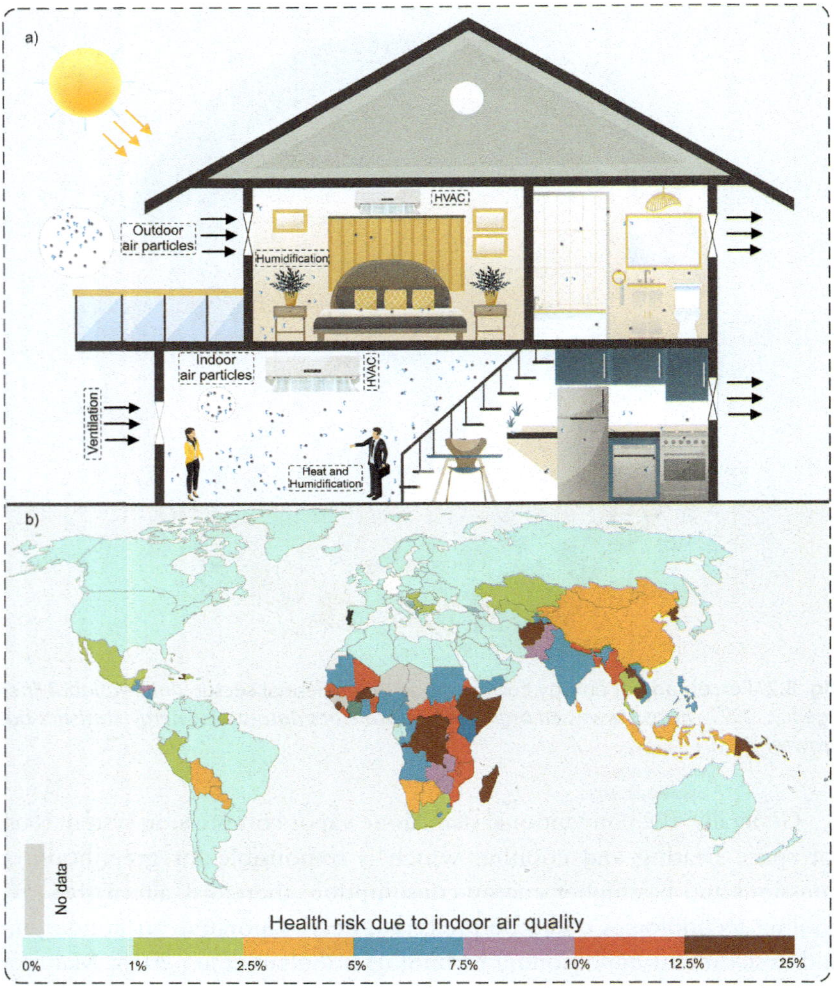

Fig. 8.3 (A) Indoor air quality (B) Health risk due to indoor air quality across the world. *(Ritchie, H., Roser, M., 2022. Our World in Data https://ourworldindata.org/indoor-air-pollution.)*

potentially recover the total energy (sensible & latent) from the stale/exhaust air stream. Still, the better configuration of the membrane for MERV will maximize the energy recovery efficiency of the system. Improvement in indoor air quality is also responsible for higher energy consumption. Stale air causes health hazards, so MERV can also improve

Fig. 8.4 Classification of the energy recovery systems.

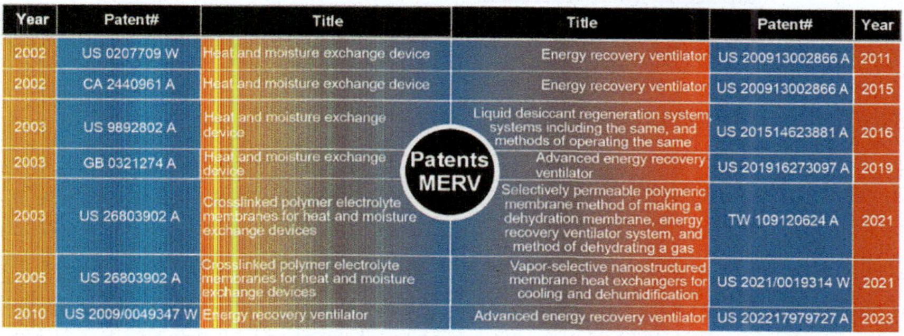

Year	Patent#	Title	Title	Patent#	Year
2002	US 0207709 W	Heat and moisture exchange device	Energy recovery ventilator	US 200913002866 A	2011
2002	CA 2440961 A	Heat and moisture exchange device	Energy recovery ventilator	US 200913002866 A	2015
2003	US 9892802 A	Heat and moisture exchange device	Liquid desiccant regeneration system, systems including the same, and methods of operating the same	US 201514623881 A	2016
2003	GB 0321274 A	Heat and moisture exchange device	Advanced energy recovery ventilator	US 201916273097 A	2019
2003	US 26803902 A	Crosslinked polymer electrolyte membranes for heat and moisture exchange devices	Selectively permeable polymeric membrane method of making a dehydration membrane, energy recovery ventilator system, and method of dehydrating a gas	TW 109120624 A	2021
2005	US 26803902 A	Crosslinked polymer electrolyte membranes for heat and moisture exchange devices	Vapor-selective nanostructured membrane heat exchangers for cooling and dehumidification	US 2021/0019314 W	2021
2010	US 2009/0049347 W	Energy recovery ventilator	Advanced energy recovery ventilator	US 202217979727 A	2023

Fig. 8.5 Comprehensive detail of patents published on MERV.

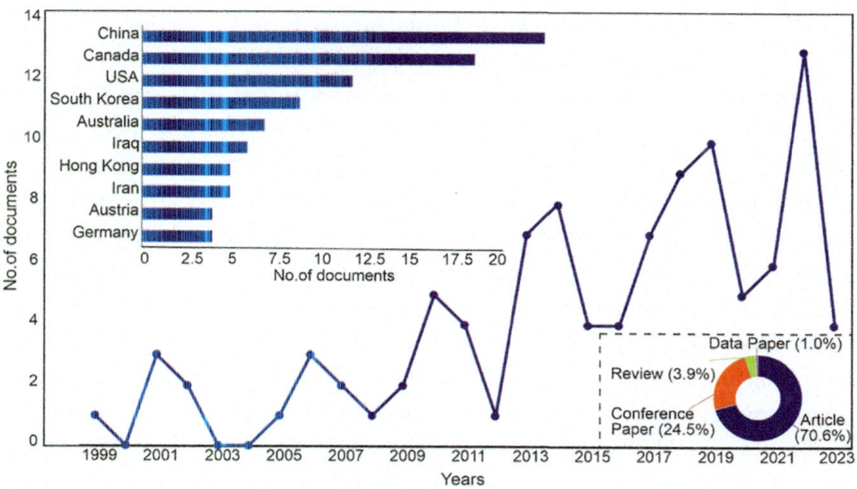

Fig. 8.6 Number of publications on MERV, publication type and leading countries on MERV.

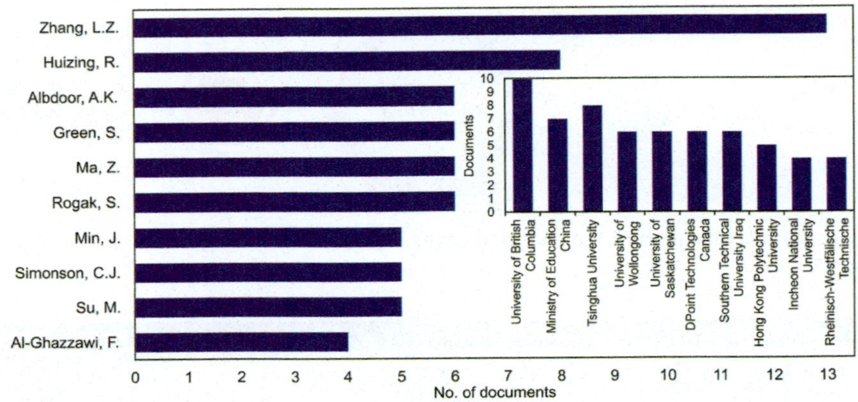

Fig. 8.7 Leading authors and universities in MERV.

indoor air quality with low energy consumption. In this regard, in the last couple of decades, MERVs have gained much attention and possess a linear rise in publications. Fig. 8.5 shows the comprehensive review details of the patents published on MERV to show the increasing interest of the world in these vehicles. Fig. 8.6 shows the number of publications in the last couple of decades, the leading countries in this technology, and the percentage of the publications. China and Canada are the leading in MERV technology, and 70% are the article type in publications. Fig. 8.7 elaborates on the leading institution and author in the world.

In literature, extensive studies can be found on MERVs. Fig. 8.8 shows the keywords used in literature exhibiting the relationship between the membrane and its other performance indicators. MERVs possess superiority as they recover both latent and sensible energy, are easy to install, have compact structures, no moving parts, easy repair and maintenance if needed, and work in extreme environments. The main issue in the adaptability of MERV in cold environments is the frost issue. Fig. 8.9 shows the summary of MERV. Membrane is the semipermeable enthalpy exchanger, which is a multifaceted structure based on the many physical properties, i.e., thermal conductivity, membrane thickness, membrane density, and water-associated properties, i.e., diffusivity, water contact angel, and selectivity. Each property affects the system's performance. Therefore, the appropriate selection membrane material is necessary to improve the system effectiveness. MERV is subclassified into three groups: 1) membrane, 2) configuration, and 3) operating condition. The manuscript reviews the MERV systems, different configurations, and membrane materials.

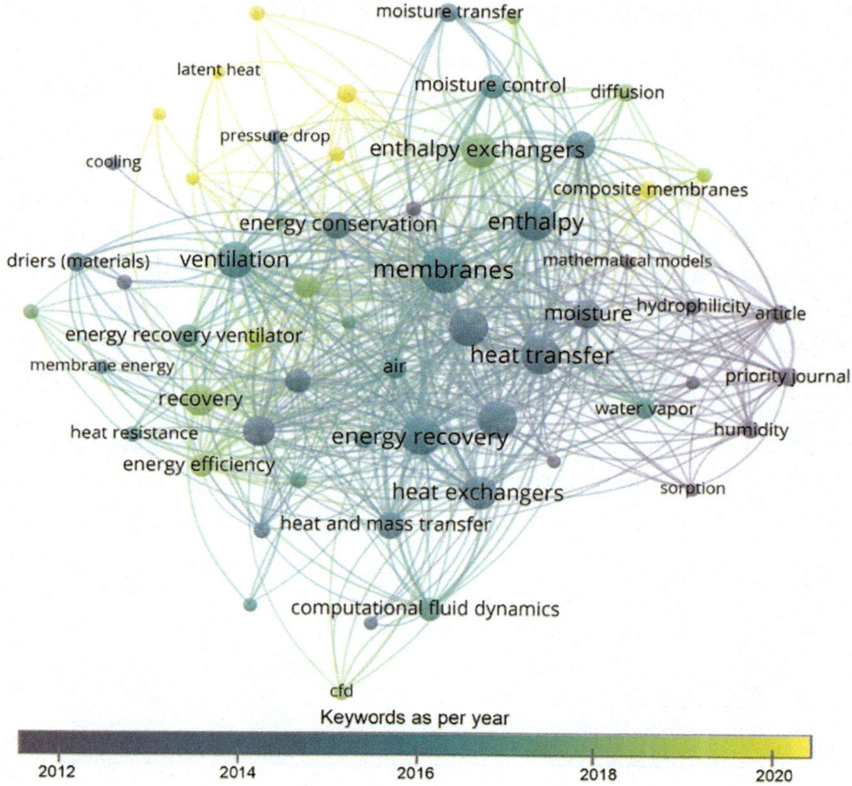

Fig. 8.8 Keywords presentation of MERVs.

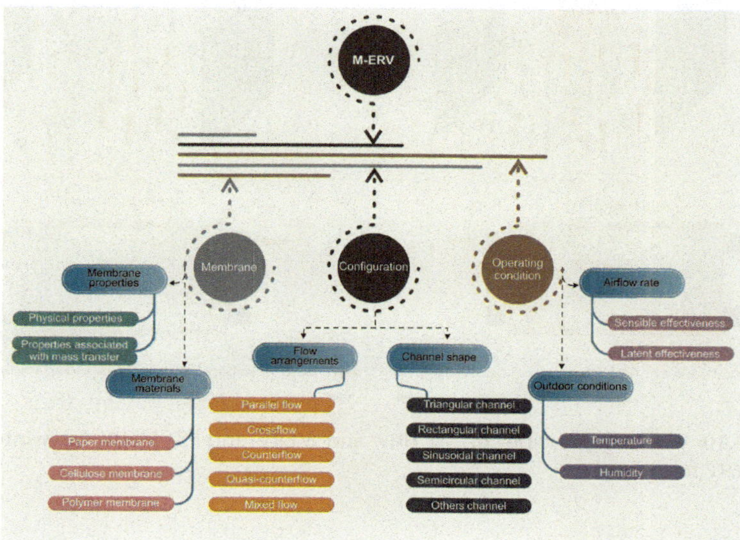

Fig. 8.9 Brief summary of MERVs.

8.2 Membrane energy recovery ventilators

AN MERV is a mechanical device that recovers the total enthalpy from the feed and sweep air streams. The basic principle of HRVs and MERVS is presented in Fig. 8.10, which is primarily based on the two-recovery phenomenon: (1) sensible energy and (2) latent energy of exhaust air. HRV can only recover sensible energy, whereas ERV recovers the total energy (latent & sensible). The MERV is classified as two air streams: feed air stream and sweep air stream. The feed air stream carries the air from the atmosphere, and the sweep air stream carries air from the room that passes through MERV and transfers their heat to each other. The preconditioned feed air stream is supplied to the room and the sweep air stream is expelled out to the atmosphere.

The performance of MERV is greatly affected by seasonal variation. Fig. 8.10 shows the working of the MERVs. In summer, the ambient air

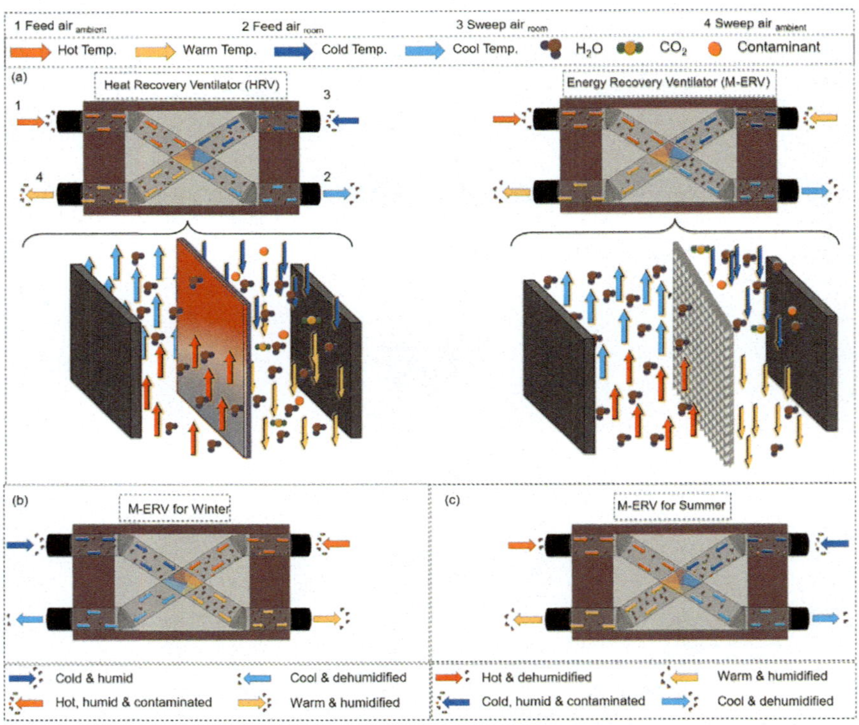

Fig. 8.10 Working principle of (A) HRV and MERV and (B) MERV for winter and (C) MERV for summer.

contains high temperatures and low humidity. The exhaust air contains low temperature and high humidity. The ambient air and the exhaust air that pass through MERV interact with each other and exchange sensible and latent heat with each other; preconditioned air is supplied to a room that contains relatively high humidity and low temperature. In winter the ambient air contains low temperature and high humidity. The exhaust air contains high temperature and low humidity. The ambient air and exhaust air stream pass through MERV, interact with each other, and exchange sensible and latent heat with each other; preconditioned air is supplied to a room that contains relatively high humidity and temperature. The detail inside of the MERV is the membrane heat exchanger. The membrane is the complex polymer porous media that allows the water vapor to adsorb through it.

8.3 Channel geometry

Channel geometry is a significant parameter in heat and mass transfer through MERVs. Various channel designs, i.e., square, triangular, sinusoidal, circular, and trapezoidal are discussed here with their impact on MERVs effectiveness. This section shows the experimental and simulation review of the membrane channels.

8.3.1 Rectangular channel

In the rectangular channel, space and height of the channel are key parameters influencing MERVs. In the literature, it is reported that the total enthalpy transfer through the MERVs increases as the channel height is short which enhances the air flow rate with the decrease in the energy transfer contact area (Min and Su, 2010). However, another study indicated that by increasing the channel height from 1.8 to 2.9 mm while maintaining the face velocity and channel thickness, the total effectiveness increased by up to 21% (Yaïci et al., 2013). Furthermore, if the channel height is 1.8 mm, it will give the maximum energy-saving potential annually (Kim, 2016). Other structural parameters, such as hydraulic diameter, are also important; as the hydraulic diameter decreases total energy recovery increases (Al-Waked et al., 2015). Conclusively, at constant air velocity and channel height, both effectiveness and resistance decrease; hence, optimal channel height is required to maximize effectiveness while minimizing resistance.

8.3.2 Sinusoidal wave channel

In the sinusoidal wave channel, the most significant parameters are amplitude, wavelength ratio, and included angle. In literature, two amplitudes, 5 mm and 10 mm were tested and concluded that a small amplitude is preferred (Islamoglu and Parmaksizoglu, 2003); whereas, if the included angle varies from 0 to 160° the friction coefficient increases two orders of the amplitude and Nu increases from 3 to 5. Therefore, the optimal concluded angle would be better to enhance the heat transfer and decrease the resistance (Ciofalo et al., 1996; Mehrabian et al., 2000). Furthermore, with the increase in the Re and amplitude-wavelength ratio, the Nu and coefficient of fraction increase. The heat transfer enhancement was insignificant in the smaller amplitude-wavelength ratio channel, whereas in high Re and amplitude-wavelength ratio, it showed considerable heat transfer (Wang and Chen, 2002).

8.3.3 Triangular channel

In the triangular channel, the most significant parameter is the apex angle; it is a widely employed channel in MERVs (Beattie et al., 2018; Liang et al., 2017; Zhang and Chen, 2011). It is observed that under the same operating condition if the apex angle increases the performance of MERVs also increases. With the increase in the apex angle from 0–90°, the permeance of air, resistance, and pressure drop increase due to enhancement in the contact area and turbulence effect. Furthermore, by changing the apex angle from 45–90°, the energy consumption was reduced up to 80 to 88% at a constant face velocity of 2.5 m/s in parallel corrugations (Scott et al., 2000). In another study, it is found that cross-corrugations develop complex flow patterns (Zhang and Chen, 2011). The author concluded that if the apex angle varies from 90–120°, it shows the maximum heat transfer, but the coefficient of friction was found to be approximately half at 120° (Liu and Niu, 2015). In another study, it is reported that the apex angle of 150° shows the best heat and mass transfer (Sharif et al., 2018).

8.3.4 Novel channel

Other novel membrane channels including trapezoidal, curved, and circular, are also discussed. The trapezoidal channel is investigated with varying sidewall angles of 45° and 30° in crossflow MERVs. The results showed at 30° sidewall angle there was better heat transfer and less frictional resistance. If the sidewall angle is less, it means the channel has a big base termed as aspect ratio, which mostly affects the connective frictional resistance (Liu and Niu, 2015). Furthermore, in another study, a comprehensive evaluation of different channels was exhibited. The results indicated that the circular channel shows the best heat and mass transfer rates than triangular, curved, and rectangular channels. The main reason was the largest contact surface area than other channels (Jafarizave et al., 2019). Conclusively, based on the properties of the channel mentioned above an optimized channel shape would result in higher effectiveness with minimum cost.

8.4 Flow-based classification of MERVs

The flow in the MERVs is a much more significant parameter affecting the performance. Broadly, the flow is classified as; crossflow, counterflow, quasi-counterflow, and z-shaped flow. → Table 8.1 shows the classification of the MERV system based on the flow. → Fig. 8.11 shows the sensible effectiveness of parallel flow, crossflow, counterflow, and quasi-counterflow. → Fig. 8.12 shows the latent effectiveness of parallel flow, crossflow, counterflow, and quasi-counterflow. Conclusively, as the flow increases, the effectiveness decreases, so the minimum flow in necessary for better heat and mass transfer between the air streams.

8.5 Operating parameters

Operating condition is mainly based on the outdoor air state and flow rate. Table 8.2 shows the impact of the state of outdoor air, i.e., temperature, humidity and flow rate on the MERV system. Table 8.2 elaborates that as the flow rate increases the sensible, latent and total system effectiveness decreased. If the temperature is increased the total and sensible effectiveness in increased and latent effectiveness in decreased. Lastly, if the humidity increases and temperature and flow rate are constant than the sensible and latent effectiveness increases but the total effectiveness may increase, or decrease based on other factors, including the type of membrane material.

8.6 Membrane

The membrane is the energy exchanger and plays a vital role in the heat and mass transfer through MERV; therefore, the membrane materials are the key performance indicators of the MERV system. Membrane is the layer of selective material between the two feed and exhaust air streams. In MERVs

Table 8.1 Flow based classification of MERVs.

Sr. No.	Category and schematic	Findings and conclusions	Reference
1	Crossflow MERV 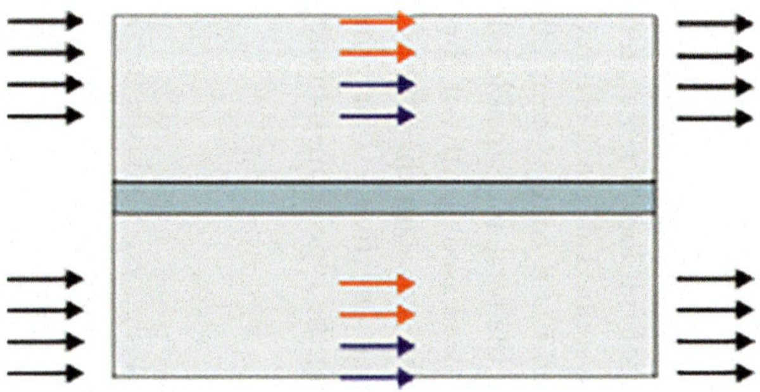	• <50% effectiveness • Easy in installation • Greatly affected by flow velocity	Yaïci et al. (2013); Salafi et al. (2021); Al-Waked et al. (2015); Min et al. (2012); Asasian-Kolur et al. (2022)
2	Crossflow MERV 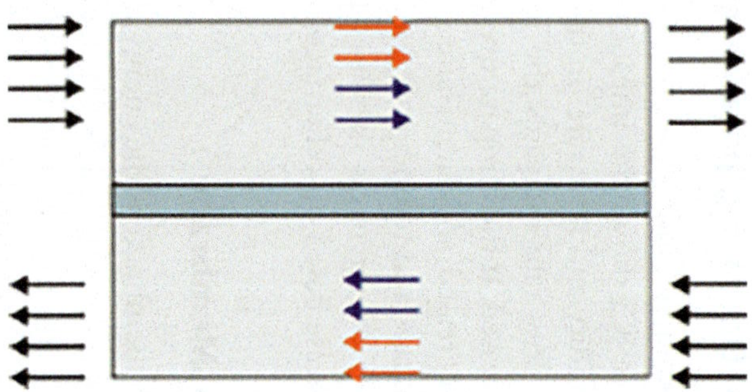	• 50 to 70% effectiveness • The efficiency of the crossflow is greater than the counter flow • Convenient in installation	Salafi et al. (2021); Asasian-Kolur et al. (2022)

3	Counterflow MERV 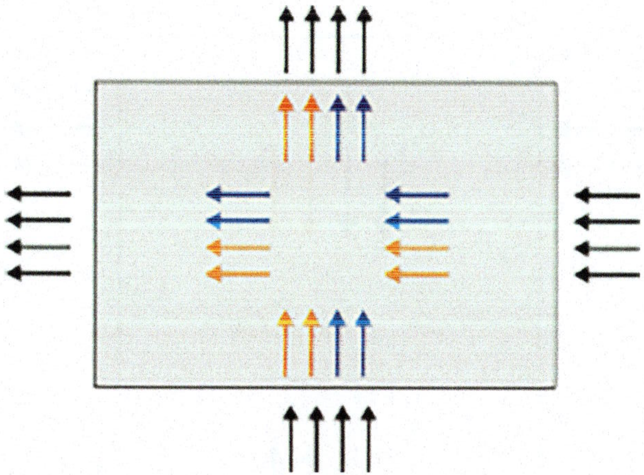	• Theoretical effectiveness is considered 100% but its effectiveness is less as compared to the cross flow • Difficult to install and handle	Zhang et al. (2011); Asasian-Kolur et al. (2022); Zender–Świercz (2021); Koester et al. (2017)
4	Quasi counterflow MERVs 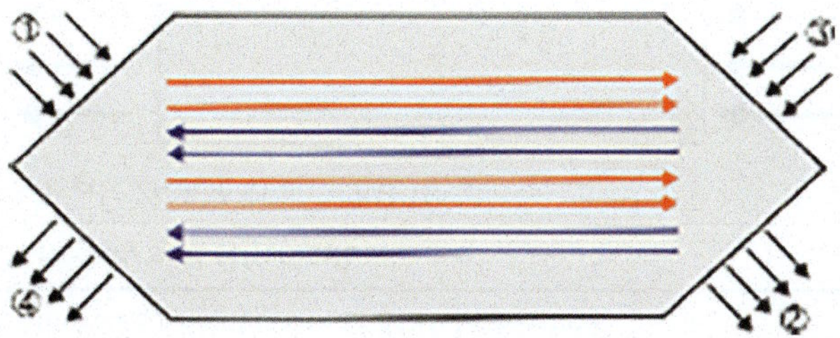	• Mixed flow configuration • Combined flow cross flow and counter flow • Sensible and latent effectiveness is In between the pure cross and counterflow configurations • Convenient to separate inlet and outlet at higher effectiveness	Nasif et al. (2010); Zhang (2010); Liu et al. (2009b); Asasian-Kolur et al. (2022)

Continued

Table 8.1 Flow based classification of MERVs—cont'd.

Sr. No.	Category and schematic	Findings and conclusions	Reference
5	Mixed/ Z-shaped flow MERVs 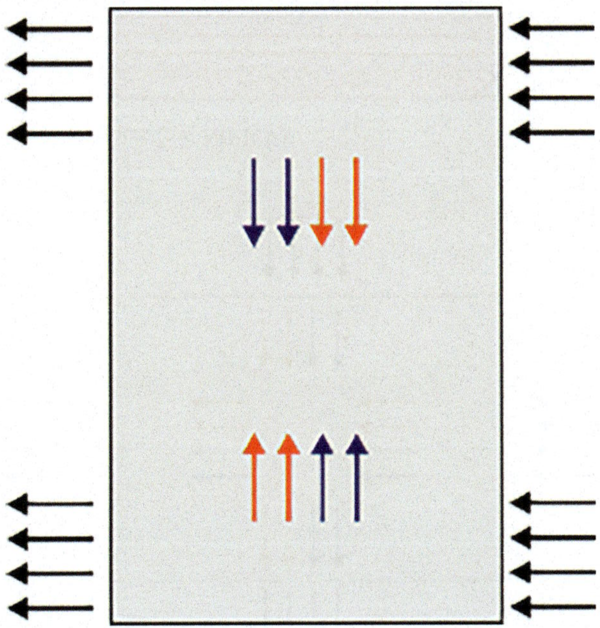	• Mixed flow configuration • Effectiveness of MERVs increases by 14.6% • High latent effectiveness as compared to sensible effectiveness • More efficient than crossflow	Asasian-Kolur et al. (2022); Albdoor et al. (2020)

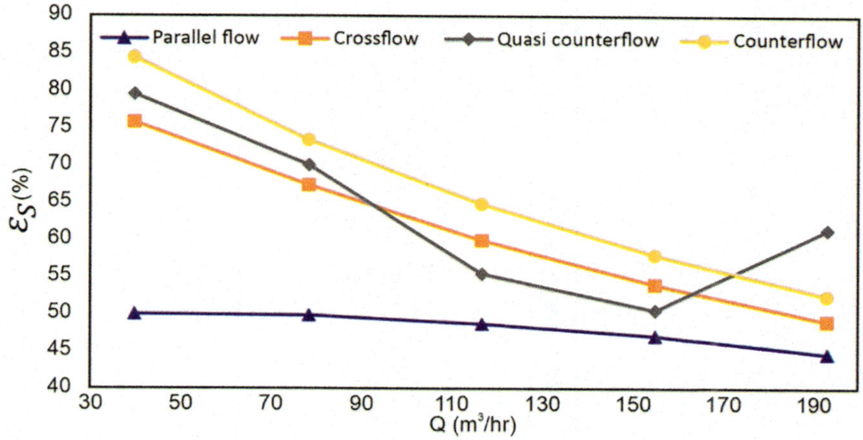

Fig. 8.11 Sensible effectiveness for different air flow configurations.

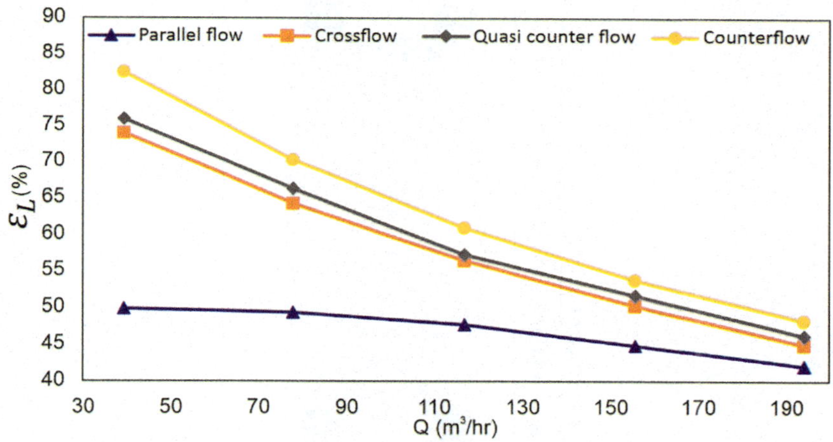

Fig. 8.12 Latent effectiveness for different air flow configurations.

the overall performance is based on the membrane heat exchanger its properties. The properties of the membrane are classified as physical and membrane-associated properties. Physical properties are pore size, porosity, thermal conductivity, membrane thickness, and modulus of elasticity. Membrane-associated properties are moisture diffusivity, vapor contact angle, sorption capacity, and isothermal moisture diffusion constant. Though MERV is not a new technology, it still attracted much attraction. So, the number of materials is tested with its all its pros and cons. Since the material is classified as conventional, advanced, and emerging membranes.

Table 8.2 Impact of outdoor air on the performance of MERVs.

Outdoor air state			Operating condition of MERV	Effectiveness			Ref.
Q	**T**	**w**	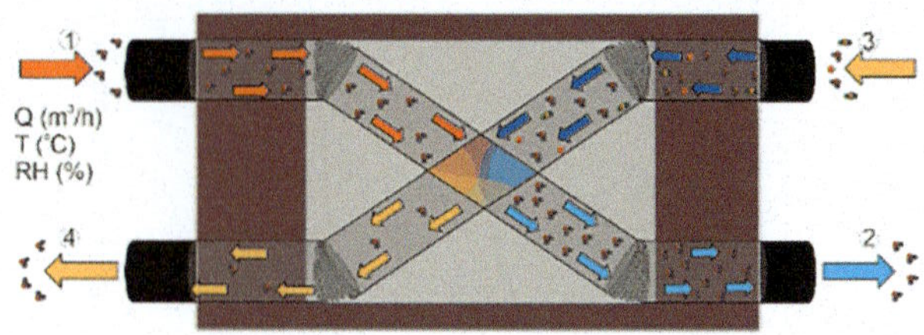	ε_s	ε_L	ε_T	Yaïci et al. (2013); Min and Ming (2010); Engarnevis et al. (2018); Kassai and Al-Hyari (2019); Liu et al. (2016)
(m^3/h)	(°C)	(kg_a/kg_w)		(-)	(-)	(-)	
↑	==	==		↓	↓	↓	
==	↑	==		↑	↓	↑	
==	==	↑		↑	↑	↑	

8.6.1 Membrane properties

8.6.1.1 Physical properties

Pore size refers to the nominal diameter of the membrane pores (Liu et al., 2019). Pore size is a significant component that affects the mass transfer through the membrane (Alkhudhiri et al., 2012). Based on the pore size, the membrane is classified as dense and porous. The pore size for the dense and porous membranes are 0.1 nm and 100 nm, respectively (Liu et al., 2016). The large pore size enhances the mass transfer through the membrane and influences the indoor air quality because it allows hazardous contaiminants into the conditioned space. Therefore, the optimal pore size is necessary for the MERVs.

Porosity is the ratio of the volume of the membrane pores to the total volume of the membrane (Alkhudhiri et al., 2012; Abdel-Salam et al., 2014). Porosity influences the mass flux and thermal efficiency of the membrane energy exchanger (Al-Turki et al., 2021). Higher porosity can potentially increase the mass flux but reduces the conductive heat dissipation (Alkhudhiri et al., 2012). Certain mechanical properties adversely affect the porosity, i.e., flow.

Thermal conductivity is the ability of membrane material to transfer heat. Usually, the membrane material has low thermal conductivity, ranging from 0.12 to 0.33 W/m.k. (Zhang, 2009b; Al-Waked et al., 2021). Moreover, when the membrane layer is thin then the total thermal resistance is high which makes it insignificant to the thermal resistance of the membrane, therefore the thermal conductivity has a minor impact on the sensible energy. Because of the moisture resistance, the thermal conductivity is also neglected due to its minor impact on latent energy (Al-Waked and Nasif, 2019; Vickers, 2017). → Fig. 8.13 shows the impact of thermal conductivity on enthalpy effectiveness. Conclusively, the membrane conductivity could be neglected because it accounts for less than half of the total heat transfer in the membrane energy exchanger system.

Thickness of the membrane material is one of the most significant properties that impact the sensible and latent heat recovery of the MERV system. Increasing thickness is responsible for the reduction in the surface contact area and number of channels. When the thickness of the membrane is close to zero sensible and latent effectiveness increases with negligible heat conduction and can be ignored (Liu et al., 2016; Vickers, 2017; Zhang et al., 2000). Furthermore, as the thickness of the membrane decreases the heat and mass transfer area also decreases which can be encountered using a supporting structure to accomplish optimal aspect ratio (Paul et al., 2019).

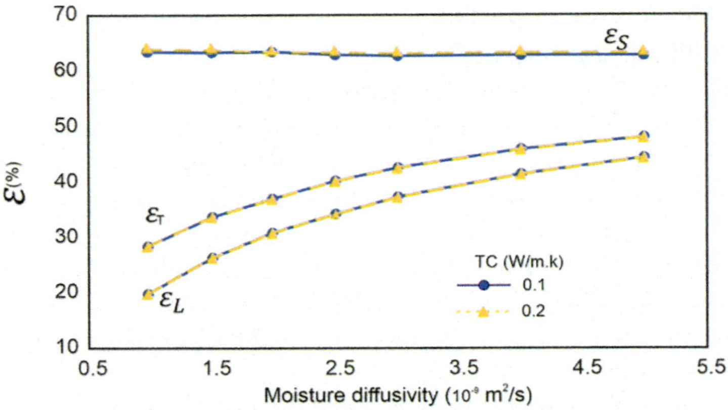

Fig. 8.13 Enthalpy effectiveness for different air flow configurations against moisture diffusivity.

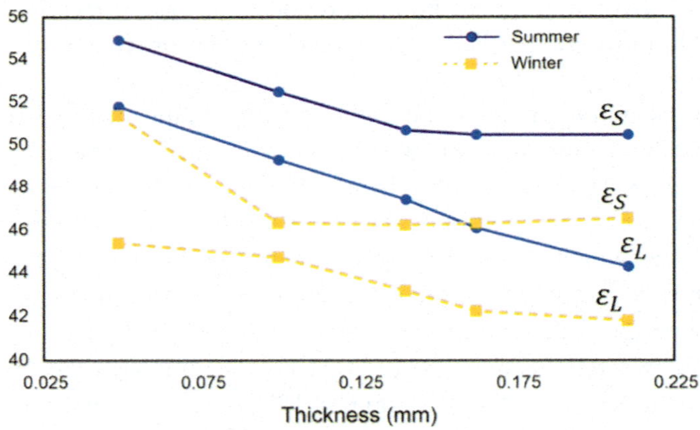

Fig. 8.14 Enthalpy effectiveness for different air flow configurations against thickness.

The membrane thickness acts differently as per outdoor conditions, i.e., as the thickness increases (0.1–0.2 mm), the sensible effectiveness does not vary in summer but slightly varies in winter (Yaïci et al., 2013). Fig. 8.14 shows that the membrane thickness increases the system effectiveness in both summer and winter decreases. Consequently, a small thickness of the membrane could be a better option to enhance the heat and mass transfer with a better optimal aspect ratio.

A renowned modulus shows the ratio of stress to strain within the elastic limit (Scott, 2000). It is the most critical parameter in the design of the

MERVs to reduce membrane deflection (Ge et al., 2014). It strongly relates to the pressure drop of the MERV channels whereas the pressure drop could also be possible due to the difference in pressure between two contiguous membrane channels (Alonso et al., 2017). There could be a significant membrane deflection by applying moderate pressure drop across both sides of the membrane (Larson et al., 2008). Membrane thickness and modulus of elasticity are inversely proportional to the membrane deflection, which indicates the significance of the optimal modulus of elasticity to avoid membrane deflection. In addition, membrane deflection results in variable channel space and uneven flow distribution (Abdel-Salam et al., 2017). This will reduce the overall system effectiveness with the lower convective heat and mass transfer coefficient (Larson et al., 2008; Zhang, 2009a). In this regard, to improve the system performance bulge cell test is used to measure the deflection and stress-strain diagram of the membrane (Ge et al., 2014; Larson et al., 2008). Therefore, the membrane deflection could be reduced using the support structure in the channel and using high elastic modulus membrane.

8.6.1.2 Mass transfer-associated properties

Mass diffusivity/permeability is the most crucial property in MERVs' performance. It is the amount of the vapors/mass that diffuses through the unit surface area in the unit time (Liu et al., 2019). As the moisture diffusivity increases the mass transfer through the membrane increases. Though there are many methods to measure mass diffusivity, the wet/dry method is considered as the standard method detailed in the ASTM E96 standard (Ge et al., 2014; Huang, 2008; McCullough et al., 2003; ASTM International, 2013). The outdoor parameters (temperature and humidity) influence the mass diffusivity in the membrane (Ge et al., 2014). Though in many studies reported in the literature that the mass diffusivity is accounted as the constant or averaged value (Yaïci et al., 2013; Koester et al., 2017; Albdoor et al., 2020; Zhang, 2009b; Woods, 2014; Zaw et al., 2013), on the other hand, for the high accuracy the variable mass diffusivity with respect to the outdoor parameters is also reported in the literature (Albdoor et al., 2020; Nasif et al., 2012; Nasif and Al-Waked, 2014; Nasif, 2015). Mass diffusivity resistance is the reciprocal of mass diffusivity. Mass diffusivity resistance is the ability of a membrane to resist permeation of the mass through it which accounts ~65%–90% of the total mass transfer resistance (Min and Su, 2010). Mass diffusivity influences the total effectiveness of the MERVs, which signifies the mass diffusivity.

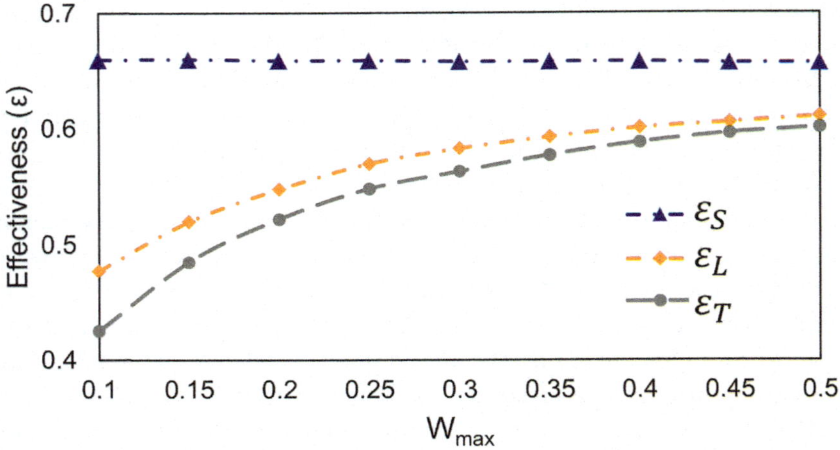

Fig. 8.15 Sensible, latent, and total effectiveness and maximum moisture uptake.

Maximum moisture water uptake is the ability of the membrane to absorb the maximum moisture in it. As the moisture uptake in the membrane increases the sensible effectiveness is slightly affected but the latent effectiveness is significantly affected (Min and Su, 2010; Min et al., 2011). Fig. 8.15 shows the sensible effectiveness is slightly changed as the moisture uptake increases but the latent and total effectiveness increases as the moisture uptake in membrane increases. Furthermore, the moisture uptake varies from 0.1–0.5 kg/kg which enhances 14% and 18% latent and total effectiveness respectively (Shen and Min, 2020).

As the isothermal constant is a significant property exhibited by the membrane material. As the C increases the mass resistance in the membrane is increased and then progressively decreased (Min and Su, 2010; Min et al., 2011). The change in the mass resistances with C is closely related to the adsorptive curve. As the C increases the sensible effectiveness remained unchanged whereas the total and latent effectiveness increased and then decreased. If the C=1 it shows the maximum effectiveness of MERVs (Shen and Min, 2020). Fig. 8.16 Justifies the maximum moisture uptake id isothermal constant is unity.

Water contact angle (α) is the measure degree of the wetness of mass/vapor droplet with the membrane surface and classifies the membrane as hydrophilic and hydrophobic. If the water vapor contact angle is greater than 90° shows the hydrophobic membrane and <90° indicates the hydrophilic membrane (Zhang et al., 2008; Albdoor et al., 2020; Zhang et al., 2012). Fig. 8.17 shows the water contact angle of different membranes.

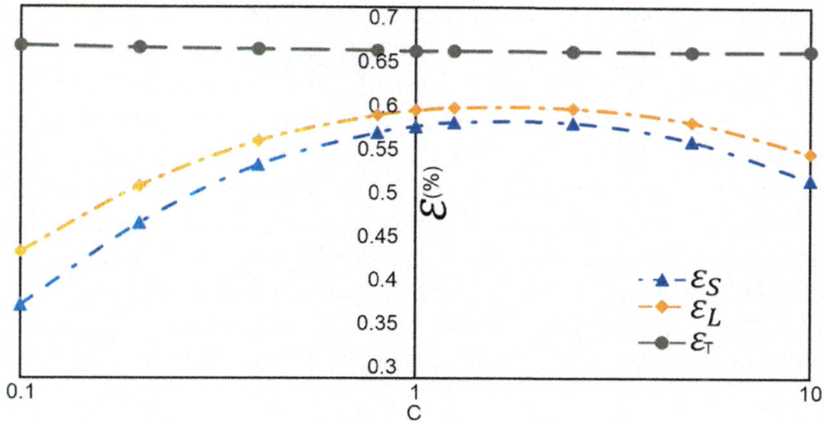

Fig. 8.16 Sensible, latent, and total effectiveness and isothermal constant.

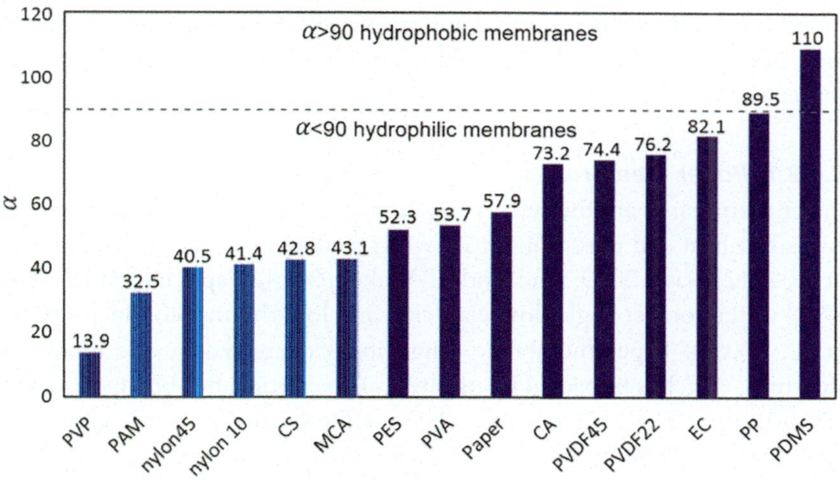

Fig. 8.17 Water contact angle of different membranes.

Furthermore, if the water contact angle is <90° indicates the better wettability of the membrane. The water vapor contact angle is also related to the physical properties of the membrane, i.e., porosity, and surface roughness (Zhang et al., 2008). Therefore, the lower the value of the water contact angle means the better mass transfer through the membrane.

Membrane selectivity is the ability of mass transfer with the crossing air streams (Liu et al., 2019). Higher selectivity is also important to maintain indoor air quality as it reduces the crossing of the contamination and allows better mass transfer in MERVs. It is inversely relates to the pore size which means as the pore size increases membrane selectivity decrease (Woods, 2014) therefore, it could be concluded that the as higher membrane selectivity is mandatory to enhance mass transfer and reduce health hazards.

The tortuosity factor is the ratio of the length of the pathway through which the mass/vapor diffuses into the porous membrane to the length of the straight pathway (Alkhudhiri et al., 2012). The touristy factor is dependent on the pore geometry. The pore geometry in the membrane varies which makes the tortuosity factor complex, however the tortuosity factor is assumed as 2. Higher tortuosity factor would result in lower mass flux through the membrane (El-Bourawi et al., 2006; Lawson and Lloyd, 1997; Bandini et al., 1997; Phattaranawik et al., 2003; Khayet et al., 2004).

8.6.2 Conventional materials

Conventional membrane materials, i.e., paper, cellulose and polymeric are discussed based on their physical and mass-associated properties. Tables 8.3 and 8.4 show the physical and mass transfer-associated properties, respectively.

8.6.2.1 Paper membranes

Paper membranes are the widely used inexpensive hydrophilic material that is used to heat and mass transfer between feed and sweep air streams (Nasif et al., 2012; Nasif, 2015; Nasif and Al-Waked, 2014). Paper membranes have <90° water contact angle, low elasticity, and low thermal conductivity (Liu et al., 2009a). Paper membrane longevity is decreased by the membrane structure, which is impacted by its physical properties after being exposed to air moisture. This will enhance the membrane deformation and reduce effectiveness. Typically, the paper membrane diffusivity varies from 10^{-11} to 10^{-7} m^2/s (Zhang et al., 2008, 2010; Zhang, 2016). Furthermore, by changing the chemical structure of the membrane, i.e., LiCl impregnated on the paper membrane the mass-associated properties also can be improved. As such the paper membrane longevity cannot be enhanced but it can be replaced because it is inexpensive and easily available (Liu et al., 2009a).

Various types of kraft paper with varying physical properties are also employed to test MERVs effectiveness. Two paper membranes KP45 and KP60 are compared having 0.0027, and 0.0029 porosity and 78, and 98 μm thickness, respectively. The results demonstrate that employing Mylar as a heat transfer surface, KP60, regardless of thickness, has a potential to save 78% more energy than KP45 (Liu et al., 2009a).

Table 8.3 Physical properties of paper, cellulose, and polymeric membranes.

Sr. No.	Physical properties						
	Name	Porosity (-)	Pore size (μm)	Thickness (mm)	Thermal conductivity (W/m.K)	Density (kg/m^3)	Ref.
Paper membranes							
1	KP-45 g	0.0027	–	78	–	–	Nasif et al. (2012); Nasif (2015)
2	KP-60 g	0.0029	–	98	0.12	–	Al-Waked et al. (2021); Nasif (2015)
3	KP-75 g	–	–	130	–	–	Nasif and Al-Waked (2014); Nasif et al. (2013)
4	ER paper	–	–	96	0.31	–	Zhang (2016)
5	Treated paper	–	–	–	0.18	–	Zhang et al. (2010)
6	Plate paper	-	-	100	0.44	–	Zhang (2009b)
7	LiCl treated paper	–	–	42	0.13	–	Lee et al. (2012); Lee et al. (2013a)
8	Paper membrane	–	–	55	0.44	–	Zhang et al. (2008)
Cellulose membranes							
9	AC	–	–	$5x10^{-3}$	0.41	760	ZHANG et al. (2008)
10	Cellulose	–	0.22	$1.5x10^{-3}$	–	–	Min et al. (2010)
Polymeric membranes							
11	PVDF	–	0.22	$1x10^{-1}$	–	–	Min et al. (2010)
12	PES	–	0.22	$1x10^{-1}$	–	–	Min et al. (2010)
13	PP	0.41	0.22	$3.2x10^{-2}$	0.16	370	Liu et al. (2016)
14	PDMS	–	–	$5.5x10^{-2}$	–	–	Lekshminarayanan et al. (2020)
15	PE	–	0.068	$1.05x10^{-1}$	–	–	Engarnevis et al. (2020)
16	Nylon	–	0.45	$1x10^{-1}$	–	–	Albdoor et al. (2020)

Table 8.4 Properties associated with the mass transfer of paper, cellulose, and polymeric membranes.

Sr. No.	Properties associated with mass transfer				
	Name	Membrane nature	Moisture diffusivity (m^2/s)	Permeability (mol/m s Pa)	Ref.
Paper membranes					
1	KP-45 g	–	–	–	Nasif et al. (2012); Nasif (2015)
2	KP-60 g	–	–	–	Al-Waked et al. (2021); Nasif (2015)
3	KP-75 g	–	–	–	Nasif and Al-Waked (2014); Nasif et al. (2013)
4	ER paper	–	$1.78x10^{-7}$	–	Zhang (2016)
5	Treated paper	–	$5x10^{-7}$	–	Zhang et al. (2010)
6	Plate paper	–	$3.5x10^{-11}$	–	Zhang (2009b)
7	LiCl treated paper	–	$3x10^{-11}$	–	Lee et al. (2012); Lee et al. (2013a)
8	Paper membrane	–	$6.08x10^{-12}$	–	Zhang et al. (2008)
Cellulose membranes					
9	AC	Hydrophilic	$1.05x10^{-11}$	–	Zhang et al. (2008)
10	Cellulose	–	$6.72x10^{-7}$	–	Min et al. (2010)
Polymeric membranes					
11	PVDF	–	$1.65x10^{-6}$	–	Min et al. (2010)
12	PES	–	$6.11x10^{-7}$	–	Min et al. (2010)
13	PP	Hydrophobic	$1.6x10^{-12}$	–	Liu et al. (2016)
14	PDMS	–	–	$1.2x10^{-11}$	Lekshminarayanan et al. (2020)
15	PE	–	–	$6.86x10^{-1o}$	Engarnevis et al. (2020)
16	Nylon	Hydrophilic	$1.5x10^{-6}$	–	Albdoor et al. (2020)

It is important to describe that the SEM images of paper, PSS, PES, Cellulose, MC, AC, LiCl-impregnated, PVDF, PSF membranes, and likewise can be found from the cited literature (Min et al., 2008).

8.6.2.2 Cellulose membrane

Cellulose membranes are one of the most common types of membrane used for the heat and mass transfer in MERVs, which is superior in vapor permeability and shows better performance in total energy recovery than paper membranes (Zhang et al., 2008; Liu et al., 2009b; Zhang, 2009b; Zhang, 2006; Li et al., 2015; Zhang et al., 2011). The novel membrane CSLM, which is composed of two PVDF and one cellulose acetate layer, has been reported in the literature. According to the findings, this novel membrane results in a water vapor diffusivity that is twice as high as that of the solid acetate membrane (Liu et al., 2009b). Zhang et al. (2011) developed the single-step asymmetric membrane which is prepared by using deionized water as an additive, acetic acid solvent, and porous skin medium with the coagulant medium exchange. The study concluded that the cross-corrugated structure with a one-step asymmetric membrane shows 20% higher sensible effectiveness and 40% higher latent effectiveness.

8.6.2.3 Polymeric membranes

Polymeric membranes are the semi permeable widely employed for heat and mass transfer through the MERVs. Polymeric membranes are inexpensive, superior in physical properties, and easily reproducible (Qu et al., 2018). The properties of the membrane-associated with the mass transfer are much significant in the membrane selection. The appropriate fabrication of the membrane will be helpful to control the structural properties of the membrane. The fabrication process includes film casting, aerosol deposition, in situ polymerization, ion-assisted deposition, ion exchange, dip coating, hydrothermal synthesis, spray coating, electrospinning and phase inversion layer-by-layer deposition sputtering, and etching (Metz et al., 2005). The higher permeability and selectivity exhibit the higher enthalpy effectiveness of the MERVs. → Fig. 8.18 shows the permeability and selectivity of various polymeric membranes at 30°C.

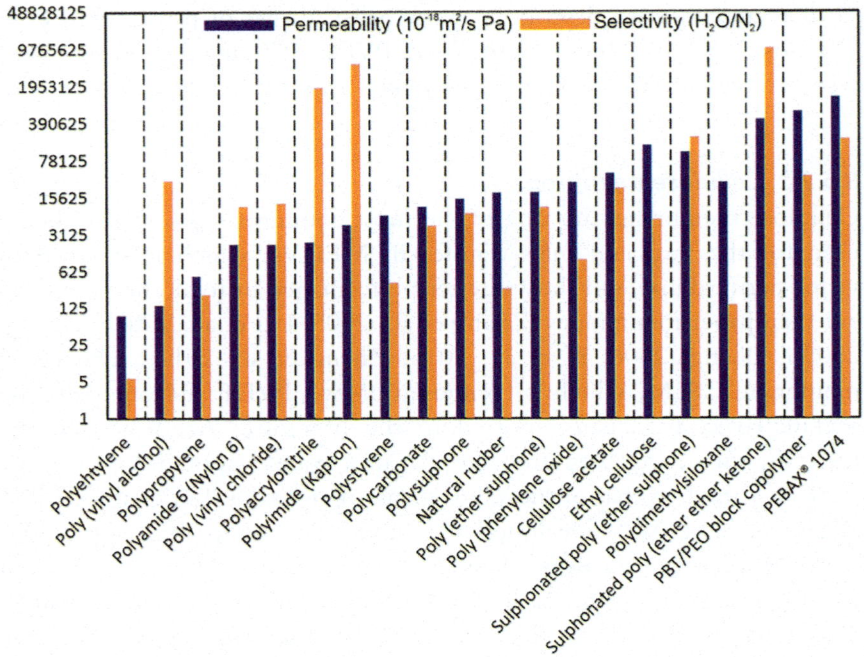

Fig. 8.18 Permeability and selectivity of polymeric membranes.

8.6.3 Advanced materials

8.6.3.1 Ceramic membranes

Ceramic membranes are microporous structure having hydrophilic nature due to the hydroxyl group and are made of titania, silica, zirconia, and alumina (Curcio et al., 2010). Usually, the ceramic membrane can be used at high temperatures because ceramic membranes are thermally stable i.e., silica-alumina (Asaeda et al., 1986). The ceramic membrane is not easily prone to microorganisms, contaminants, and abrasion in dehumidification applications. This makes ceramic membranes superior but still, these are just undervalued ones and tested only on a lab scale due to a lack of technical expertise (Zhang et al., 2020; Zhang, 2006; Issaoui and Limousy, 2019). Ceramic membrane has varying physical properties, i.e., porosity having an asymmetric structure. The surface roughness of the membrane could be reduced by adding the micropores (2–50 nm) that are employed to the intermediate layer. The mechanical stability of the membrane can be increased by macropores (> 50 nm) (Xiao et al., 2021). Furthermore, nanopores present in the membrane could reduce the mass flux or mass transport in the MERVs. Nanopore membranes are expensive, so it is difficult to commercialize them (Xiong et al., 2017).

8.6.3.2 Composite membranes

Composite membranes are designed to enhance the hydrophilic characteristics of the membrane through modification in the structural and chemical properties of the membrane. Usually, membranes are modified to enhance moisture diffusivity of the membrane which typically varies from (10^{-9} to 10^{-13} m^2/s). The methods used for hydrophilic modification are surface coating and mixing with hydrophilic materials, grafting of the membrane surface with hydroxyl, and a sulfonic acid group. Surface modification causes a few structural changes that affect mass transfer and makes the system complex (Jin et al., 2019). Therefore, the composite membrane has relatively high permeability and strength compared to other membranes and has attracted much attention recently. In the literature, it was reported that the low-cost polymer-based composite membrane has a support structure named polyamide to control mass transfer (Kim, 2020).

8.6.3.3 Zeolite membranes

Zeolite membranes are microporous structures having hydrophilic nature that can be used for heat and mass transfer applications. Zeolite membrane consists of the inorganic alumina silicate, and silico-alumino-phosphate (Caro and Noack, 2010; Kanoğlu et al., 2004; Morigami et al., 2001; Sato et al., 2008). In addition, due to the hydrophilic nature of the zeolite membrane and alteration in properties due to the water adsorption it can be used in other pervaporation separations of the aqueous solution. Furthermore, the high selectivity of the zeolite membrane is reported in zeolite A (Linde Type A: LTA) and pentasil zeolite (MFI-type ZSM-5: Zeolite Socony Mobil-5) membranes. Usually, the zeolite membrane is used as a supporting material in the composite membranes, and zeolite on the ceramic membranes is also used in the vapor separation processes. It signifies the application of the zeolite membrane (Al-Alili et al., 2015; Okamoto et al., 2001).

8.6.4 Emerging materials

8.6.4.1 Liquid membranes

Liquid membrane can also be employed to recover the heat and mass from the feed and sweep air stream instead of solid membranes. The application of the liquid membrane differs because they require a separate system to transfer heat and mass as compared to the solid membranes (Liu et al., 2019; Ronghui et al., 2020). The moisture is transferred to the liquid side by the difference in air/gas vapor pressure (Krause et al., 2020). Moreover, the characteristics of desiccant material can be changed by adding, i.e., nanoparticles, and surfactants. However, the complexity and sensitivity of this technology, i.e., low heat capacity and corrosive of agents make it unreliable (Liu et al., 2019).

In recent years liquid membranes attracted much attention because of superior mass-associated properties i.e., mass diffusivity is $\sim 10^{-9}$ m^2/s (Zhang, 2006). Furthermore, the liquid membrane can be selected if it has low viscosity, less vapor pressure than a liquid, and high hydrophilicity (Ito, 2000; Kudasheva et al., 2016). Liquid membranes can be classified as organic solvent, ionic liquids, and aqueous solutions of inorganic salts. Various organic materials such as polyethylene glycol, and tri-ethylene glycol with an aqueous solution i.e., lithium chloride are employed in the liquid membrane (Liu et al., 2019; Ito, 2000). There are many other organic solids with the aqueous solution that can be used as a liquid membrane, but the applicability of a liquid membrane has not been investigated up till now (Bouzenada et al., 2014; Gurubalan et al., 2017). In this regard, the ionic liquid membrane has attracted much attention (Qu et al., 2018; Ito, 2000; Kudasheva et al., 2016; Scovazzo, 2010; Zang et al., 2011) due to its low flammability and vapor pressure, high polarity, and have low melting point, therefore, it could be potential alternatives membrane. Furthermore, except hydrophilicity nature of ionic liquid membrane, the other characteristics of the membrane i.e., less biodegradable, inexpensive, and less toxic but the major disadvantage is leaching (Singh et al., 2014; Francesco et al., 2011; Park and Jeong, 2020). In this regard, Gebreyohannes et al. (2020) proposed a novel liquid membrane for ionic amino acids by encapsulation as carbonaceous sub-microcapsules and dispersion in a thin polydimethylsiloxane (PDMS) coating layer of hollow fibers. This novel membrane was much more capable of reducing the leaching issue and effective entrapment of ionic liquids. Furthermore, the results show that the selectivity and vapor permeability of this novel membrane 4500 for water over N_2 and 10,600 GPU (gas permeation unit $1GPU = 10^{-6}$ $cm^3/$ $(s.cm^2.cm\ Hg)$).

8.6.4.2 Graphene oxide paper-based membranes

Graphene oxide paper-based membrane has been investigated for its mass-associated properties, i.e., water permeability, adsorption capacity, and membrane selectivity (Shin et al., 2016; Liu et al., 2017). In literature, it is reported that at 80% and ~31°C outdoor parameters 2D graphene oxide membrane shows higher selectivity (N_2 > 104) and permeability (1.82 × 10^5 barrel). Furthermore, in another comparative study graphene oxide framework with uncross linking and glyoxal cross linking. The results showed that the glyoxal cross linking was ~10^6 high dehumidification factor and was proven more effective (Hung et al., 2020).

8.6.4.3 Metal organic frameworks-based membranes

Recently, metal organic framework (MOF) based membranes have gained much attention air to air energy recovery and other dehumidification applications. In literature, for isothermal membrane-based dehumidification ionic liquids encapsulated by MOF were proposed and claimed to eradicate the poor stability by direct encapsulation of ionic liquids to support structure. The results indicated that the membrane shows the maximum water vapor permeability (2.36 × 10^{-4} mol/m^2.s.Pa) and separation factor of 1560 H_2O/N_2 in dehumidification application (Park and Jeong, 2020). The thin layer of MOF is applied to the PVDF membrane to enhance mass transport and hydrophilic properties (Albdoor et al., 2021). Furthermore, various MOF absorbent fillers were employed in mass removal, and the results indicated 2.8 times more water vapor permeance with a selectivity of 542 (Ingole et al., 2018).

8.7 Adaptability constraints

Though the MERVs have attracted much attention as a low cost, environment friendly, adaptability in extreme weather conditions, it could not break the industrial barrier. One of the main reasons is the implementation of the ecofriendly/ carbon emission reduction laws. There are less people familiar with MERV technology including the architects, developers, and end users which is responsible of the the lack of skilled labor. The membrane fouling due to heavy dust storm minimizes its applicability it high dust storm regions. The other constraints in the MERV technology is the need of the close space and have minimum applicability at small scale but have a more applicability at centralized air condition system.

8.8 Conclusions

The study represents the comprehensive review of membrane-based energy recovery ventilators. The different types of channel configuration, flow configuration, operating parameters, membrane properties, membrane types and adaptability constraints of MERV discussed. Novel channel geometries improve the heat and mass transfer but still the optimized channel configuration results in improved system effectiveness. Counterflow MERV exhibits better system performance, but it is difficult to install i.e., inlet and outlet adjustments but mixed/z-shaped flow improves the system effectiveness ~15% and is easy to install. In operating MERV system if the flow rate increases than system performance decreases. Membrane properties are the key factors in MERV performance. Some properties are more influential to the system's performance and some possess minimum impact. High modulus of elasticity is required to minimize the membrane deflection and reduce the pressure drop to enhance the MERV effectiveness. Moisture diffusivity resistance impacts mass permeance through membrane and reduces system effectiveness. Polymeric membrane and paper membrane are widely employed because they are inexpensive, easy to handle, reproducible and good in heat and mass permanence. Ceramic membrane is superior in heat transfer as it is thermally stable. MOF membrane possesses 2.8 times more mass permeance.

Acknowledgment

The United Kingdom Government funds this research (and/or) pilot activity through the Project, APP47457, titled "Super–efficient Sustainable Cooling Solution for All Applications (S2Cool)" under the Ayrton Challenge Programme of the UK Research and Innovation (UKRI); however, the views expressed do not necessarily reflect the UK Government's official policies. The Project, APP47457, is implemented/led by Northumbria University, UK, together with partners to develop sustainable cooling solution for several applications.

Funding statement

The United Kingdom Government funds this research (and/or) pilot activity through the Project, APP47457, titled "Super–efficient Sustainable Cooling Solution for All Applications (S2Cool)" under the Ayrton Challenge Programme of the UK Research and Innovation (UKRI); however, the views expressed do not necessarily reflect the UK Government's official policies. The Project: APP47457 is implemented/led by Northumbria University, UK, together with partners to develop sustainable cooling solution for several applications.

CRediT authorship contribution statement

Muhammad Asif was involved in the conceptualization, methodology, software, formal analysis, investigation, writing of the original draft. **Muhammad Sultan, Zahid**

Mahmood Khan, and **Muhammad Wakil Shahzad** were involved in the conceptualization, methodology, validation, resources, visualization, supervision, project administration, Writing—Review Editing, Funding acquisition. **Faheem Hassan Akhtar, Hadeed Ashraf, Naveed Husnain, Uzair Sajjad, Muhammad Farooq, and Muhammad Yasar Javaid** were involved in the conceptualization, methodology, software, formal analysis, investigation, writing (review and editing).

Data availability statement

The datasets used and/or analyzed during the current study are available from the corresponding authors upon request.

Conflicts of interest

The authors declare no conflict of interest.

Declaration of generative AI and AI-assisted technologies in the writing process

No generative AI/AI-assisted technology is used in this paper.

References

Abdel-Salam, M.R.H., Besant, R.W., Simonson, C.J., 2017. Performance testing of 2-fluid and 3-fluid liquid-to-air membrane energy exchangers for HVAC applications in cold–dry climates. Int. J. Heat Mass Transf. 00179310. 106, 558–569. Available from: http://www.journals.elsevier.com/international-journal-of-heat-and-mass-transfer/. doi:10.1016/j.ijheatmasstransfer.2016.09.024.

Abdel-Salam, M.R.H., Ge, G., Fauchoux, M., Besant, R.W., Simonson, C.J., 2014. State-of-the-art in liquid-to-air membrane energy exchangers (LAMEEs): a comprehensive review. Renew. Sustain. Energy Rev. 13640321. 39, 700–728. doi:10.1016/j.rser.2014.07.022.

Al-Alili, A., Hwang, Y., Radermacher, R., 2015. Performance of a desiccant wheel cycle utilizing new zeolite material: experimental investigation. Energy 03605442. 81, 137–145. Available from: https://www.elsevier.com/inca/publications/store/4/8/3/. doi:10.1016/j.energy.2014.11.084.

Al-Turki, Y.A., Mebarek-Oudina, F., Ahmadian, A., Baleanu, D., 2021. Flat sheet direct contact membrane distillation desalination system using temperature-dependent correlations: thermal efficiency via a multi-parameter sensitivity analysis based on Monte Carlo method. J. Therm. Anal. Calorim. 15882926. 144 (6), 2641–2652. Available from: https://www.springer.com/journal/10973. doi:10.1007/s10973-020-10503-6.

Al-Waked, R.F., Nasif, M.S., 2019. Air to air energy recovery from HVAC systems under different membrane materials. Univ. J. Mech. Eng. 23323361. 7 (2), 37–49. Available from: http://www.hrpub.org/download/20190330/UJME2-15112786.pdf. doi:10.13189/ujme.2019.070202.

Al-Waked, R., Mostafa, D.B., Nasif, M.S., 2021. Performance of energy recovery ventilators under different climatic regions. Energy Eff. 15706478. 14 (1). Available from: http://www.springer.com/environment/journal/12053. doi:10.1007/s12053-020-09917-w.

Al-Waked, R., Nasif, M.S., Morrison, G., Behnia, M., 2015. CFD simulation of air to air enthalpy heat exchanger: variable membrane moisture resistance. Appl. Therm. Eng. 13594311. 84, 301–309. Available from: http://www.journals.elsevier.com/applied-thermal-engineering/. doi:10.1016/j.applthermaleng.2015.03.067.

Albdoor, A.K., Zhenjun, M., Cooper, P., 2020. Experimental investigation and performance evaluation of a mixed-flow air to air membrane enthalpy exchanger with different configurations. Appl. Therm. Eng. 13594311. 166, 114682. doi:10.1016/j.applthermaleng.2019.114682.

Albdoor, A.K., Zhenjun, M., Cooper, P., Al-Ghazzawi, F., Liu, J., Richardson, C., Wagner, P., 2021. Air-to-air enthalpy exchangers: membrane modification using metal-organic frameworks, characterisation and performance assessment. J. Clean. Prod. 09596526. 293, 126157. doi:10.1016/j.jclepro.2021.126157.

Alkhudhiri, A., Darwish, N., Hilal, N., 2012. Membrane distillation: a comprehensive review. Desalination. 00119164. 287, 2–18. doi:10.1016/j.desal.2011.08.027.

Alonso, M.J., Mathisen, H.M., Aarnes, S., Liu, P., 2017. Performance of a lab-scale membrane-based energy exchanger. Appl. Therm. Eng. 13594311. 111, 1244–1254. Available from: http://www.journals.elsevier.com/applied-thermal-engineering/. doi:10.1016/j.applthermaleng.2015.11.119.

Arrow, K.J., 2007. Global climate change: a challenge to policy. Econ. Voice. 1553-3832. 4 (3). doi:10.2202/1553-3832.1270.

Asaeda, A., Du, L.D., Ikeda, K., 1986. Experimental studies of dehumidification of air by an improved ceramic membrane. J. Chem. Eng. Jpn. 00219592. 19 (3), 238–240. doi:10.1252/jcej.19.238.

Asasian-Kolur, N., Sharifian, S., Haddadi, B., Pourhoseinian, M., Shekarbaghani, Z.M., Harasek, M., 2022. Membrane-based enthalpy exchangers for coincident sensible and latent heat recovery. Energy Convers. Manag. 01968904. 253, 115144. doi:10.1016/j.enconman.2021.115144.

ASTM International, 2013. ASTME96/E96M-10: Standard Test Methods for Water Vapor Transmission of Materials. ASTM International. doi:10.1520/E0096_E0096M-10.

Bandini, S., Saavedra, A., Sarti, G.C., 1997. Vacuum membrane distillation: experiments and modeling. AIChE J. 00011541. 43 (2), 398–408. Available from: https://www.interscience.wiley.com. doi:10.1002/aic.690430213.

Beattie, C., Fazio, P., Zmeureanu, R., Rao, J., 2018. Experimental study of air-to-air heat exchangers for use in Arctic housing. Appl. Therm. Eng. 13594311. 129, 1281–1291. Available from: http://www.journals.elsevier.com/applied-thermal-engineering/. doi:10.1016/j.applthermaleng.2017.10.112.

Bouzenada, S., Kaabi, A.N., Fraikin, L., Léonard, A., 2014. Experimental Study on dehumidification/regeneration of Liquid Desiccant: liBr Solution. Procedia Comput. Sci. 18770509. 32, 673–680. doi:10.1016/j.procs.2014.05.476.

Caro, J., Noack, M., 2010. Zeolite membranes–status and prospective. In: Ernst, S. (Ed.), Advances in Nanoporous Materials. Elsevier, pp. 1–96. doi:10.1016/S1878-7959(09)00101-7.

Ciofalo, M., Stasiek, J., Collins, M.W., 1996. Investigation of flow and heat transfer in corrugated passages—II. Numerical simulations. Int. J. Heat Mass Transf. 00179310. 39 (1), 165–192. doi:10.1016/s0017-9310(96)85014-9.

Curcio, E., Di Profio, G., Drioli, E., 2010. Membrane Distillation and Osmotic Distillation, Vol. **4**. Elsevier BV, pp. 1–20. doi:10.1016/b978-0-08-093250-7.00015-3.

El-Bourawi, M.S., Ding, Z., Ma, R., Khayet, M., 2006. A framework for better understanding membrane distillation separation process. J. Membr. Sci. 03767388. 285 (1-2), 4–29. doi:10.1016/j.memsci.2006.08.002.

Engarnevis, A., Huizing, R., Green, S., Rogak, S., 2018. Heat and mass transfer modeling in enthalpy exchangers using asymmetric composite membranes. J. Membr. Sci. 18733123. 556, 248–262. Available from: https://www.elsevier.com/locate/memsci. doi:10.1016/j.memsci.2018.03.007.

Engarnevis, A., Huizing, R., Green, S., Rogak, S., 2020. Fouling of composite water vapor transport membranes by aerosol nanoparticles. J. Membr. Sci. 03767388. 614, 118440. doi:10.1016/j.memsci.2020.118440.

European Commission, 2020. In focus: energy efficiency in buildings. Available from: https://ec.europa.eu/info/news/focus-energy-efficiency-buildings-2020-lut-17_en.

Francesco, F.D., Calisi, N., Creatini, M., Melai, B., Salvo, P., Chiappe, C., 2011. Water sorption by anhydrous ionic liquids. Green Chem. 14639270. 13 (7), 1712–1717. doi:10.1039/c1gc15080d.

Ge, G., Mahmood, G.I., Moghaddam, D.G., Simonson, C.J., Besant, R.W., Hanson, S., Erb, B., Gibson, P.W., 2014. Material properties and measurements for semi-permeable membranes used in energy exchangers. J. Membr. Sci. 18733123. 453, 328–336. doi:10.1016/j.memsci.2013.11.013.

Gebreyohannes, A.Y., Upadhyaya, L., Silva, L.P., Falca, G., Carvalho, P.J., Nunes, S.P., 2020. Hollow fibers with encapsulated green amino acid-based ionic liquids for dehydration. ACS Sustain. Chem. Eng. 21680485. 8 (48), 17763–17771. Available from: http://pubs.acs.org/journal/ascecg. doi:10.1021/acssuschemeng.0c06001.

Global CCS Institute, 2021. The Global status of CCS . Available from: https://www.globalccsinstitute.com/publications/global-status-of-ccs-2021/.

Gurubalan, A., Maiya, M.P., Tiwari, S., 2017. Performance characterization of membrane dehumidifier with desiccants in flat-plate arrangement. Energy Build. 03787788. 156, 151–162. doi:10.1016/j.enbuild.2017.09.040.

Hansen, J., Sato, M., Ruedy, R., Lo, K., Lea, D.W., Medina-Elizade, M., 2006. Global temperature change. Proc. Natl. Acad. Sci. U. S. A. 00278424. 103 (39), 14288–14293. doi:10.1073/pnas.0606291103.

Huang, J., 2008. A device for characterizing water vapor permeability of polymer membranes. Int. J. Polym. Anal. Charact. 1023-666X. 13 (1), 37–48. doi:10.1080/10236660701760262.

Hung, W.S., Lai, Y.L., Lee, P.H., Chiao, Y.H., Sengupta, A., Sivakumar, M., Lee, K.R., Lai, J.Y., 2020. Tuneable interlayer spacing self-assembling on graphene oxide-framework membrane for enhance air dehumidification. Sep. Purif. Technol. 18733794. 239. Available from: http://www.journals.elsevier.com/separation-and-purification-technology/. doi:10.1016/j.seppur.2019.116499.

Ingole, P.G., Sohail, M., Abou-Elanwar, A.M., Baig, M.I., Jeon, J.D., Choi, W.K., Kim, H., Lee, H.K., 2018. Water vapor separation from flue gas using MOF incorporated thin film nanocomposite hollow fiber membranes. Chem. Eng. J. 13858947. 334, 2450–2458. Available from: https://www.elsevier.com/inca/publications/store/6/0/1/2/7/3/index.htt. doi:10.1016/j.cej.2017.11.123.

International Energy Agency, 2021. Energy statistics data browser. Available from: https://www.iea.org/data-and-statistics/data-tools/energy-statistics-data-browser.

Islamoglu, Y., Parmaksizoglu, C., 2003. The effect of channel height on the enhanced heat transfer characteristics in a corrugated heat exchanger channel. Appl. Therm. Eng. 23 (8), 979–987. doi:10.1016/S1359-4311(03)00029-2.

Issaoui, M., Limousy, L., 2019. Low-cost ceramic membranes: synthesis, classifications, and applications. Comptes Rendus Chimie 16310748. 22 (2-3), 175–187. Available from: http://www.elsevier.com/journals/comptes-rendus-chimie/1631-0748. doi:10.1016/j.crci.2018.09.014.

Ito, A., 2000. Dehumidification of air by a hygroscopic liquid membrane supported on surface of a hydrophobic microporous membrane. J. Membr. Sci. 175 (1), 35–42. doi:10.1016/S0376-7388(00)00404-X.

Jafarizave, M., Khaleghi, A., Rezakazemi, M., 2019. Development of CFD model for membrane-based energy recovery ventilators. Chem. Eng. Res. Des. 02638762. 145, 226–234. doi:10.1016/j.cherd.2019.03.019.

Jin, Y., Dan, H., Lin, Y., Shi, L., 2019. Hydrophilic modification of polyvinylidene fluoride membrane by blending amphiphilic copolymer via thermally induced phase separation. Polym. Adv. Technol. 1042-7147. 30 (1), 110–119. doi:10.1002/pat.4449.

Kanoğlu, M., Özdinç Çarpınlıoğlu, M., Yıldırım, M., 2004. Energy and exergy analyses of an experimental open-cycle desiccant cooling system. Appl. Therm. Eng. 13594311. 24 (5-6), 919–932. doi:10.1016/j.applthermaleng.2003.10.003.

Kassai, M., Al-Hyari, L., 2019. Investigation of ventilation energy recovery with polymer membrane material-based counter-flow energy exchanger for nearly zero-energy buildings. Energies. 1996-1073. 12 (9), 1727. doi:10.3390/en12091727.

Khayet, M., Khulbe, K.C., Matsuura, T., 2004. Characterization of membranes for membrane distillation by atomic force microscopy and estimation of their water vapor transfer coefficients in vacuum membrane distillation process. J. Membr. Sci. 238 (1-2), 199–211. doi:10.1016/j.memsci.2004.03.036.

Kim, N.-H., 2016. Effect of the channel height of the enthalpy exchanger on energy saving by enthalpy recovery ventilator. Int. J. Air-Cond. Refrig. 2010-1325. 24 (03), 1650020. doi:10.1142/s2010132516500206.

Kim, N.H., 2020. Moisture diffusion coefficients of a PE/PA composite membrane in an enthalpy exchanger for heat and moisture recovery. J. Mech. Sci. Technol. 19763824. 34 (12), 5295–5302. Available from: http://www.springerlink.com/content/1738-494X. doi:10.1007/s12206-020-1130-3.

Kingma, B., Lichtenbelt, W.V.M., 2015. Energy consumption in buildings and female thermal demand. Nat. Clim. Change 17586798. 5 (12), 1054–1056. Available from: http://www.nature.com/nclimate/index.html. doi:10.1038/nclimate2741.

Koester, S., Falkenberg, M., Logemann, M., Wessling, M., 2017. Modeling heat and mass transfer in cross-counterflow enthalpy exchangers. J. Membr. Sci. 03767388. 525, 68–76. doi:10.1016/j.memsci.2016.10.030.

Krause, R., Rosenbaum, H., Ralf Herzog, A.V.S., 2020. Applications with semipermeable membranes for humidity recovery in HVAC systems. MATEC Web Conf. 2261-236X. 324, 03003. doi:10.1051/matecconf/202032403003.

Kudasheva, A., Kamiya, T., Hirota, Y., Ito, A., 2016. Dehumidification of air using liquid membranes with ionic liquids. J. Membr. Sci. 18733123. 499, 379–385. Available from: https://www.elsevier.com/locate/memsci. doi:10.1016/j.memsci.2015.10.069.

Larson, M.D., Besant, R.W., Simonson, C.J., 2008. The effect of membrane deflections on flow rate in crossflow air-to-air exchangers. HVAC&R Res. 1078-9669. 14 (2), 275–288. doi:10.1080/10789669.2008.10391008.

Lawson, K.W., Lloyd, D.R., 1997. Membrane distillation. J. Membr. Sci. 124 (1), 1–25. doi:10.1016/S0376-7388(96)00236-0.

Lee, E.J., Lee, J.P., Kim, N.H., 2013a. Moisture transfer characteristics of a LiCl-impregnated paper membrane used for an enthalpy exchanger. J. Mech. Sci. Technol. 1738494X. 27 (5), 1527–1537. doi:10.1007/s12206-013-0333-2.

Lee, E.J., Lee, J.P., Sim, H.M., Kim, N.H., 2012. Modeling and verification of heat and moisture transfer in an enthalpy exchanger made of paper membrane. Int. J. Air-Cond. Refrig. 20 (3), 1250015. doi:10.1142/s2010132512500150.

Lekshminarayanan, G., Croal, M., Maisonneuve, J., 2020. Recovering latent and sensible energy from building exhaust with membrane-based energy recovery ventilation. Sci.

Technol. Built Environ. 2374474X. 26 (7), 1000–1012. Available from: http://www.tandfonline.com/loi/uhvc21. doi:10.1080/23744731.2020.1761712.

Li, Z.X., Zhong, T.S., Niu, J.L., Xiao, F., Zhang, L.Z., 2015. Conjugate heat and mass transfer in a total heat exchanger with cross-corrugated triangular ducts and one-step made asymmetric membranes. Int. J. Heat Mass Transf. 00179310. 84, 390–400. Available from: http://www.journals.elsevier.com/international-journal-of-heat-and-mass-transfer/. doi:10.1016/j.ijheatmasstransfer.2015.01.032.

Liang, C., Tong, X., Lei, T., Zhenxing, L., Guoshan, W., 2017. Optimal design of an air-to-air heat exchanger with cross-corrugated triangular ducts using a particle swarm optimization algorithm. Appl. Sci. 2076-3417. 7 (6), 554. doi:10.3390/app7060554.

Liu, P., Alonso, M.J., Mathisen, H.M., Simonson, C., 2016. Performance of a quasi-counter-flow air-to-air membrane energy exchanger in cold climates. Energy Build. 03787788. 119, 129–142. doi:10.1016/j.enbuild.2016.03.010.

Liu, R., Gong, T., Zhang, K., Lee, C., 2017. Graphene oxide papers with high water adsorption capacity for air dehumidification. Sci. Rep. 20452322. 7 (1). Available from: https://www.nature.com/srep/index.html. doi:10.1038/s41598-017-09777-y.

Liu, S., Riffat, S., Zhao, X., Yuan, Y., 2009b. Impact of adsorbent finishing and absorbent filming on energy exchange efficiency of an air-to-air cellulose fibre heat & mass exchanger. Build. Environ. 03601323. 44 (9), 1803–1809. doi:10.1016/j.buildenv.2008.12.004.

Liu, S., Zhao, X., Riffat, S., Yuan, Y., 2009a. Comparative study of hydrophilic materials for air-to-air heat/mass exchanger. Int. J. Low-Carbon Technol. 1748-1317. 4 (2), 120–130. doi:10.1093/ijlct/ctp012.

Liu, X.P., Niu, J.L., 2015. Effects of geometrical parameters on the thermohydraulic characteristics of periodic cross-corrugated channels. Int. J. Heat Mass Transf. 00179310. 84, 542–549. doi:10.1016/j.ijheatmasstransfer.2015.01.046.

Liu, X., Qu, M., Liu, X., Wang, L., 2019. Membrane-based liquid desiccant air dehumidification: a comprehensive review on materials, components, systems and performances. Renew. Sustain. Energy Rev. 18790690. 110, 444–466. Available from: https://www.journals.elsevier.com/renewable-and-sustainable-energy-reviews. doi:10.1016/j.rser.2019.04.018.

Mahmood, M.H., Sultan, M., Miyazaki, T., Koyama, S., Maisotsenko, V.S., 2016. Overview of the Maisotsenko cycle – a way towards dew point evaporative cooling. Renew. Sustain. Energy Rev. 18790690. 66, 537–555. Available from: https://www.journals.elsevier.com/renewable-and-sustainable-energy-reviews. doi:10.1016/j.rser.2016.08.022.

McCullough, E.A., Kwon, M., Shim, H., 2003. A comparison of standard methods for measuring water vapour premeability of fabrics. Meas. Sci. Technol. 09570233. 14 (8), -1402–1408. Available from: http://www.iop.org/EJ/journal/0957-0233. doi:10.1088/0957-0233/14/8/328.

Mehrabian, M.A., Poulter, R., Quarini, G.L., 2000. Hydrodynamic and thermal characteristics of corrugated channels: experimental approach. Exp. Heat Transf. 15210480. 13 (3), 223–234. doi:10.1080/08916150050174904.

Metz, S.J., De Ven, W.J.C.V., Potreck, J., Mulder, M.H.V., Wessling, M., 2005. Transport of water vapor and inert gas mixtures through highly selective and highly permeable polymer membranes. J. Membr. Sci. 03767388. 251 (1-2), 29–41. doi:10.1016/j.memsci.2004.08.036.

Min, J.C., Su, M., Wang, L.N., 2012. Experimental and theoretical investigations of membrane-based energy recovery ventilator performance. Int. J. Air-Cond. Refrig. 20 (1), 1150004. doi:10.1142/s2010132511500040.

Min, J., Liu, X., Hu, T., 2008. Experimental studies of transport properties of water vapor through membranes. Refrig. Air Cond 1–8. https://docs.lib.purdue.edu/iracc/989/.

Min, J., Ming, S., 2010. Performance analysis of a membrane-based enthalpy exchanger: effects of the membrane properties on the exchanger performance. J. Membr. Sci. 348 (1–2), 376–382. doi:10.1016/j.memsci.2009.11.032.
Min, J., Su, M., 2010. Performance analysis of a membrane-based energy recovery ventilator: effects of membrane spacing and thickness on the ventilator performance. Appl. Therm. Eng. 13594311. 30 (8-9), 991–997. doi:10.1016/j.applthermaleng.2010.01.010.
Min, J., Su, M., 2011. Performance analysis of a membrane-based energy recovery ventilator: effects of outdoor air state. Appl. Therm. Eng. 13594311. 31 (17-18), 4036–4043. doi:10.1016/j.applthermaleng.2011.08.006.
Min, J., Teng, H., Liu, X., 2010. Evaluation of moisture diffusivities in various membranes. J. Membr. Sci. 03767388. 357 (1-2), 185–191. doi:10.1016/j.memsci.2010.04.019.
Min, J., Teng, H., Song, Y., 2011. Experimental and numerical investigations of moisture permeation through membranes. J. Membr. Sci. 03767388. 367 (1-2), 174–181. doi:10.1016/j.memsci.2010.10.064.
Morigami, Y., Kondo, M., Abe, J., Kita, H., Okamoto, K., 2001. The first large-scale pervaporation plant using tubular-type module with zeolite NaA membrane. Sep. Purif. Technol. 13835866. 25 (1-3), 251–260. doi:10.1016/S1383-5866(01)00109-5.
Nasif, M., AL-Waked, R., Morrison, G., Behnia, M., 2010. Membrane heat exchanger in HVAC energy recovery systems, systems energy analysis. Energy Build. 03787788. 42 (10), 1833–1840. doi:10.1016/j.enbuild.2010.05.020.
Nasif, M.S., 2015. Effect of utilizing different permeable material in air-to-air fixed plate energy recovery heat exchangeron energy saving. ARPN J. Eng. Appl. Sci. 18196608. 10 (21), 10153–10158. Available from: http://www.arpnjournals.org/jeas/research_papers/rp_2015/jeas_1115_3041.pdf.
Nasif, M.S., Al-Waked, R., 2014. Seasonal weather conditions effect on energy consumption and CO_2 emission for air conditioning systems coupled with enthalpy energy recovery heat exchanger. APCBEE Procedia. 22126708. 10, 42–48. doi:10.1016/j.apcbee.2014.10.012.
Nasif, M.S., Al-Waked, R., Behnia, M., Morrison, G., 2012. Modeling of air to air enthalpy heat exchanger. Heat Transf. Eng. 15210537. 33 (12), 1010–1023. doi:10.1080/01457632.2012.659616.
Nasif, M.S., Al-Waked, R., Behnia, M., Morrison, G., 2013. Air to air fixed plate enthalpy heat exchanger, performance variation and energy analysis. J. Mech. Sci. Technol. 1738494X. 27 (11), 3541–3551. doi:10.1007/s12206-013-0872-6.
Noor, S., Ashraf, H., Sultan, M., Khan, Z.M., 2020. Evaporative cooling options for building air-conditioning: a comprehensive study for climatic conditions of multan (Pakistan). Energies 19961073. 13 (12). Available from: https://www.mdpi.com/1996-1073/13/12/3061. doi:10.3390/en13123061.
Okamoto, K.I., Kita, H., Horii, K., Tanaka, K., Kondo, M., 2001. Zeolite NaA membrane: preparation, single-gas permeation, and pervaporation and vapor permeation of water/organic liquid mixtures. Ind. Eng. Chem. Res. 08885885. 40 (1), 163–175. Available from: http://pubs.acs.org/journal/iecred. doi:10.1021/ie0006007.
Park, S., Jeong, H.K., 2020. Highly H2O permeable ionic liquid encapsulated metal-organic framework membranes for energy-efficient air-dehumidification. J. Mater. Chem. A 20507496. 8 (44), 23645–23653. Available from: http://pubs.rsc.org/en/journals/journal/ta. doi:10.1039/d0ta07780a.
Paul, B.K., Kawula, S., Song, C., 2019. A manufacturing process design for producing a membrane-based energy recovery ventilator with high aspect ratio support ribs. J. Manuf. Syst. 02786125. 52, 242–252. Available from: http://www.elsevier.com. doi:10.1016/j.jmsy.2019.04.004.
Pérez-Lombard, L., Ortiz, J., Pout, C., 2008. A review on buildings energy consumption information. Energy Build. 03787788. 40 (3), 394–398. doi:10.1016/j.enbuild.2007.03.007.

Phattaranawik, J., Jiraratananon, R., Fane, A.G., 2003. Effect of pore size distribution and air flux on mass transport in direct contact membrane distillation. J. Membr. Sci. 03767388. 215 (1-2), 75–85. Available from: https://www.elsevier.com/locate/memsci. doi:10.1016/S0376-7388(02)00603-8.

Qu, M., Abdelaziz, O., Gao, Z., Yin, H., 2018. Isothermal membrane-based air dehumidification: a comprehensive review. Renew. Sustain. Energy Rev. 18790690. 82, 4060–4069. Available from: https://www.journals.elsevier.com/renewable-and-sustainable-energy-reviews. doi:10.1016/j.rser.2017.10.067.

Ritchie, H., Roser, M., 2022. Our World in Data https://ourworldindata.org/indoor-air-pollution.

Ronghui, Q., Dong, C., Zhang, L.-Z., 2020. A review of liquid desiccant air dehumidification: from system to material manipulations. Energy Build. 03787788. 215, 109897. doi:10.1016/j.enbuild.2020.109897.

Salafi, M., Asasian-Kolur, N., Sharifian, S., Ghadimi, A., 2021. A flat-plate spiral-channeled membrane heat exchanger for methane dehumidification: comparison of kraft paper and thin-film composite membrane. Int. J. Therm. Sci. 12900729. 167, 107046. doi:10.1016/j.ijthermalsci.2021.107046.

Santamouris, M., Papanikolaou, N., Livada, I., Koronakis, I., Georgakis, C., Argiriou, A., Assimakopoulos, D.N., 2001. On the impact of urban climate on the energy consumption of buildings. Sol. Energy. 0038092X. 70 (3), 201–216. doi:10.1016/s0038-092x(00)00095-5.

Sato, K., Sugimoto, K., Nakane, T., 2008. Separation of ethanol/ethyl acetate mixture by pervaporation at 100-130 °C through NaY zeolite membrane for industrial purpose. Microporous Mesoporous Mater. 13871811. 115 (1-2), 170–175. doi:10.1016/j.micromeso.2007.10.054.

Scott, K., Mahmood, A.J., Jachuck, R.J., Hu, B., 2000. Intensified membrane filtration with corrugated membranes. J. Membr. Sci. 03767388. 173 (1), 1–16. doi:10.1016/s0376-7388(00)00327-6.

Scott, N.H., 2000. Area modulus of elasticity: definition and properties. J. Elast. 58 (3), 269–275. doi:10.1023/A:1007675928019.

Scovazzo, P., 2010. Testing and evaluation of room temperature ionic liquid (RTIL) membranes for gas dehumidification. J. Membr. Sci. 03767388. 355 (1-2), 7–17. doi:10.1016/j.memsci.2010.02.067.

Sharif, A., Ameel, B., T'Jollyn, I., Lecompte, S., De Paepe, M., 2018. Comparative performance assessment of plate heat exchangers with triangular corrugation. Appl. Therm. Eng. 13594311. 141, 186–199. Available from: http://www.journals.elsevier.com/applied-thermal-engineering/. doi:10.1016/j.applthermaleng.2018.05.111.

Shen, Z.J., Min, J.C., 2020. Outdoor air state that may cause moisture condensation in membrane-type total heat exchangers using different membrane materials. Energy Rep. 23524847. 6, 227–233. doi:10.1016/j.egyr.2019.08.049.

Shin, Y., Liu, W., Schwenzer, B., Manandhar, S., Chase-Woods, D., Engelhard, M.H., Devanathan, R., Fifield, L.S., Bennett, W.D., Ginovska, B., Gotthold, D.W., 2016. Graphene oxide membranes with high permeability and selectivity for dehumidification of air. Carbon 00086223. 106, 164–170. Available from: http://www.journals.elsevier.com/carbon/. doi:10.1016/j.carbon.2016.05.023.

Singh, M.P., Singh, R.K., Chandra, S., 2014. Ionic liquids confined in porous matrices: physicochemical properties and applications. Prog. Mater. Sci. 00796425. 64, 73–120. doi:10.1016/j.pmatsci.2014.03.001.

Sultan, M., El-Sharkawy, I.I., Miyazaki, T., Saha, B.B., Koyama, S., 2015. An overview of solid desiccant dehumidification and air conditioning systems. Renew. Sustain. Energy Rev. 18790690. 46, 16–29. Available from: https://www.journals.elsevier.com/renewable-and-sustainable-energy-reviews. doi:10.1016/j.rser.2015.02.038.

Vickers, N.J., 2017. Animal communication: When I'm calling you, will you answer too? Curr. Biol. 18790445. 27 (14), R713. Available from: http://www.elsevier.com/journals/current-biology/0960-9822. doi:10.1016/j.cub.2017.05.064.
Wang, -C.-C., Chen, C.-K., 2002. Forced convection in a wavy-wall channel. Int. J. Heat Mass Transf. 00179310. 45 (12), 2587–2595. doi:10.1016/s0017-9310(01)00335-0.
Woods, J., 2014. Membrane processes for heating, ventilation, and air conditioning. Renew. Sustain. Energy Rev. 13640321. 33, 290–304. Available from: https://www.journals.elsevier.com/renewable-and-sustainable-energy-reviews. doi:10.1016/j.rser.2014.01.092.
Xiao, L., Yang, M., Yuan, W.Z., Huang, S.M., 2021. Macroporous ceramic membrane condenser for water and heat recovery from flue gas. Appl. Therm. Eng. 13594311. 186. Available from: http://www.journals.elsevier.com/applied-thermal-engineering/. doi:10.1016/j.applthermaleng.2020.116512.
Xiong, Y., Tan, H., Wang, Y., Xu, W., Mikulčić, H., Duić, N., 2017. Pilot-scale study on water and latent heat recovery from flue gas using fluorine plastic heat exchangers. J. Clean. Prod. 09596526. 161, 1416–1422. Available from: https://www.journals.elsevier.com/journal-of-cleaner-production. doi:10.1016/j.jclepro.2017.06.081.
Yaïci, W., Ghorab, M., Entchev, E., 2013. Numerical analysis of heat and energy recovery ventilators performance based on CFD for detailed design. Appl. Therm. Eng. 13594311. 51 (1-2), 770–780. doi:10.1016/j.applthermaleng.2012.10.003.
Zaw, K., Safizadeh, M.R., Luther, J., Ng, K.C., 2013. Analysis of a membrane based air-dehumidification unit for air conditioning in tropical climates. Appl. Therm. Eng. 13594311. 59 (1-2), 370–379. doi:10.1016/j.applthermaleng.2013.05.029.
Zender–Świercz, E., 2021. A review of heat recovery in ventilation. Energies. 1996-1073. 14 (6), 1759. doi:10.3390/en14061759.
Zhang, J., Zhaohao, L., Zhang, H., Chen, H., Gao, D., 2020. Numerical study on recovering moisture and heat from flue gas by means of a macroporous ceramic membrane module. Energy 207, 118230. doi:10.1016/j.energy.2020.118230.
Zhang, L.-Z., Chen, Z.-Y., 2011. Convective heat transfer in cross-corrugated triangular ducts under uniform heat flux boundary conditions. Int. J. Heat Mass Transf. 00179310. 54 (1-3), 597–605. doi:10.1016/j.ijheatmasstransfer.2010.09.010.
Zhang, L., 2006. Fabrication of a lithium chloride solution based composite supported liquid membrane and its moisture permeation analysis. J. Membr. Sci. 03767388. 276 (1-2), 91–100. doi:10.1016/j.memsci.2005.09.035.
Zhang, L., Liang, C., Pei, L., 2008. Heat and moisture transfer in application scale parallel-plates enthalpy exchangers with novel membrane materials. J. Membr. Sci. 03767388. 325 (2), 672–682. doi:10.1016/j.memsci.2008.08.041.
Zhang, L.Z., 2006. Evaluation of moisture diffusivity in hydrophilic polymer membranes: a new approach. J. Membr. Sci. 03767388. 269 (1-2), 75–83. doi:10.1016/j.memsci.2005.06.020.
Zhang, L.Z., 2009a. Flow maldistribution and performance deteriorations in membrane-based heat and mass exchangers. J. Heat Transf. 15288943. 131 (11), 1–7. Available from: http://asmedl.aip.org/getpdf/servlet/GetPDFServlet?filetype=pdf&id=JHTRAO000131000011111801000001&idtype=cvips&prog=normal. doi:10.1115/1.3154832China.
Zhang, L.Z., 2009b. Heat and mass transfer in plate-fin enthalpy exchangers with different plate and fin materials. Int. J. Heat Mass Transf. 00179310. 52 (11-12), 2704–2713. doi:10.1016/j.ijheatmasstransfer.2008.12.014.
Zhang, L.Z., 2010. Heat and mass transfer in a quasi-counter flow membrane-based total heat exchanger. Int. J. Heat Mass Transf. 00179310. 53 (23-24), 5478–5486. doi:10.1016/j.ijheatmasstransfer.2010.07.009.
Zhang, L.Z., 2016. A reliability-based optimization of membrane-type total heat exchangers under uncertain design parameters. Energy 03605442. 101, 390–401.

Available from: https://www.elsevier.com/inca/publications/store/4/8/3/. doi:10.1016/j.energy.2016.02.032.
Zhang, L.Z., Liang, C.H., Pei, L.X., 2010. Conjugate heat and mass transfer in membrane-formed channels in all entry regions. Int. J. Heat Mass Transf. 00179310. 53 (5-6), 815–824. doi:10.1016/j.ijheatmasstransfer.2009.11.043.
Zhang, L.Z., Wang, Y.Y., Wang, C.L., Xiang, H., 2008. Synthesis and characterization of a PVA/LiCl blend membrane for air dehumidification. J. Membr. Sci. 03767388. 308 (1-2), 198–206. doi:10.1016/j.memsci.2007.09.056.
Zhang, L.Z., Zhang, X.R., Miao, Q.Z., Pei, L.X., 2012. Selective permeation of moisture and VOCs through polymer membranes used in total heat exchangers for indoor air ventilation. Indoor Air 16000668. 22 (4), 321–330. Available from: http://onlinelibrary.wiley.com/journal/10.1111/(ISSN)1600-0668. doi:10.1111/j.1600-0668.2011.00762.xChina.
Zhang, X.R., Zhang, L.Z., Liu, H.M., Pei, L.X., 2011. One-step fabrication and analysis of an asymmetric cellulose acetate membrane for heat and moisture recovery. J. Membr. Sci. 03767388. 366 (1-2), 158–165. doi:10.1016/j.memsci.2010.09.054.
Zhang, Y., Jiang, Y., Zhang, L.Z., Deng, Y., Jin, Z., 2000. Analysis of thermal performance and energy savings of membrane based heat recovery ventilator. Energy. 03605442. 25 (6), 515–527. doi:10.1016/S0360-5442(99)00087-0.

CHAPTER 9

Investigating energy-efficient configurations of membrane energy recovery ventilators using CFD

Muhammad Asif[a], Muhammad Sultan[a], Zahid Mahmood Khan[a], Muhammad Wakil Shahzad[b], Faheem Hassan Akhtar[c], Hadeed Ashraf[a], Naveed Husnain[d], Uzair Sajjad[e], Muhammad Farooq[f], and Muhammad Yasar Javaid[g]

[a]Department of Agricultural Engineering, Bahauddin Zakariya University, Multan, Pakistan
[b]Department of Mechanical and Construction Engineering, Northumbria University, Newcastle Upon Tyne, United Kingdom
[c]Department of Chemistry and Chemical Engineering, Lahore University of Management Sciences (LUMS), Lahore, Pakistan
[d]Department of Mechanical Engineering, Bahauddin Zakariya University, Multan, Pakistan
[e]Interdisciplinary Research Center for Sustainable Energy Systems (IRC-SES), King Fahd University of Petroleum and Minerals, Dhahran, Saudi Arabia
[f]Department of Mechanical Engineering, University of Engineering and Technology, Lahore, Pakistan
[g]Department of Mechanical Engineering and Technology, Government College University Faisalabad, Faisalabad, Pakistan

9.1 Introduction

The exponential rise in global population is expected to approach 9.7 billion, thereby increasing energy demand by about 47%, particularly in developing Asian countries, by the end of 2050 (International Energy Agency, World Energy Balances, 2021). Domestic and building sector energy consumption has been steadily increasing, reaching 36% of total global energy consumption (United Nations Environment Programme, 2021). In tropical climates, >50% of energy is used in heating, ventilation, and air conditioning (HVAC) systems in the building sector (Chua et al., 2013; Aleem et al., 2020). Fig. 9.1A shows the HVAC system breakdown of energy consumption in the building sector, reproduced from the global status report for buildings and construction (2021). The energy consumption includes ventilation (34%), cooling (27%), heating (17%), pumps (16%), and cooling towers (6%). Fig. 9.1B exhibits the global energy consumption breakdown in the building sector. As a technique for conserving energy in the buildings, green

Sustainable Hybrid Cooling and Dehumidification Strategies for Controlled Environments
https://doi.org/10.1016/B978-0-443-48555-8.00007-4

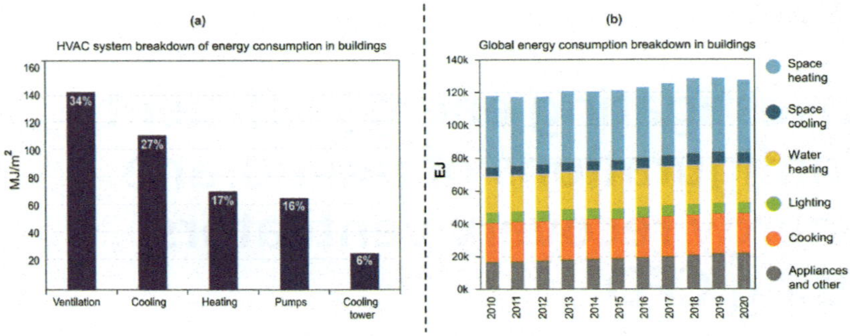

Fig. 9.1 Energy consumption in buildings: (A) HVAC air conditioning, and (B) Global energy consumption. *(Reproduced from United Nations Environment Programme, 2021. Global status report for buildings and construction: towards a zero-emission, efficient and resilient buildings and construction sector. Nairobi.)*

energy generation is also being used as biogas, but the energy usage in the buildings is the source of the CO_2 generation due to a lack of clean energy (Khan et al., 2022; Sultan et al., 2018; Zhang et al., 2021). Conventional HVAC systems are vapor-compression-based and use refrigerants such as hydrochlorofluorocarbons and hydrofluorocarbons, thereby consuming excessive amounts of energy and degrading the environment.

In the last couple of decades, green air conditioning has attracted much more attention than conventional cooling systems, i.e., vapor compression air conditioning systems (VAC). The green air conditioning system is evaporative cooling, i.e., Maisotsenko cycle-based evaporative cooling, indirect evaporative cooling, direct evaporative cooling, and) desiccant air conditioning. The green air conditioning systems can effectively control the (Sultan et al., 2016, 2015, 2021; Mahmood et al., 2016; Raza et al., 2020; Noor et al., 2020, 2021). The membrane-based energy recovery ventilator (MERV) could be a significant parameter in energy reduction in buildings. As per the operating standard, the ERV is classified into three types: energy recovery wheel, plate ERV, and MERV (Bonilla, 1962; Nakonieczny, 2006; Zhang et al., 2000). MERV is the emerging concept in energy reduction in buildings. The working mechanism of the MERV is discussed below.

9.2 Membrane-based energy recovery ventilators

MERVs are employed with air conditioning systems to improve indoor air quality and energy conservation within a predetermined space. It works by transferring enthalpies between the incoming and outgoing air streams, reducing the amount of energy needed while maintaining thermal comfort. The membrane refers to a thin, permeable material that separates the two air streams, typically made of a polymeric material with high heat and moisture-transfer properties. MERVs can be installed in residential or commercial buildings to improve air quality and minimize energy costs.

In building HVAC, MERV systems are widely used. Commonly, ERVs are used in commercial buildings. However, most of the latent and sensible energy is wasted from the buildings/indoor-conditioned spaces as exhaust air. The specific proportion of latent and sensible energy in the exhaust air can be extracted through the MERV systems. Fig. 9.2A shows the membrane that acts as the energy exchanger, which is dependent on the enthalpy difference between the ambient air and exhaust air. Fig. 9.2B shows the MERV system in which feed air is an air stream that carries air from the atmosphere to the building/conditioned space, whereas the sweep air stream carries the air from the building/conditioned space to the atmosphere. In winter, the supply air is preheated and humidified, whereas in summer, the supply air is precooled and dehumidified. MERV can recover 60%–65% of the energy from the exhaust air (ASHRAE, 2020), whereas the global annual consumption of electricity was around 516,595 and 412,627 ktoe in the residential sector and commercial/public sector, respectively (International Energy Agency, 2019). Indoor air essentially needs to be replaced with a fresh air load to maintain the threshold level

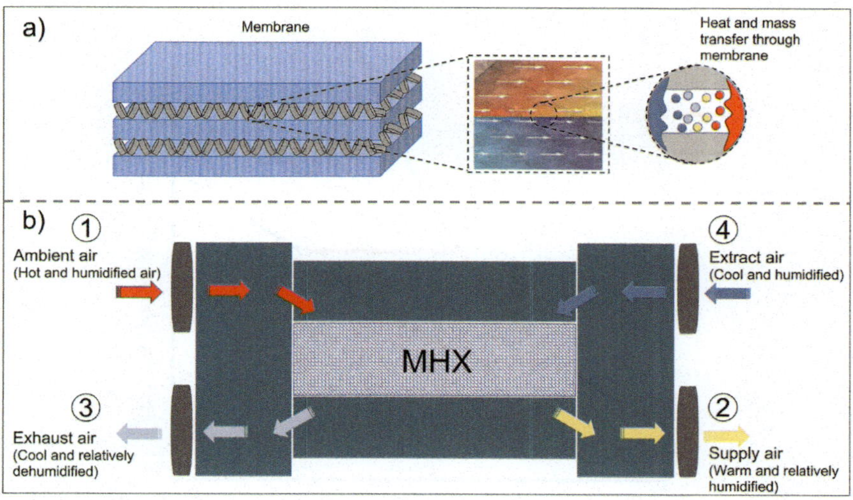

Fig. 9.2 Schematic diagrams of (A) membrane enthalpy exchanger, and (B) MERV system.

of indoor air (Daisey et al., 2003; Kreider and Rabl, 1994). The MERV can effectively decrease the energy demand in the sense of fresh air load without scarifying indoor air quality (Zhang et al., 2021; Zhong and Kang, 2009; Fisk and Turiel, 1983; Drost, 1993; Papakostas and Kiosis, 2015).

As per the flow configuration, the MERV is classified as the crossflow, counterflow, parallel flow, and multiple-pass heat exchanger. The performance of the ERV is analyzed in a number of studies, whereas on the basis of effectiveness, the counterflow system is preferred (Phillips et al., 1989; ASHRAE, 2020; Zhong and Kang, 2009; Zhang and Niu, 2001; Papakostas and Kiosis, 2015; Nizovtsev et al., 2016; Al-Waked et al., 2013; Fouih et al., 2012). Fig. 9.3 exhibits the psychrometric framework of the conventional VAC system integrated with the MERV that reduces energy consumption.

During the last couple of decades, several studies have been carried out on the performance and design improvement of MERVs. Zhang (2007, 2010, 2016) conducted various studies and explored the total enthalpy transfer of membrane enthalpy exchanger and its optimization. The authors conducted research related to the membrane spacing, membrane thickness, moisture diffusivity, sorption, thermal conductivity, outdoor air state, and its effect on the performance of MERVs. The study concluded that if the

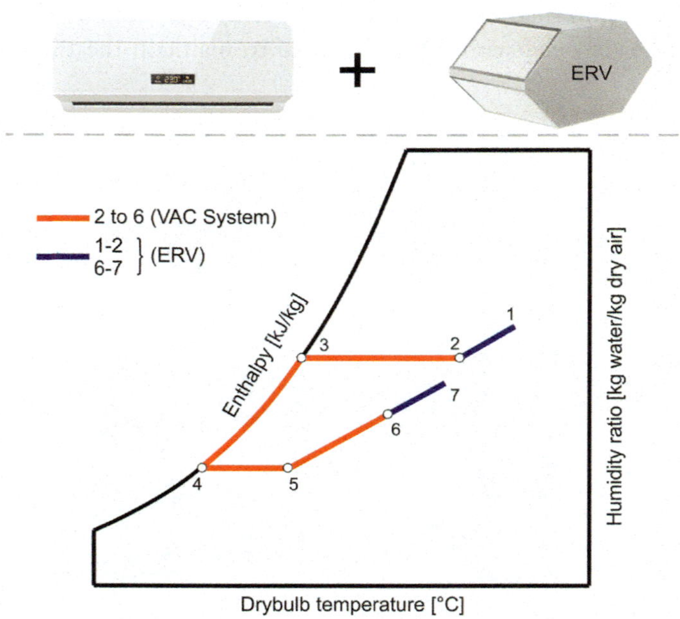

Fig. 9.3 Psychrometric representation of the VCAC with ERV.

structural parameters of the membrane (thickness and channel height) are increased, then the enthalpy effectiveness is decreased and vice versa. Furthermore, sensible effectiveness was less affected by the ambient condition as compared to latent effectiveness (Min and Su, 2010a, 2010b, 2011). The authors investigated the annual energy consumption and enthalpy effectiveness of membrane heat exchanger (MHX) and concluded that MHX can effectively reduce the 8% load in contrast to conventional air conditioning systems (Nasif and Al-Waked, 2014). The performance of any system is mainly dependent upon the materials' properties. In the case of the MHX, the moisture permeability, moisture diffusivity, and sorption capacity play a major role in the system effectiveness. The authors studied the mass permeation through the asymmetric membrane (Liu et al., 2001), whereas in another study the authors studied the mass permeation through the alumina composite membrane and concluded that the moisture diffusivity, depends upon the vapor pressure across the membrane side (Aranda et al., 1995). The authors compared the cellulose membrane, PVDF, and PES membranes in the study and concluded that the PVDF shows the maximum moisture diffusivity whereas in the other relevant study, he also concluded the humid air flow rate and membrane are the functions of the sorption capacity of the membrane (Min and Wang, 2013; Min et al., 2010, 2011). The authors studied the mass permeation through the polyimide membrane and concluded that the higher flow rates are essential to remove the concentration polarization effects of membranes having higher permeabilities (Chen et al., 2011). The authors reported that the mass diffusivity increases as the temperature increases up to ~60% humidity and then decreases due to the water clustering (Lee et al., 2013).

The performance of the MERV is reciprocal to the change in ambient conditions. The author studied seasonal variation of the MERV with the indoor setpoints (Zhou et al., 2007), whereas in another study the authors studied the air-to-air recovery ventilator and concluded that the AARV is potentially applicable for the building's air conditioning with higher recovery potential (Zhong and Kang, 2009). The author studied the effects of the outdoor environment on the effectiveness of the MERV and concluded that in summer as the weighted coefficients of the sensible efficiency go down as the ambient temperature goes down, heat goes down and vice versa for latent recovery efficiency (Liu et al., 2010). In another study, the author studied the seasonal effect on MERV and concluded that the MERV is superior in the summer season (Nasif and Al-Waked, 2014). Recently, the authors developed the regression model for the MERVs and concluded that the model is more likely applicable in summer than winter (Gao et al., 2021; Qiu et al., 2019), whereas in another relevant study, the author analyzed the condensation in an energy recovery ventilator (ERV). The study concludes that 4°C and 78% outdoor relative humidity and 50%–60% and 25°C room temperature and humidity could cause condensation in winter (Abadi et al., 2022).

The geometrical/structural parameters, i.e., membrane structure, length, apex angle, and height of MHX, affect the performance of MERV. The authors analyzed the effect of apex angle and aspect ratio on thermohydraulic parameters in a corrugated triangular duct. They concluded that the aspect ratio relatively strongly influences the flow frictional loss and heat and pressure drop brought about by the apex angle (Liu and Niu, 2015). Balanced and unbalanced flow conditions were analyzed by authors on cross-corrugated triangular duct and concluded that the unbalanced flow is helpful in optimizing the MERV to enhance enthalpy transfer (Sebai et al., 2014). The author evaluated the three-dimensional model for MHX on the air channel establishment. Furthermore, the comparison between various obstacles, circular, triangular, and rectangular, show that circular MHX is the maximum mean effective (Sabek et al., 2018). The author investigated the effect of the ambient air state on the asymmetric composite membrane MHX with triangular duct. The study concluded that the total energy effectiveness is greatly affected by the outdoor air relative humidity more than temperature (Engarnevis et al., 2018). Jafarizave et al. (2019) studied the rectangular, triangular, and curvature-type MHX and concluded that the curvature-type MERV shows the minimum pressure drop, whereas in another study, the authors studied the geometrical effect on thermodynamic properties and concluded that the apex angle from 90 to 120 shows the maximum heat transfer into the cross-corrugated triangular MERV (Liu and Niu, 2015).

According to the above studies, it is obvious that the geometrical parameters, i.e., the configurations of the membrane, affect the performance of MERV, and appropriate geometry is mandatory to improve the performance of the MERV. Therefore, this proposed study focuses on the Computational fluid dynamics (CFD), heat, and mass transfer analysis of different configurations of MERV systems. In this study, novel membrane channel structures, including rectangular, triangular, and semicircular configurations with counterflow, were used as these proposed membrane channels. The structural, physical, and operating parameters of all proposed configurations are the same.

9.3 Methodology

In the present study, rectangular, triangular, and semicircular geometry configurations of MERV are compared by COMSOL Multiphysics software based on process air pressure variation and enthalpy transfer. Fig. 9.4 shows the geometrical representation of the proposed configurations. The shape, structural configuration, and relative arrangement of MERV components are considered key performance indicators of the MERV system. Appropriate geometry leads toward the optimal design and thereby high performance of the MERV system. Fig. 9.5 represents the generalized

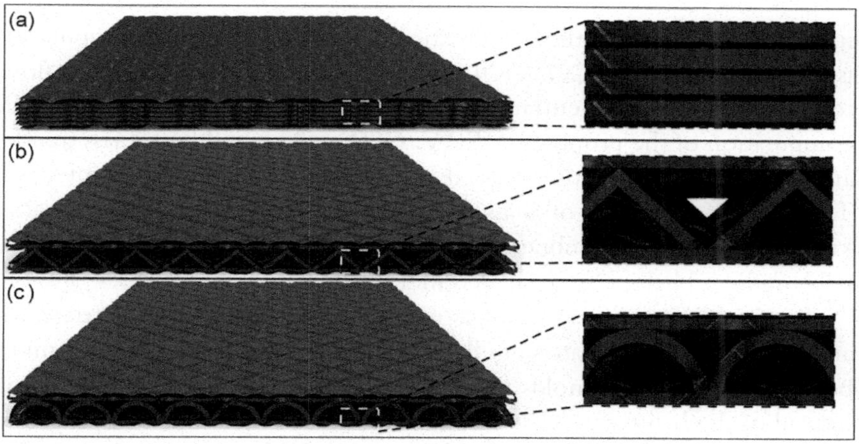

Fig. 9.4 Membrane configurations used in this study: (A) rectangular, (B) triangular, and (C) semicircular configurations.

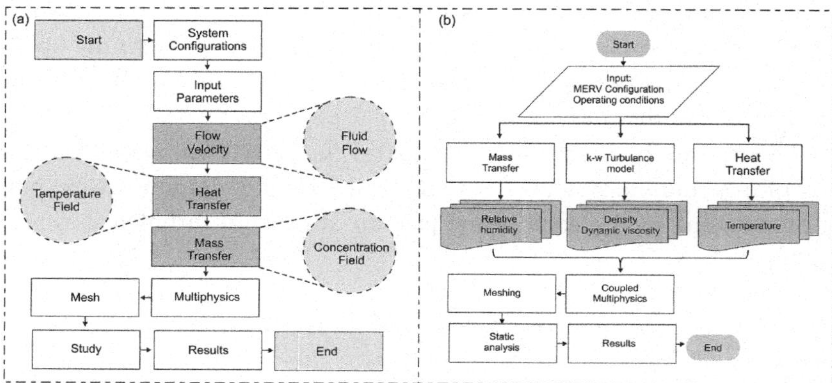

Fig. 9.5 Flow of (A) research-executing strategy of this study, and (B) research methodology considerations.

Table 9.1 Membrane properties used for MERVs in this study.

Parameters	Symbol	Units	Values
Mass diffusivity in membrane	D_{vm}	m^2/s	2.82×10^{-5}
Mass diffusivity in air streams	D_{va}	m^2/s	3.77×10^{-6}
Kinematic viscosity	η	m^2/s	15.89×10^{-6}
Membrane's density	D	kg/m^3	856
Thermal conductivity	K	W/(m.K)	0.127
Channel length	Y	mm	496
Channel height	H	mm	271

modeling algorithm for the computational process. The model computation is carried out by the COMSOL Multiphysics. Thereby, the fluid flow, temperature, and concentration field are simultaneously solved for the computation of the proposed MERV. CFD analysis and heat/mass transfer analyses are consequently coupled in COMSOL Multiphysics interface. The numerical values of mass diffusivity in membrane & air streams, kinematic viscosity, membrane density and thermal conductivity, and channel length and height are provided in Table 9.1.

Process airflow from the ducts of complex geometries (i.e., triangular and semicircular) fluctuates rapidly due to variation in the Reynolds number. Thereby, the Reynolds turbulent model is selected for computation instead of the laminar model. The Navier–Stokes equation is employed to compute the fluid flow. The Navier–Stokes and continuity equations are summations of momentum and mass conservation. The assumptions for the model computation are listed as (1) the model is steady state, (2) the fluid is incompressible, (3) the heat conduction between the air streams is minimum, and (4) the vapor diffusion is negligible. Table 9.2 shows the operating parameters of the study, whereas Fig. 9.5B shows the model used and the variables in the study.

The continuity equation (Li et al., 2015) helps to describe the conservation of mass in fluid systems and is essential for understanding a wide range of phenomena such as fluid flow, pressure drop, and turbulence as given by (Eq. 1).

$$\frac{\partial \rho u}{\partial x} + \frac{\partial \rho u}{\partial y} + \frac{\partial \rho u}{\partial z} = 0 \tag{1}$$

where ρ, u shows .and fluid density in kg/m^3 and velocity of flow in m/s.

Table 9.2 Operating parameters and air state used in the study.

Parameters/Conditions	Symbol	Units	Values
Ambient temperature	T_a	K	313.15
Extract temperature	T_e	K	298.15
Ambient relative humidity	RH_a	–	0.30
Extract relative humidity	RH_e	–	0.50
Reynolds number	R_e	–	100–6000

Navier–Stokes equation (Li et al., 2015) is used to study the behavior of fluid flow given as (Eq. 2).

$$\frac{\partial}{\partial x_j}\left(\rho u_i u_j\right) = -\left(\frac{\partial P}{\partial x_i}\right) + \frac{\partial}{\partial j}\left(\tau_{ij} + \tau_{ij}^t\right) \tag{2a}$$

$$\tau_{ij} = \mu\left(\frac{\partial u_i}{\partial x_j} + \frac{\partial u_i}{\partial x_j}\right) \tag{2b}$$

$$\tau_{ij}^t = -\rho\overline{u_i u_j} \tag{2c}$$

where, μshows dynamic viscosity (Pa-s), P indicates pressure (Pa), and τ shows shear stress (N/m^2). Energy relationship to describe the energy conservation in a fluid system given as (Eq. 3).

$$\frac{\partial}{\partial x_j}\left(\rho C_p u_j T\right) = \frac{\partial}{\partial x_j}\left(h_j + h_j^t\right) \tag{3a}$$

$$h_j = \frac{\mu C_p}{pr}\left(\frac{\partial T}{\partial x_j}\right) \tag{3b}$$

$$h_j^t = -\rho C_p \overline{u_j' T'} \tag{3c}$$

where, C_p shows specific heat in (J/kg K) and h is heat flux (kW/m^2).

Conservation of mass flux is presented in Eq. 4, available in literature from

$$\frac{\partial}{\partial x_j}\left(\rho u_j X\right) = \frac{\partial}{\partial x_j}\left(q_j + q_j^t\right) \tag{4a}$$

$$q_j = D_{\mathrm{va}}\left(\frac{\partial X}{\partial x_j}\right) \tag{4b}$$

$$q_j^t = -\rho \overline{u_j{}' X'} \tag{4c}$$

where, D_{va} is mass diffusivity of dry air (m^2/s), X is humidity ratio (kg/kg), and q is mass diffusion flux (kg/ms^2). Fig. 9.5B represents the models used for the computational process by the COMSOL Multiphysics interface software. The k-w turbulent model is coupled with the heat and mass transfer model simultaneously and computed for MERV. Heat and mass transfer in feed and sweep air streams are represented by Eq. 5 from Engarnevis et al. (2018) and Jafarizave et al. (2019). The moisture transport in the membrane is estimated by (Eq. 6) (Jafarizave et al., 2019).

$$\rho_s C_{ps} v_s d\left(\frac{\partial T_s}{\partial x}\right) + 2h_s(T_s + T_{ms}) = 0 \tag{5a}$$

$$v_s d\left(\frac{\partial Y_s}{\partial x}\right) + 2k_s(X_{ws} + X_{wms}) = 0 \tag{5b}$$

$$\rho_e C_{pe} v_e d\left(\frac{\partial T_e}{\partial x}\right) + 2h_e(T_e + T_{me}) = 0 \tag{5c}$$

$$v_e d\left(\frac{\partial Y_e}{\partial x}\right) + 2k_e(X_{we} + X_{wme}) = 0 \tag{5d}$$

$$m_w = -D_{wm}\rho_m\left(\frac{X_{ms} - X_{me}}{t}\right) \tag{6}$$

where, X_{ms} is water/mass uptake in membrane (kg/kg), v is flow velocity (m/s), h and k are convective heat and mass transfer coefficients respectively. The K-w model is widely used due to its simplicity and higher precision as compared to the other RANS models, particularly in simulating flows with separated regions or turbulence-driven recirculation. is given as Eq. 7 available in the literature (Manual, 2023).

$$u_j \frac{\partial k}{\partial x_j} = \tau_{ij}\left(\frac{\partial u_i}{\partial x_j}\right) - \beta^* k\omega + \frac{\partial}{\partial x_j}\left[(\nu + (\sigma^* \nu_T))\left(\frac{\partial k}{\partial x_j}\right)\right] \tag{7a}$$

$$u_j \frac{\partial \omega}{\partial x_j} = \alpha \frac{\omega}{k} \tau_{ij} \left(\frac{\partial u_i}{\partial x_j} \right) - \beta \omega^2 + \frac{\partial}{\partial x_j} \left[(v + \sigma v_T) \left(\frac{\partial \omega}{\partial x_j} \right) \right] + 2\sigma_{\omega^2} \frac{1}{\omega} \left(\frac{\partial k}{\partial x_j} \right) \left(\frac{\partial \omega}{\partial x_j} \right) \tag{7b}$$

The constants of the model is given as $\alpha, \sigma_k^*, \sigma_w, \beta_o, \beta_o^*, k_v$ and B13/25, 1/2, 1/2, 9/125, 9/100, 0.41, and 5.2, respectively. The integrated heat and mass transfer model with the K-w model shows the mass transfer within the air streams is given by Eqs. 8 and 9 available in literature (Manual, 2023).

$$M_v u \left(\frac{\partial C_v}{\partial x_j} \right) + \left(\frac{\partial g_w}{\partial x_j} \right) = G \tag{8a}$$

$$g_w = -M_v D_{va} \left(\frac{\partial c_v}{\partial x_j} \right) \tag{8b}$$

$$C_v = \text{RHC}_{\text{sat}} \tag{8c}$$

$$g_{\text{Ic}} = 0 \tag{8d}$$

where, C_v shows vapor concentration, M_v molar mass per water vapor in (kg/mol), D_{va} is vapor diffusion coefficients in air, C_{sat} is saturation concentration of water vapor, RHshows the relative humidity. Heat transfer between the air streams is given by Eq. 9 available in the literature.

$$\rho C_p.\text{u} \left(\frac{\partial T}{\partial x_j} \right) + \left(\frac{\partial q}{\partial x_j} \right) = Q \tag{9a}$$

$$q = -k \left(\frac{\partial T}{\partial x_j} \right) \tag{9b}$$

where, C_p is specific heat of air at constant pressure (J/kg.k), q is conductive heat flux (W/m^2), and Q is heat source point (W/m^3). Thermodynamics of the moist air is given by Eq. 10 available in the literature.

$$0 \leq \text{RH} \leq 1, c_v = c_v (T, P_A, x_{\text{vap}}) \tag{10a}$$

$$x_{\text{vap}} = \frac{M_v}{M_a} \frac{\text{RH} p_{\text{sat}}(\text{TRH})}{\text{PRH} - \text{RHP}_{\text{sat}}(\text{TRH})} \tag{10b}$$

where, P_A is absolute pressure (Pa), x_{vap} is moisture content, M_a is the molar mass of the dry air (kg/mol), $P_{sat}$is saturation vapor pressure. When the temperature and relative humidity of the sweep and feed air stream in MERV are known, the enthalpy effectiveness can be calculated easily. The mean values of the ambient, supply, extract, and exhaust streams are used for computation. The flowrate is assumed to be the constant. The sensible, latent, and total performance of MERV is given by Eq. 11, available in the literature (Jyothiprakash et al., 2019; Kho et al., 2017).

$$\varepsilon_s = \frac{\left(T^a_{amb} - T^a_s\right) - \left(T^a_{ext} - T^a_e\right)}{2\left(T^a_{amb} - T^a_s\right)} \tag{11a}$$

$$\varepsilon_L = \frac{\left(X^a_{amb} - X^a_s\right) - \left(X^a_{ext} - X^a_e\right)}{2\left(X^a_{amb} - X^a_s\right)} \tag{11b}$$

$$\varepsilon_T = \frac{\left(h^a_{amb} - h^a_s\right) - \left(h^a_{ext} - h^a_e\right)}{2\left(h^a_{amb} - h^a_s\right)} \tag{11b}$$

where, T_{amb}, T_s, T_{ext}, and are ambient, supply, extract, and exhaust temperature (K), respectively. X_{amb}, X_s, X_{ext}, and X_e are ambient, supply, extract, and exhaust humidity ratio (kg_w/kg_{DA}), respectively. h_{amb}, h_s, h_{ext}, and h_e are ambient, supply, extract, and exhaust enthalpy (kJ/kg_{DA}), respectively. All other symbols are available in the nomenclature.

9.4 Results and discussion

Annual energy consumption in building and service sectors is about 35 million tera joules (TJ). MERV could save about half of the energy used in these sectors. Fig. 9.6 shows the global energy-saving potential in building and service sectors with MERV as compared to without MERV. The performance of MERV for energy recovery is primarily dependent on configuration of the system, e.g., rectangular, triangular, and semicircular. Each configuration on the system level performance was investigated from the viewpoints of heat and mass transfer characteristics, i.e., pressure drop, temperature, relative humidity, and sensible as well as the latent effectiveness. Therefore, the study investigated heat and mass transfer characteristics of the MERV in the COMSOL Multiphysics. The pressure drop is referred to as the pressure profile if the pressure across the sides of MERV (feed and

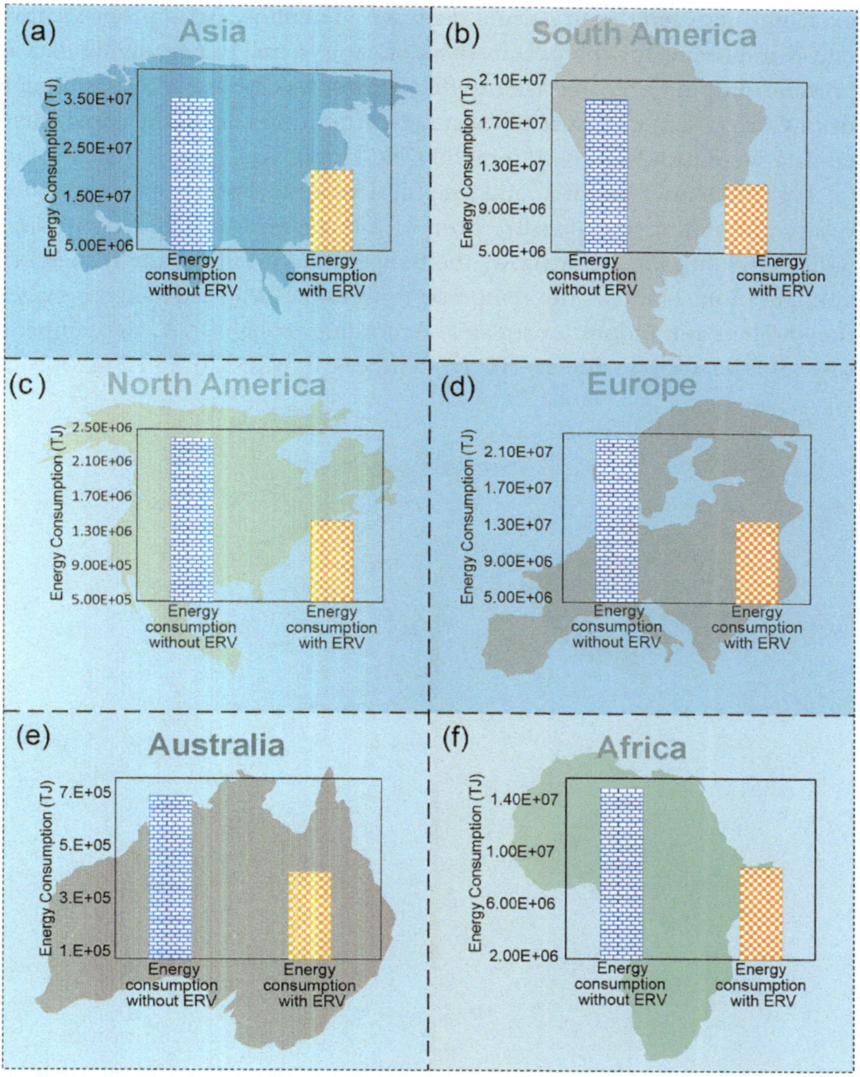

Fig. 9.6 Energy consumption in (A) Asia, (B) South America, (C) North America, (D) Europe, (E) Australia, and (F) Africa with and without MERV. *(Source: International Energy Agency, World Energy Balances, 2021. https://www.iea.org/data-and-statistics/data-tools/energy-statistics-data-browser.)*

sweep) is more than the pressure drop; it is also increased and vice versa. An increase in flow rate increases the pressure drop in the MERV. While the sensible and latent effectiveness of the MERV system decreases with an increase in flow rate, it increases as higher flow rates and the enthalpy exchange between the air streams decreases. Therefore, an appropriate flow rate is important for the effectiveness of the system. Experimental data of pressure drop sensible and latent effectiveness was calibrated by using data from Core (2023) to ensure the accuracy. After calibration, the experimental data were processed with the CFD-based results.

Fig. 9.7 shows pictorial and graphical representations of temperature, relative humidity, and pressure drop of a rectangular configuration-based MERV system. Fig. 9.7A shows the temperature profile of the rectangular configuration. It shows the temperature (sensible heat) exchange between the ambient and indoor air streams. According to Fig. 9.7A, the temperature on the supply side is 302 K, whereas it was 313.15 K in outside

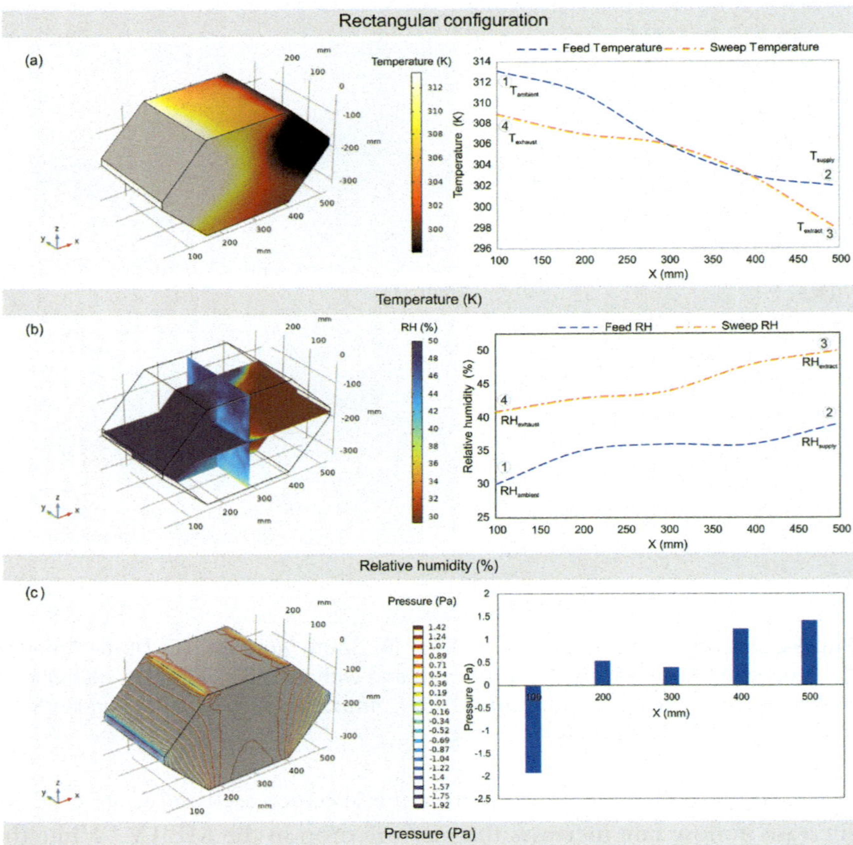

Fig. 9.7 Performance profiles of (A) temperature, (B) relative humidity, and (C) pressure drop for the rectangular configuration of the MERV system.

conditions (i.e., ambient conditions). The exhaust temperature was 309 K, as it was 298.15 K inside. In addition, heat transfer occurs between two air streams, resulting in a lower temperature on the supply outlet side as compared to the ambient inlet side. Fig. 9.7B shows profiles of relative humidity (latent heat) of the rectangular configuration-based MERV system. According to the results, the relative humidity ranges from 30% to 50%. According to Fig. 9.7B, the relative humidity at the supply side is 39%, whereas it was 30% ambient/outside. At exhaust, the humidity was 41%, whereas it was 50% inside the conditioned space. The relative humidity shows the latent heat recovery. Fig. 9.7C shows the pressure distribution in the rectangular configuration-based MERV system. According to Fig. 9.7C, profiles of pressure drop range from −1.92 to 1.42 Pa, while near the inside and outside surfaces of the system, it was less than −1 Pa and −1.92 to 1.42 Pa, respectively. However, it shows the excessive pressure drop was observed in at the corners and near boundary walls because of higher friction of air with the surface.

Fig. 9.8 shows pictorial and graphical representation of temperature, relative humidity, and pressure drop of the triangular configuration-based MERV system. Fig. 9.8A shows the temperature profile of the triangular configuration. It shows the temperature exchange between the ambient and indoor air streams. According to Fig. 9.8A, the temperature on the supply side is 301 K, whereas it was 313.15 K in outside conditions (i.e., ambient conditions). The exhaust temperature is 311 K, as it was 298.15 K inside. In addition, heat transfer occurs between two air streams, resulting in a lower temperature on the supply outlet side as compared to the ambient inlet side. Fig. 9.8B shows profiles of relative humidity (i.e., moisture transfer) by the rectangular configuration-based MERV system. According to Fig. 9.8B, the relative humidity at the supply side is 41%, whereas it was 30% ambient/outside. At exhaust, the humidity is 37%, whereas it was 50% inside the conditioned space. The relative humidity shows the latent heat recovery. The mass/humidity exchange between the air streams improves the latent heat transfer to the supply side. Fig. 9.8C shows the pressure distribution in the triangular configuration-based MERV system. According to Fig. 9.8C, profiles of pressure drop range from −0.67 to 4.35 Pa, while near the inside and outside surfaces of the system, pressure distribution exhibits more variability due to the wall friction and reduces system effectiveness.

Fig. 9.9 shows pictorial and graphical representation of temperature, relative humidity, and pressure drop of a semicircular configuration-based MERV system. Fig. 9.9A shows the temperature profile of the semicircular

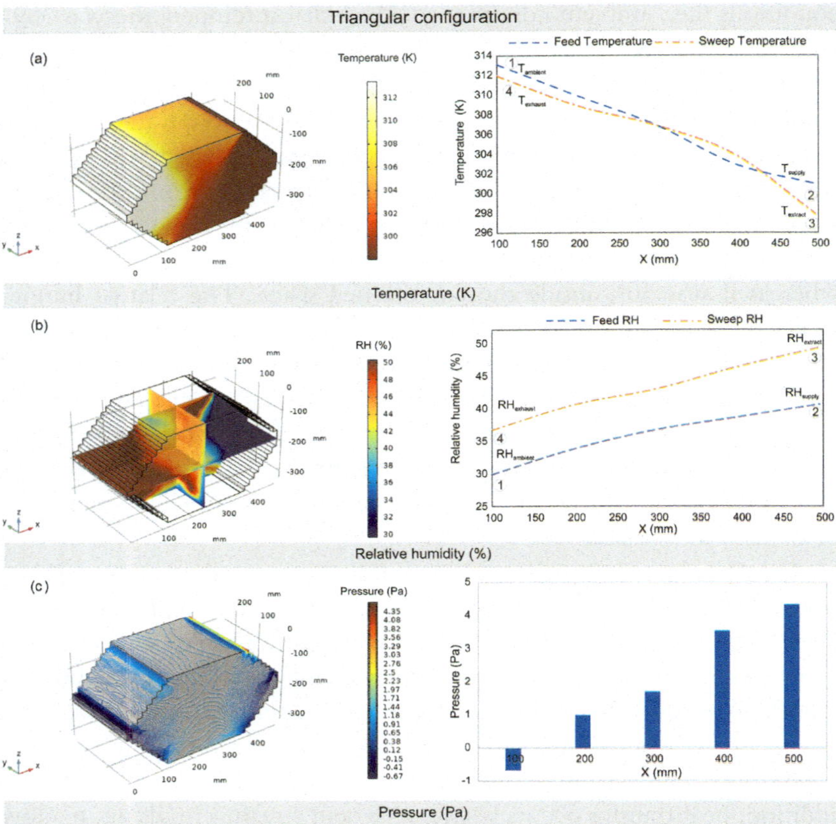

Fig. 9.8 Performance profiles of (A) temperature, (B) relative humidity, and (C) pressure drop for the triangular configuration of the MERV system.

configuration. It shows the temperature exchange between the ambient and indoor air streams. According to Fig. 9.9A, the temperature at supply side is 299 K, whereas it was 313.15 K on outside conditions (i.e., ambient conditions). The exhaust temperature was 312 K, whereas it was 298.15 K inside. In addition, heat transfer occurs between two air streams, resulting in a lower temperature on the supply outlet side as compared to the ambient inlet side. Fig. 9.9B shows profiles of relative humidity (i.e., moisture transfer) by the semicircular configuration-based MERV system. According to Fig. 9.9B, the relative humidity at the supply side is 46%, whereas it was 30% ambient/outside. At exhaust, the humidity is 35%, whereas it was 50% inside the conditioned space. The relative humidity shows the latent heat recovery. The mass/humidity exchange between the air streams improves

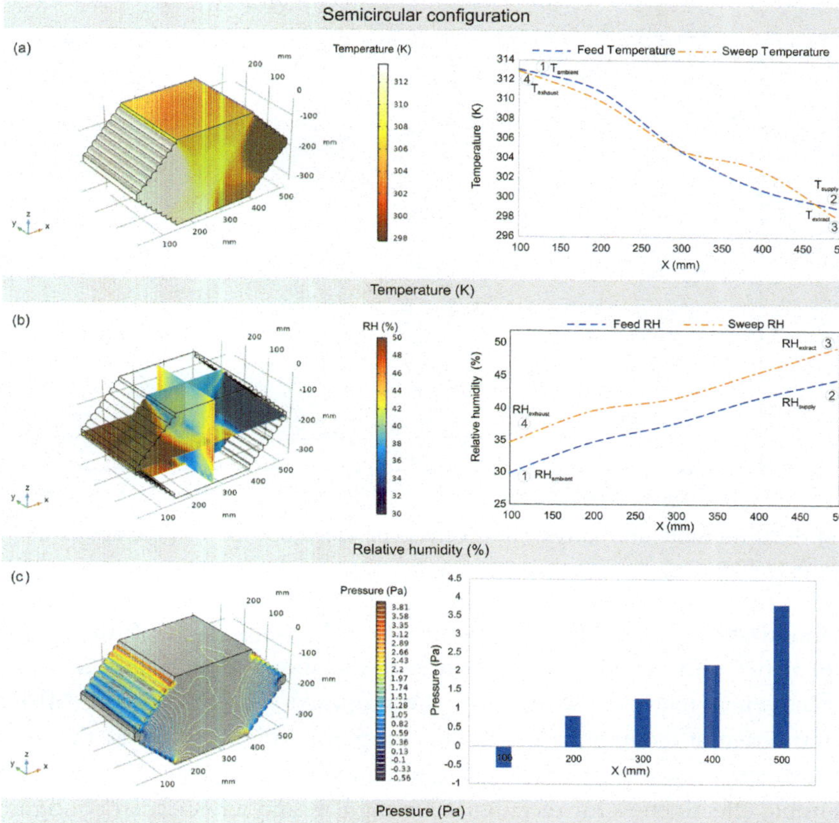

Fig. 9.9 Performance profiles of (A) temperature, (B) relative humidity, and (C) pressure drop for the semicircular configuration of the MERV system.

the latent heat transfer to the supply side as compared to the outdoor air state. Fig. 9.9C shows the pressure distribution in the semicircular configuration-based MERV system. According to Fig. 9.9C, profiles of pressure variability with the space range from −0.56 to 3.81 Pa, while near the inside and outside surfaces of the system, pressure distribution exhibits more variability due to the wall friction and reduces system effectiveness.

Keeping in view the outcome of the study, it can be concluded that the performance of all configurations on the MERV system is mainly influenced by heat and mass transfer characteristics such as pressure drop, temperature, and relative humidity. Higher pressure drops cause a decrease in the system's effectiveness. Conversely, higher temperature and relative humidity recovery result in higher system effectiveness. The semicircular

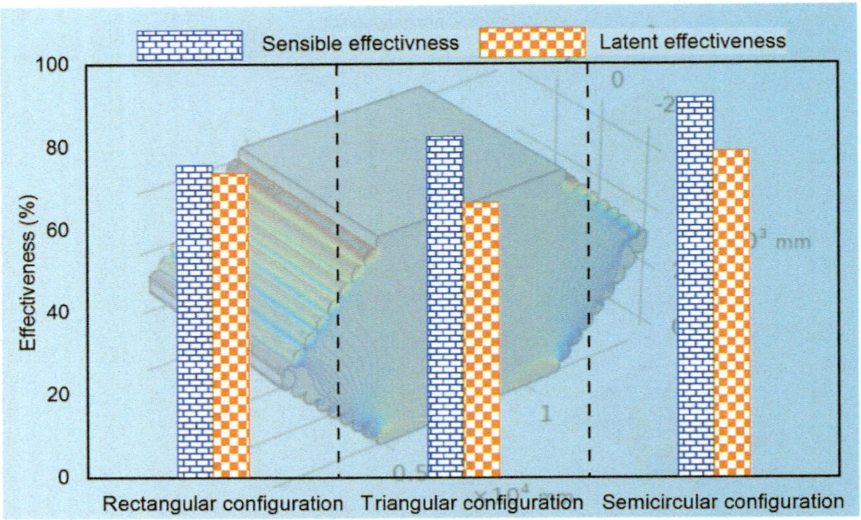

Fig. 9.10 Sensible and latent effectiveness of the proposed configurations of the MERV system.

configuration-based MERV system showed a minimum pressure drop. For better understanding or validation of the above-mentioned results, the effectiveness method was used for the proposed configurations of MERV irrespective of the membrane material properties. Fig. 9.10 shows the sensible effectiveness of proposed configurations for the MERV system. The sensible effectiveness for rectangular, triangular, and semicircular configuration was 76.85%, 82.59%, and 90.47%, respectively. Similarly, the latent effectiveness for rectangular, triangular, and semicircular configurations was 75.26%, 66.67%, and 79.05%, respectively. As per the proposed configuration mentioned above, the simulation results showed that the semicircular configurations of the membrane shows the maximum enthalpy exchange between the air streams. Therefore, based on pressure drop, sensible, and latent effectiveness, it was concluded that the semicircular-based MERV system could be a potential option for energy recovery.

9.5 Conclusions

The present study aims to investigate rectangular, triangular, and semicircular configurations of membrane integration into ERVs. A CFD-based low Reynolds number k-w turbulence model available in COMSOL

Multiphysics is used. Heat and mass transfer through the air streams is studied and verified using the effectiveness method of all proposed configurations. The outdoor and indoor air states used in the analysis are 313.15 K and 30%, 50%, and 298.15 K, respectively. The temperature at the supply side of MERV in rectangular, triangular, and semicircular configurations is given as such, respectively. The relative humidity profile at the supply side of MERV in rectangular, triangular, and semicircular configurations varies from 39%, 41%, and 46%, respectively. The temperatures at the exhaust side are 309 K, 311 K, and 312 K, respectively. The relative humidity at exhaust side is 41%, 37%, and 35%, respectively. The pressure distribution of rectangular, triangular, and semicircular configurations of the MERV ranges from −1.92 to 1.42 Pa, −0.76 to 4.35 Pa, and −0.56 to 3.81 Pa, respectively. The latent and sensible effectiveness of the rectangular, triangular, and semicircular configurations are 75.26%, 66.67%, 79.05%, 76.85%, 82.59%, and 90.47%. Conclusively, the semicircular configuration of MERV is recommended because of better heat and mass transfer, sensible and latent effectiveness, and less pressure drop as compared to triangular and rectangular configurations.

Acknowledgment

The United Kingdom Government funds this research (and/or) pilot activity through the Project, APP47457, titled "Super-efficient Sustainable Cooling Solution for All Applications (S2Cool)," under the Ayrton Challenge Programme of the UK Research and Innovation (UKRI); however, the views expressed do not necessarily reflect the UK Government's official policies. The Project, APP47457, is implemented/led by Northumbria University, UK, together with partners to develop sustainable cooling solution for several applications.

Funding statement

The United Kingdom Government funds this research (and/or) pilot activity through the Project, APP47457, titled "Super–efficient Sustainable Cooling Solution for All Applications (S2Cool)," under the Ayrton Challenge Programme of the UK Research and Innovation (UKRI); however, the views expressed do not necessarily reflect the UK Government's official policies. The Project, APP47457, is implemented/led by Northumbria University, UK, together with partners to develop sustainable cooling solution for several applications.

Conflicts of interest

The authors declare no conflict of interest.

Declaration of generative AI and AI-assisted technologies in the writing process

No generative AI/AI-assisted technology is used in this paper.

CRediT authorship contribution statement

Muhammad Asif was involved in the conceptualization, methodology, software, formal analysis, investigation, writing of the original draft. **Muhammad Sultan, Zahid Mahmood Khan**, and **Muhammad Wakil Shahzad** were involved in the conceptualization, methodology, validation, resources, visualization, supervision, project administration, writing (Review and editing), funding acquisition. **Faheem Hassan Akhtar, Hadeed Ashraf, Naveed Husnain, Uzair Sajjad, Muhammad Farooq, and Muhammad Yasar Javaid** were involved in the conceptualization, methodology, software, formal analysis, investigation, writing (review and editing).

References

Abadi, I.R., Aminian, B., Nasr, M.R., Huizing, R., Green, S., Rogak, S., 2022. Experimental investigation of condensation in energy recovery ventilators. Energy Build. 03787788. 256. Available from: https://www.journals.elsevier.com/energy-and-buildings. doi:10.1016/j.enbuild.2021.111732.

Al-Waked, R., Nasif, M.S., Morrison, G., Behnia, M., 2013. CFD simulation of air to air enthalpy heat exchanger. Energy Convers. Manag. 01968904. 74, 377–385. doi:10.1016/j.enconman.2013.05.038.

Aleem, M., Hussain, G., Sultan, M., Miyazaki, T., Mahmood, M.H., Sabir, M.I., Nasir, A., Shabir, F., Khan, Z.M., 2020. Experimental investigation of desiccant dehumidification cooling system for climatic conditions of multan (Pakistan). Energies. 19961073. 13 (21). Available from: https://www.mdpi.com/1996-1073/13/21/5530. doi:10.3390/en13215530.

Aranda, P., Chen, W.J., Martin, C.R., 1995. Water transport across polystyrenesulfonate/alumina composite membranes. J. Membr. Sci. 03767388. 99 (2), 185–195. doi:10.1016/0376-7388(94)00214-J.

ASHRAE, 2020. ASHRAE Handbook Heating, Ventilating, and Air-Conditioning: Systems and Equipment. American Society of Heating, Refrigerating and Air-Conditioning Engineers, Georgia, United States.

Bonilla, C.F., 1962. Thermowheel rotary air-to-air heat exchanger: engineering characteristics and economics. Arch. Environ. Health 4, 285–294. doi:10.1080/00039896.1962.10663155.

Chen, G.Q., Scholes, C.A., Qiao, G.G., Kentish, S.E., 2011. Water vapor permeation in polyimide membranes. J. Membr. Sci. 18733123. 379 (1-2), 479–487. doi:10.1016/j.memsci.2011.06.023.

Chua, K.J., Chou, S.K., Yang, W.M., Yan, J., 2013. Achieving better energy-efficient air conditioning—a review of technologies and strategies. Applied Energy. 03062619. 104, 87–104. doi:10.1016/j.apenergy.2012.10.037.

Core, 2023. Memmbrane based energy recovery ventilator (MERV). Available from: https://core.life/eu/product-solutions/.

Daisey, J.M., Angell, W.J., Apte, M.G., 2003. Indoor air quality, ventilation and health symptoms in schools: an analysis of existing information. Indoor Air. 09056947. 13 (1), 53–64. doi:10.1034/j.1600-0668.2003.00153.x.

Drost, M.K., 1993. Air-to-air heat exchanger performance. Energy Build. 03787788. 19 (3), 215–220. doi:10.1016/0378-7788(93)90029-t.

Engarnevis, A., Huizing, R., Green, S., Rogak, S., 2018. Heat and mass transfer modeling in enthalpy exchangers using asymmetric composite membranes. J. Membr. Sci. 18733123. 556, 248–262. Available from: www.elsevier.com/locate/memsci. doi:10.1016/j.memsci.2018.03.007.

Fisk, W.J., Turiel, I., 1983. Residential air-to-air heat exchangers: performance energy savings, and economics. Energy Build. 03787788. 5 (3), 197–211. doi:10.1016/0378-7788(83)90005-1.

Fouih, Y.E., Stabat, P., Rivière, P., Hoang, P., Archambault, V., 2012. Adequacy of air-to-air heat recovery ventilation system applied in low energy buildings. Energy Build. 03787788. 54, 29–39. doi:10.1016/j.enbuild.2012.08.008.

Gao, H., Zhenhai, L., Qiu, S., Yang, B., Shan, L., Wen, Y., 2021. Energy exchange efficiency prediction from non-linear regression for membrane-based energy-recovery ventilator cores. Appl. Therm. Eng. 13594311. 197, 117353. doi:10.1016/j.applthermaleng.2021.117353.

International Energy Agency, World Energy Balances, 2021. https://www.iea.org/data-and-statistics/data-tools/energy-statistics-data-browser.

International Energy Agency, 2019. Electricity information: overview. Available from: https://www.iea.org/reports/electricity-information-overview.

Jafarizave, M., Khaleghi, A., Rezakazemi, M., 2019. Development of CFD model for membrane-based energy recovery ventilators. Chem. Eng. Res. Des. 02638762. 145, 226–234. doi:10.1016/j.cherd.2019.03.019.

Jyothiprakash, K.H., Harshith, J., Abhimanyu Sharan, K.N.S., Krishnegowda, Y.T., 2019. Thermodynamic optimization of three-fluid cross-flow heat exchanger using GA and PSO heuristics. Therm. Sci. Eng. Prog. 24519049. 11, 289–301. doi:10.1016/j.tsep.2019.04.009.

Khan, M.U., Rehman, M.M.U., Sultan, M., Rehman, T.U., Sajjad, U., Yousaf, M., Ali, H.M., Bashir, M.A., Akram, M.W., Ahmad, M., Asif, M., 2022. Key prospects and major development of hydrogen and bioethanol production. Int. J. Hydrogen Energy. 03603199. 47 (62), 26265–26283. Available from: http://www.journals.elsevier.com/international-journal-of-hydrogen-energy/. doi:10.1016/j.ijhydene.2022.06.224.

Kho, N., Huynha, H.L., Soh, Y.C., Cai, W., 2017. Performance characterization of the membrane-based energy recovery system. Procedia Eng. 18777058. 214, 50–58. doi:10.1016/j.proeng.2017.09.822.

Kreider, J.F., Rabl, A., 1994. Heating and Cooling of Buildings: Design for Efficiency. McGraw-Hill, New York.

Lee, E.-J., Lee, J.-P., Kim, N.-H., 2013. Moisture transfer characteristics of a LiCl-impregnated paper membrane used for an enthalpy exchanger. J. Mech. Sci. Technol. 1738-494X. 27 (5), 1527–1537. doi:10.1007/s12206-013-0333-2.

Li, Z.X., Zhong, T.S., Niu, J.L., Xiao, F., Zhang, L.Z., 2015. Conjugate heat and mass transfer in a total heat exchanger with cross-corrugated triangular ducts and one-step made asymmetric membranes. Int. J. Heat Mass Transf. 00179310. 84, 390–400. Available from: http://www.journals.elsevier.com/international-journal-of-heat-and-mass-transfer/. doi:10.1016/j.ijheatmasstransfer.2015.01.032.

Liu, J., Li, W., Liu, J., Wang, B., 2010. Efficiency of energy recovery ventilator with various weathers and its energy saving performance in a residential apartment. Energy Build. 03787788. 42 (1), 43–49. doi:10.1016/j.enbuild.2009.07.009.

Liu, L., Chen, Y., Shuguang, L., Deng, M., 2001. The effect of a support layer on the permeability of water vapor in asymmetric composite membranes. Sep. Sci. Technol. 0149-6395. 36 (16), 3701–3720. doi:10.1081/SS-100108357.

Liu, X.P., Niu, J.L., 2015. Effects of geometrical parameters on the thermohydraulic characteristics of periodic cross-corrugated channels. Int. J. Heat Mass Transf. 00179310. 84, 542–549. doi:10.1016/j.ijheatmasstransfer.2015.01.046.

Mahmood, M.H., Sultan, M., Miyazaki, T., Koyama, S., Maisotsenko, V.S., 2016. Overview of the Maisotsenko cycle—a way towards dew point evaporative cooling. Renew. Sustain. Energy Rev. 18790690. 66, 537–555. Available from: https://www.journals.elsevier.com/renewable-and-sustainable-energy-reviews. doi:10.1016/j.rser.2016.08.022.

Manual, R., 2023. COMSOL Multiphysics Reference Manual. https://doc.comsol.com/6.1/docserver/#!/com.comsol.help.cfd/cfd_ug_fluidflow_noniso.07.05.html.

Manual, R., 2023. K-ω Turbulence Model. https://doc.comsol.com/5.5/doc/com.comsol.help.cfd/cfd_ug_fluidflow_single.06.090.html.

Min, J., Hu, T., Liu, X., 2010. Evaluation of moisture diffusivities in various membranes. J. Membr. Sci. 03767388. 357 (1-2), 185–191. doi:10.1016/j.memsci.2010.04.019.

Min, J., Hu, T., Song, Y., 2011. Experimental and numerical investigations of moisture permeation through membranes. J. Membr. Sci. 03767388. 367 (1-2), 174–181. doi:10.1016/j.memsci.2010.10.064.

Min, J., Su, M., 2010a. Performance analysis of a membrane-based energy recovery ventilator: effects of membrane spacing and thickness on the ventilator performance. Appl. Therm. Eng. 13594311. 30 (8-9), 991–997. doi:10.1016/j.applthermaleng.2010.01.010.

Min, J., Su, M., 2010b. Performance analysis of a membrane-based enthalpy exchanger: effects of the membrane properties on the exchanger performance. J. Membr. Sci. 03767388. 348 (1-2), 376–382. doi:10.1016/j.memsci.2009.11.032.

Min, J., Su, M., 2011. Performance analysis of a membrane-based energy recovery ventilator: effects of outdoor air state. Appl. Therm. Eng. 13594311. 31 (17-18), 4036–4043. doi:10.1016/j.applthermaleng.2011.08.006.

Min, J., Wang, L., 2013. Coupled heat and mass transfer during moisture exchange across a membrane. J. Membr. Sci. 03767388. 430, 150–157. doi:10.1016/j.memsci.2012.12.018.

Nakonieczny, K., 2006. Numerical modeling of cross-flow plate-fin air-to-air heat exchanger under unsteady flow conditions. Numer. Heat Transf. A-Appl. 1040-7782. 49 (1), 1–24. doi:10.1080/10407780500302265.

Nasif, M.S., Al-Waked, R., 2014. Seasonal weather conditions effect on energy consumption and CO_2 emission for air conditioning systems coupled with enthalpy energy recovery heat exchanger. APCBEE Proc. 22126708. 10, 42–48. doi:10.1016/j.apcbee.2014.10.012.

Nizovtsev, M.I., Borodulin, V.Y., Letushko, V.N., Zakharov, A.A., 2016. Analysis of the efficiency of air-to-air heat exchanger with a periodic change in the flow direction. Appl. Therm. Eng. 13594311. 93, 113–121. Available from: http://www.journals.elsevier.com/applied-thermal-engineering/. doi:10.1016/j.applthermaleng.2015.09.029.

Noor, S., Ashraf, H., Hussain, G., Sultan, M., Miyazaki, T., Shakoor, A., Mahmood, M.H., Riaz, M., 2021. Spatiotemporal investigation of evaporative cooling options for greenhouse air-conditioning application in Pakistan. Fresenius Environ. Bull. 30 (3), 2858–2870.

Noor, S., Ashraf, H., Sultan, M., Miyazaki, T., Mahmood, M.H., Khan, Z.M., 2020. IEICES. 24341436. 110–115. Investigation of direct and indirect evaporative cooling options for greenhouse air conditioning in multan (Pakistan)https://www.tj.kyushu-u.ac.jp/en/igses/ieices/. doi:10.5109/4102474.

Papakostas, K.T., Kiosis, G.C., 2015. Heat recovery in an air-conditioning system with air-to-air heat exchanger. J. Sustain. Energy. 1478646X. 34 (3-4), 221–231. Available from: http://www.tandfonline.com/toc/gsol20/current. doi:10.1080/14786451.2013.879139.

Phillips, E.G., Chant, R.E., Bradley, L.C., Fisher, D.R., 1989. Comparison of freezing control strategies for residential air-to-air heat recovery ventilators. ASHRAE Trans. 95, 1–6. Available from: https://www.aivc.org/sites/default/files/airbase_3541.pdf.

Qiu, S., Li, S., Wang, F., Wen, Y., Li, Z., Li, Z., Guo, J., 2019. An energy exchange efficiency prediction approach based on multivariate polynomial regression for membrane-based air-to-air energy recovery ventilator core. Build. Environ. 03601323. 149, 490–500. Available from: http://www.elsevier.com/inca/publications/store/2/9/6/index.htt. doi:10.1016/j.buildenv.2018.12.052.

Raza, H.M.U., Ashraf, H., Shahzad, K., Sultan, M., Miyazaki, T., Usman, M., Shamshiri, R. R., Zhou, Y., Ahmad, R., 2020. Investigating applicability of evaporative cooling systems for thermal comfort of poultry birds in Pakistan. Appl. Sci. (Switzerland). 20763417. 10 (13). Available from: https://res.mdpi.com/d_attachment/applsci/applsci-10-04445/article_deploy/applsci-10-04445.pdf. doi:10.3390/app10134445.

Sabek, S., Tiss, F., Chouikh, R., Guizani, A., 2018. Numerical investigation of heat and mass transfer in partially blocked membrane based heat exchanger: effects of obstacles forms. Appl. Therm. Eng. 13594311. 130, 211–220. doi:10.1016/j.applthermaleng.2017.11.019.

Sebai, R., Chouikh, R., Guizani, A., 2014. Cross-flow membrane-based enthalpy exchanger balanced and unbalanced flow. Energy Convers. Manag. 01968904. 87, 19–28. doi:10.1016/j.enconman.2014.07.002.

Sultan, M., Ashraf, H., Miyazaki, T., Shamshiri, R.R., Hameed, I.A., 2021. Temperature and humidity control for the next generation greenhouses: overview of desiccant and evaporative cooling systems. IntechOpen. doi:10.5772/intechopen.97273.

Sultan, M., El-Sharkawy, I.I., Miyazaki, T., Saha, B.B., Koyama, S., 2015. An overview of solid desiccant dehumidification and air conditioning systems. Renew. Sustain. Energy Rev. 18790690. 46, 16–29. Available from: https://www.journals.elsevier.com/renewable-and-sustainable-energy-reviews. doi:10.1016/j.rser.2015.02.038.

Sultan, M., Miyazaki, T., Koyama, S., 2018. Optimization of adsorption isotherm types for desiccant air-conditioning applications. Renew. Energy. 18790682. 121, 441–450. Available from: http://www.journals.elsevier.com/renewable-and-sustainable-energy-reviews/. doi:10.1016/j.renene.2018.01.045.

Sultan, M., Miyazaki, T., Saha, B.B., Koyama, S., 2016. Steady-state investigation of water vapor adsorption for thermally driven adsorption based greenhouse air-conditioning system. Renew. Energy. 18790682. 86, 785–795. Available from: http://www.journals.elsevier.com/renewable-and-sustainable-energy-reviews/. doi:10.1016/j.renene.2015.09.015.

United Nations Environment Programme, 2021. 2021 Global status report for buildings and construction: towards a zero-emission, efficient and resilient buildings and construction sector. Nairobi.

Zhang, L.Z., 2007. Heat and mass transfer in a cross-flow membrane-based enthalpy exchanger under naturally formed boundary conditions. Int. J. Heat Mass Transf. 00179310. 50 (1-2), 151–162. doi:10.1016/j.ijheatmasstransfer.2006.06.025.

Zhang, L.Z., 2010. Heat and mass transfer in a quasi-counter flow membrane-based total heat exchanger. Int. J. Heat Mass Transf. 00179310. 53 (23-24), 5478–5486. doi:10.1016/j.ijheatmasstransfer.2010.07.009.

Zhang, L.Z., 2016. A reliability-based optimization of membrane-type total heat exchangers under uncertain design parameters. Energy. 03605442. 101, 390–401. Available from: www.elsevier.com/inca/publications/store/4/8/3/. doi:10.1016/j.energy.2016.02.032.

Zhang, L.Z., Niu, J.L., 2001. Energy requirements for conditioning fresh air and the long-term savings with a membrane-based energy recovery ventilator in Hong Kong. Energy. 03605442. 26 (2), 119–135. Available from: www.elsevier.com/inca/publications/store/4/8/3/. doi:10.1016/S0360-5442(00)00064-5.

Zhang, Y., Jiang, Y., Zhang, L.Z., Deng, Y., Jin, Z., 2000. Analysis of thermal performance and energy savings of membrane based heat recovery ventilator. Energy. 03605442. 25 (6), 515–527. doi:10.1016/S0360-5442(99)00087-0.

Zhang, Z., Zhou, Y., Zhao, N., Li, H., Tohniyaz, B., Mperejekumana, P., Hong, Q., Wu, R., Li, G., Sultan, M., Zayan, A.M.I., Cao, J., Ahmad, R., Dong, R., 2021. Clean heating during winter season in Northern China: a review. Renew. Sustain. Energy Rev. 18790690. 149. Available from: https://www.journals.elsevier.com/renewable-and-sustainable-energy-reviews. doi:10.1016/j.rser.2021.111339.

Zhong, K., Kang, Y., 2009. Applicability of air-to-air heat recovery ventilators in China. Appl. Therm. Eng. 13594311. 29 (5-6), 830–840. doi:10.1016/j.applthermaleng.2008.04.003.

Zhou, Y.P., Wu, J.Y., Wang, R.Z., 2007. Performance of energy recovery ventilator with various weathers and temperature set-points. Energy Build. 03787788. 39 (12), 1202–1210. doi:10.1016/j.enbuild.2006.12.010.

CHAPTER 10

Investigating membrane energy recovery ventilator-assisted evaporative cooling system in subtropical and tropical climates

Hadeed Ashraf[a,i], Muhammad Sultan[a], Muhammad Wakil Shahzad[b], Muhammad Asif[a], Nadia Riaz[a], Faheem Hassan Akhtar[c], Naveed Husnain[d], Uzair Sajjad[e], Muhammad Farooq[f], Muhammad Yasar Javaid[g], and Saliha Qamar[h]

[a]Department of Agricultural Engineering, Bahauddin Zakariya University, Multan, Pakistan
[b]Department of Mechanical and Construction Engineering, Northumbria University, Newcastle Upon Tyne, United Kingdom
[c]Department of Chemistry and Chemical Engineering, Lahore University of Management Sciences (LUMS), Lahore, Pakistan
[d]Department of Mechanical Engineering, Bahauddin Zakariya University, Multan, Pakistan
[e]Interdisciplinary Research Center for Sustainable Energy Systems (IRC-SES), King Fahd University of Petroleum and Minerals, Dhahran, Saudi Arabia
[f]Department of Mechanical Engineering, University of Engineering and Technology, Lahore, Pakistan
[g]Department of Mechanical Engineering and Technology, Government College University Faisalabad, Faisalabad, Pakistan
[h]Herff College of Engineering, The University of Memphis, Memphis, TN, United States
[i]Agricultural Engineering Institute, National Agricultural Research Centre, Islamabad, Pakistan

10.1 Introduction

Building air-conditioning (AC) systems require a large amount of energy to operate, mainly due to the use of conventional vapor-compression systems. These systems consume significant amounts of electricity, leading to a high demand for energy and a corresponding increase in greenhouse gas emissions (Chancel, 2022; Yüksel et al., 2022). The use of conventional vapor compression air conditioners has a major impact on energy consumption and contributes to a larger CO_2 footprint, as the refrigerants used in these systems are potent greenhouse gases (Kubba, 2017). On the other hand, evaporative cooling (EC) systems use less energy and have a smaller CO_2 footprint compared to conventional vapor compression systems (Purushothama, 2009). However, their effectiveness is limited by local climate and humidity levels, making them less suitable for certain regions. To mitigate the impact

https://doi.org/10.1016/B978-0-443-48555-8.00005-0

of building indoor comfort/AC systems on energy consumption and the environment, it is important to adopt energy-efficient technologies, such as air-source heat pumps, and to develop alternative cooling systems that use renewable energy sources (Fehrm et al., 2002). Membrane energy recovery ventilators (ERVs) have the potential to play a significant role in reducing the energy consumption and environmental impact of conventional AC systems (Liu et al., 2023; Woods, 2014). ERVs are devices that are installed into the AC system of the building, which helps in improving the overall energy efficiency of the building. The ERVs can be used in combination with traditional AC technology to increase energy saving and maintain indoor temperature and humidity conditions for human comfort. When ERVs are used with traditional AC technology, they recover energy (sensible and latent) from one stream of air and transfer it into the other stream as per the need of the surrounding conditions. The use of ERVs with traditional AC systems could potentially help in electricity conservation, lower electricity bills, and improve human thermal comfort. By incorporating EC and ERVs into conventional vapor compression AC systems, building owners and managers can achieve a more sustainable and energy-efficient solution for building AC and several other applications (Ashraf et al., 2021; Sultan et al., 2021).

Several studies have been conducted on the use of membrane ERVs for different applications, including exhaust energy recovery. Ashraf et al. (2022) studied in detail the applicability of membrane ERVs and their different combinations with traditional AC technologies in Multan, Pakistan. The authors conveyed that the technology could save up to 49% energy used in the AC to attain the desired comfort level. However, the authors did not assess the impact of spatial variability on the performance of the ERV-based AC system. The study also concluded the climate of some cities, including Multan of the country, to be suitable for EC systems, also supported by the findings of Noor et al. (2021). Cho et al. (2023) proposed and experimentally evaluated a novel ERV system. According to the authors, the studied system had better ventilation load reduction potential compared to existing ERVs based on experimental analysis. Liu et al. (2023) estimated the impact of latent energy recovery in a house in winter conditions. The authors evaluated the system from the point of view of design, moisture recovery capability, and airflow rate. Huang et al. (2023) optimized a phase change material-based energy recovery system for buildings in colder regions. The authors concluded that this technology could potentially save the energy required for building AC. Wang et al. (2015) studied the application of ERVs for hot-humid/colder climatic regions using EnergyPlus. The authors concluded that the proposed system configuration achieved energy saving up to 49% under certain conditions. However, the paper did not investigate the performance of the system for mild desert, humid subtropical, and hot desert climates. Such diversity among climatic conditions is typically found in Pakistan. Zhang and Niu (2001) studied the application of ERVs for the hot/humid climate of Hong Kong. According to the authors, the total energy saved during the AC process was up to 58%, whereas the sensible energy saved during the AC process was only up to 10%. Jafarinejad et al. (2019) presented a novel multistage energy recovery system based on condenser-side mixing. The authors concluded that the integrated multistage energy recovery system saved up to 22.9% of the total energy consumption. Alonso et al. (2017) experimentally investigated a lab-scale membrane ERV under the climatic conditions of Norway. The authors studied the performance of three different types of membrane materials for the system. The authors concluded that permeable membrane material showed a problem of membrane swelling under higher humidity conditions, whereas the nonpermeable membrane material resulted in condensation and frost.

The membrane ERV systems are not extensively studied for the climatic conditions of Pakistan. This study presents a spatiotemporal investigation and the potential to save energy using membrane ERVs for AC of buildings. A typical commercial building was modeled and simulated in the DesignBuilder software for heat and mass transfer analyses for three main cities. The study aims to investigate the thermodynamic performance of the proposed system from the viewpoint of human thermal comfort, energy consumption, and CO_2 emissions.

10.2 Research methodology

Fig. 10.1A illustrates the geographic map of Pakistan and selected study area locations [i.e., Islamabad (humid subtropical climate), Multan (subtropical desert climate), and Karachi (tropical climate)] across the country, keeping in view the varying climatic conditions. In certain plains across Southern Pakistan, temperature peaks up to 50°C, which results in very uncomfortable conditions to live and work in. A high range of discomfort is observed in the summer months (May to September) across Pakistan. Therefore, traditional AC technology is widely used for building AC, which has been proven to harm the environment and negatively impact the overall

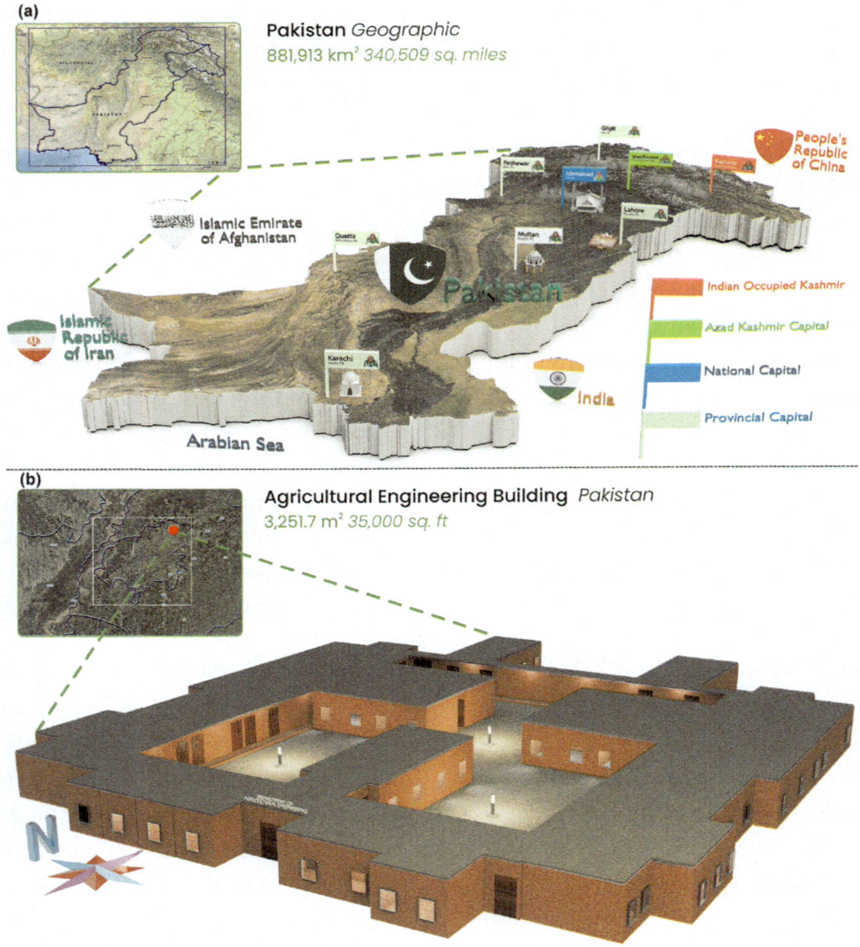

Fig. 10.1 Illustration of (A) geographic map of Pakistan and the selected study areas, and (B) layout of the building for indoor thermal comfort simulation.

energy utilization. In this regard, energy-saving technology is needed while maintaining the desired temperature and humidity conditions for the desired human comfort. In this regard, EC technology, easily available in the market, is commonly used as an alternative. However, these standalone systems, being completely dependent on outside air, sometimes fail to achieve the desired temperature and humidity conditions under certain climatic conditions (such as in monsoon season). Ergo, a sense of discomfort is observed. Instead, the Maisotsenko cycle evaporative cooling (MEC) option could be explored. This study investigates a Maisotsenko-cycle evaporative cooling-assisted vapor compression system integrated with membrane energy recovery ventilator (MEC-DX-ERV) system for AC application in three different study area locations (Multan, Islamabad, and Karachi) of Pakistan.

A nonresidential building was considered for this study. Fig. 10.1B shows the 3D model of the building studied for indoor thermal comfort simulation. Thermophysical characteristics of the studied building were incorporated into the 3D model of the building in DesignBuilder and simulated for energy use, energy saving, building indoor human thermal comfort, and CO_2 emissions.

10.2.1 Membrane ERV-assisted EC system

The working principle of the studied system, i.e., MEC-DX-ERV, is given in Fig. 10.2. Fig. 10.2A illustrates the working principle and the psychrometric representation of the MEC and DEX systems. The Maisotsenko cycle is an EC process that utilizes the principles of the age-old process of cooling, i.e., evaporation, to cool the air. In this cycle, hot air is passed over a wet membrane, causing water to evaporate and absorb sensible energy from the air. The cooled air is then returned to the room, while the water vapor is condensed back into liquid form and expelled outside. This cycle repeats continuously, producing a continuous flow of cooled air, making it an effective way to cool a building or space. The direct expansion (DX)

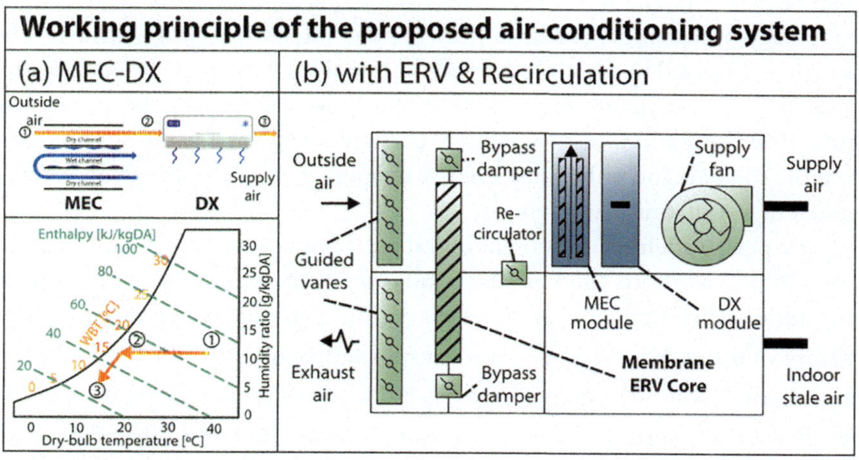

Fig. 10.2 Working principle of the (A) Maisotsenko cycle evaporative cooling (MEC) system, conventional direct expansion (DX) system, and (B) membrane energy recovery ventilator (ERV) system in an air-conditioning system.

system functions by using refrigerant gas that undergoes a perpetual sequence of compression, condensation, expansion, and evaporation. The initial step involves the compression of vaporized refrigerant, which is characterized by low pressure and low temperature, by the compressor. This compression results in the transformation of the refrigerant into a gas with high pressure and high temperature. Subsequently, the heated gas undergoes a condensation process within a condenser, during which it dissipates thermal energy to the surrounding environment and undergoes a phase change into a liquid state characterized by elevated pressure. Subsequently, the high-pressure liquid undergoes expansion via an expansion valve, leading to the formation of a low-pressure and low-temperature mixture. This mixture then undergoes evaporation within the evaporator, whereupon it absorbs heat from its surrounding environment, therefore concluding the cycle. The DX vapor compression system in AC functions by utilizing the evaporator to absorb heat from interior air, afterwards releasing this heat outside through the condenser. This process efficiently establishes a pleasant indoor atmosphere.

Fig. 10.3A illustrates the schematic of the proposed MEC-DX-ERV system providing conditioned air to the desired space. Fig. 10.3B illustrates the working principle and x-section of the membrane ERV used in the present study. The operational mechanism of a membrane ERV involves the transfer of heat and moisture between two distinct airstreams while ensuring their nonintermixing. The system comprises a sequence of alternating plates that facilitate the exchange of heat and moisture while simultaneously preserving a partition between the inlet and exiting airflow. The warm and humid inside air is crossed on one side of the plate while the cool and dry external air is directed across the opposite side of the plate. The process of heat and moisture transfer occurs through the plates, so pretreating the inlet air and enhancing energy efficiency, all the while guaranteeing efficient ventilation and savings.

As per guidelines given in literature (Engarnevis et al., 2018), the membrane substrate used in this study was a hydrophobic microporous substrate with a thickness of 100–120 + (μm), average water vapor permeability of $6.2 \times 10^{-10}\left(\frac{\text{molm}}{\text{m}^2\text{s Pa}}\right)$, water vapor diffusivity of $1.13 \times 10^{-10}\left(\frac{\text{m}^2}{\text{s}}\right)$, and thermal conductivity of $0.159\left(\frac{\text{W}}{\text{mK}}\right)$. The membrane coating used in this study was a hydrophilic polymer film with a thickness of 1–3 + (μm), average water vapor permeability of $1.8 \times 10^{-11}\left(\frac{\text{molm}}{\text{m}^2\text{s Pa}}\right)$, water vapor diffusivity of $3.3 \times 10^{-13}\left(\frac{\text{m}^2}{\text{s}}\right)$, and thermal conductivity of $0.44\left(\frac{\text{W}}{\text{mK}}\right)$.

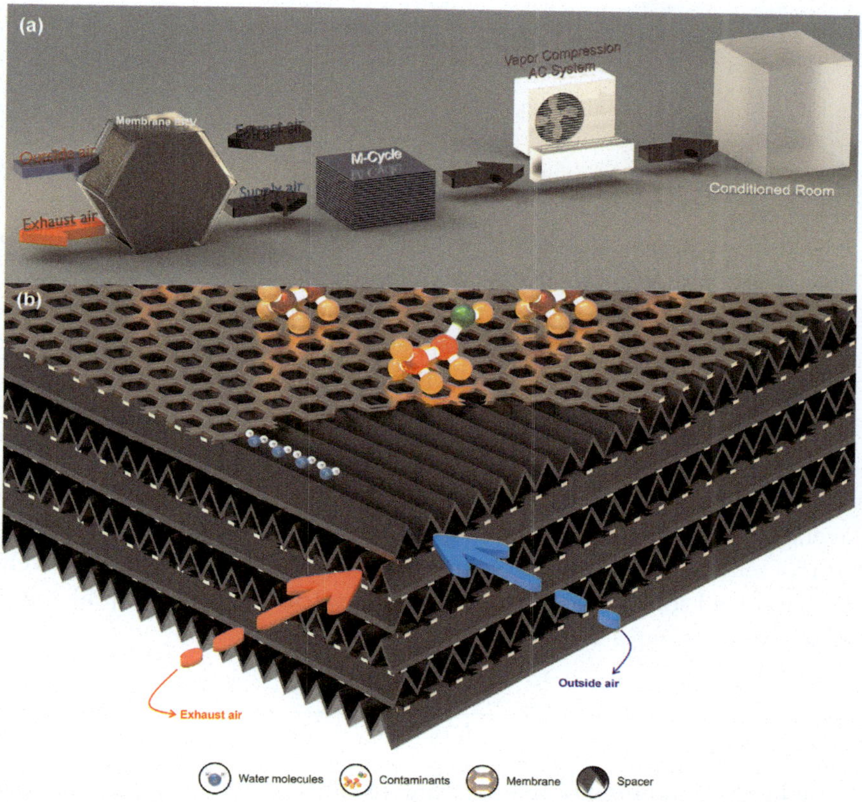

Fig. 10.3 Illustration of the (A) working principle of the MEC-DX-ERV system for a conditioned room, and (B) cross section of membrane core used in the ERV.

10.2.2 Performance assessment models

The MEC-DX-ERV AC system was thermodynamically investigated for building indoor comfort under the weather of Islamabad, Karachi, and Multan (three cities of Pakistan with different climates). To assess the performance of the proposed system, cooling capacity (Q) was the product of mass flow rate of air (m), specific heat capacity of air (C_p), and temperature difference between two air streams (ΔT) (Noor et al., 2020b; Shahzad et al., 2021; Ullah et al., 2020). Where, mass flow rate of air (m) was the product of density of air (ρ_{air}), and volumetric flow rate of air (u). Total cooling capacity (Q_{net}), total sensible energy transfer ($Q_{sensible}$), and total latent energy transfer (Q_{latent}) were calculated by (Eqs. 1–3) as per guidelines in literature (U.S. DoE, 2009).

$$Q_{net} = m(h_{SA} - h_{OA}) \tag{1}$$

$$Q_{sensible} = m(h_{SA} - h_{OA})X_{min} \tag{2}$$

$$Q_{latent} = Q_{net} - Q_{sensible} \tag{3}$$

where h_{SA} represents enthalpy of supply air, h_{OA} represents enthalpy of outside air, and X_{min} represents the minimum of supply air and outside air humidity ratio. In terms of cooling loads, area of the human body ($A_h b$) was calculated using (Eq. 4), as per guidelines (American Society of Heating and Refrigerating and Air-conditioning Engineers, Inc., 2017).

$$A_h b = 0.202 M^{0.425} H^{0.725} \tag{4}$$

where M represents mass of average human, and H represents height of average human. In addition, cooling load due to human body (Q_{hb}) was the sum of sensible and latent heats generated from skin and lungs, respectively. Cooling load sol-air temperature (T_{sa}) was calculated using (Eq. 5) (Noor et al., 2020a; Çengel and Ghajar, 2019).

$$T_{sa} = T_a + \frac{\alpha_{solar} q_{solar}}{h_o} - \frac{E\varphi(T_a^4 - T_{sur}^4)}{h_o} \tag{5}$$

where T_a represents temperature of ambient air, α_{solar} is the solar absorptivity constant, q_{solar} is the solar radiation incident on the surface, h_o is the heat transfer coefficient, E is the emissivity of the surface, φ is the solar latitude angle, and T_{sur} is the average temperature of adjacent surfaces. The cooling load from exterior surfaces (Q_{ext}) was calculated using (Eq. 6) (Noor et al., 2021).

$$Q_{ext} = h_o A_s (T_a - T_s) + \alpha_{solar} A_s q_{solar} - E A_s \varphi (T_a^4 - T_{sur}^4) \tag{6}$$

where A_s represents wall surface area and T_s is the exterior surface temperature. In addition, cooling load from fan motors (Q_m) was the ratio between the product of motor power (P_r), load factor (L_f), utilization factor (U_f), and motor efficiency (η_m). Similarly, cooling load in terms of heat load added into the conditioned space due to conduction from walls (Q_{wall}) was the product of overall heat transfer coefficient of the wall (U_{wall}), A_s, and the difference between T_{sa} and the inside temperature of the conditioned space (T_{in}). Cooling load in terms of heat load added into the space through glazing

($Q_{glazing}$) was the sum of cooling load added through the center (Q_c), edge (Q_e), and frame of glazing (Q_f), respectively, which in turn was the product of overall heat transfer coefficient of the glazing (U_g), area of the glazing (A_g) and the difference between inside and outside air temperature. Similarly, cooling load in terms of heat transfer through the floor (Q_{sp}) was the product of overall heat transfer coefficient of the floor (U_{floor}), floor perimeter (P_{floor}), and the difference between inside and outside air temperature.

Cooling load in terms of heat transfer through the conditioned space due to air exchange (ACH) was the ratio of inflow mass flowrate (m_{in}) and total volume of the building (V). In addition, infiltration cooling load was categorized into latent ($Q_{l,i}$) and sensible ($Q_{s,i}$) loads (Eqs. 7 and 8), respectively.

$$Q_{l,i} = \rho_{air} h_{fg} \mathrm{ACHV}(X_{in} - X_{out}) \tag{7}$$

$$Q_{s,i} = \rho_{air} C_p \mathrm{ACHV}(T_{in} - T_{out}) \tag{8}$$

where h_{fg} is the latent heat of evaporation, X_{in} is inside air humidity ratio, X_{in} is the outside air humidity ratio, and T_{in} & T_{out} represent the inside & outside air temperature, respectively.

In addition, performance of the proposed MEC-DX-ERV system was assessed in terms of human thermal comfort measured through predicted mean vote (PMV), given in (Eq. 9), and predicted percentage of dissatisfaction (PPD), given as (Eq. 10).

$$\begin{aligned} \mathrm{PMV} = & \left[0.303e^{(-0.036\ \mathrm{Met})} + 0.028\right](\mathrm{Met} - Q_{work}) \\ & - 3.05[5.73 - 0.007(\mathrm{Met} - Q_{work}) - P_a] - 0.42[(\mathrm{Met} - Q_{work}) - 58.15] \\ & - 0.0173\,\mathrm{Met}(5.87 - P_a) - 0.0014\,\mathrm{Met}(34 - T_{air}) - 3.96e^{-8} r_{cb} \\ & \left[(T_{cloth} + 273)^4 - (\mathrm{MRT} + 273)^4\right] - r_{cb} h_{conv}(T_{cloth} - T_{air}) \end{aligned} \tag{9a}$$

$$\begin{aligned} T_{cloth} = & 35.7 - 0.0275(\mathrm{Met} - Q_{work}) - R_c\{(\mathrm{Met} - Q_{work}) \\ & - 3.05[5.73 - 0.007(\mathrm{Met} - Q_{work}) - P_a] - 0.42[(\mathrm{Met} - Q_{work}) - 58.15] \\ & - 0.0173 Met(5.87 - P_a) - 0.0014\,\mathrm{Met}(34 - T_{air})\} \end{aligned} \tag{9b}$$

$$h_{conv} = \begin{cases} 2.38(T_{cloth} - T_{air})^{0.25} \rightarrow 2.38(T_{cloth} - T_{air})^{0.25} > 12.1\sqrt{u_{air}} \\ 12.1\sqrt{u_{air}} \rightarrow 2.38(T_{cloth} - T_{air})^{0.25} < 12.1\sqrt{u_{air}} \end{cases} \tag{9c}$$

$$r_{cb} = \begin{cases} 1.0 + 0.2\, c_i \rightarrow c_i < 0.5 \text{ clo} \\ 1.05 + 0.1\, c_i \rightarrow c_i > 0.5 \text{ clo} \end{cases} \tag{9d}$$

$$\text{PPD} = 100 - 95e^{(-0.03353\,\text{PMV}^4 - 0.2179\,\text{PMV}^2)} \tag{10}$$

where metabolic rate of human body per unit are is Met, Q_{work} is the heat transfer from human body due to work, $P_a$is the ambient air vapor pressure, T_{air} is the inside air temperature, $r_{cb}$is the ratio between clothed body and the naked body, T_{cloth} is the temperature of the clothing, mean radiant temperature was represented with MRT, h_{conv} is the heat transfer coefficient due to convection, $R_c$is the heat resistance of clothing, u_{air} is air flow velocity, and c_i is the clothing insulation.

10.3 Results

10.3.1 Climatic conditions

Results obtained from the building simulation study are presented in this section. The performance of the MEC-DX-ERV was investigated from the point of view of temperature, humidity, human thermal comfort parameters, mean radiant temperature (MRT), electricity consumption, and CO_2 emissions. The typical meteorological year (TMY) data for the three cities (i.e., Multan, Islamabad, and Karachi) is presented in Fig. 10.4. In addition, a profile of the mean minimum and maximum temperatures across Pakistan is presented in Fig. 10.5, illustrating higher temperature ranges across the middle of the year. According to Fig. 10.4A, Multan experienced the hottest summer in June with peaks around 40°C, closely followed by Karachi in July around 35°C. Winters in Pakistan are also severe in certain regions of the country. Out of the three cities investigated for this study, Islamabad experienced relatively low temperatures throughout the year with a peak in July around 35°C. According to Fig. 10.4B, Multan experienced the driest conditions in May to June with relative humidity levels ranging from 20% to 40%. Whereas this range increases up to 80%–90% in July-August, the monsoon season in the country. In case of relative humidity, Islamabad follows similar trends compared to Multan. Whereas relative humidity in Karachi varies dramatically throughout the year, with this city being so close to the coast. The temperature and humidity conditions in the selected cities are rarely the desired human thermal comfort conditions. These conditions seldom support building indoor comfort; hence, building

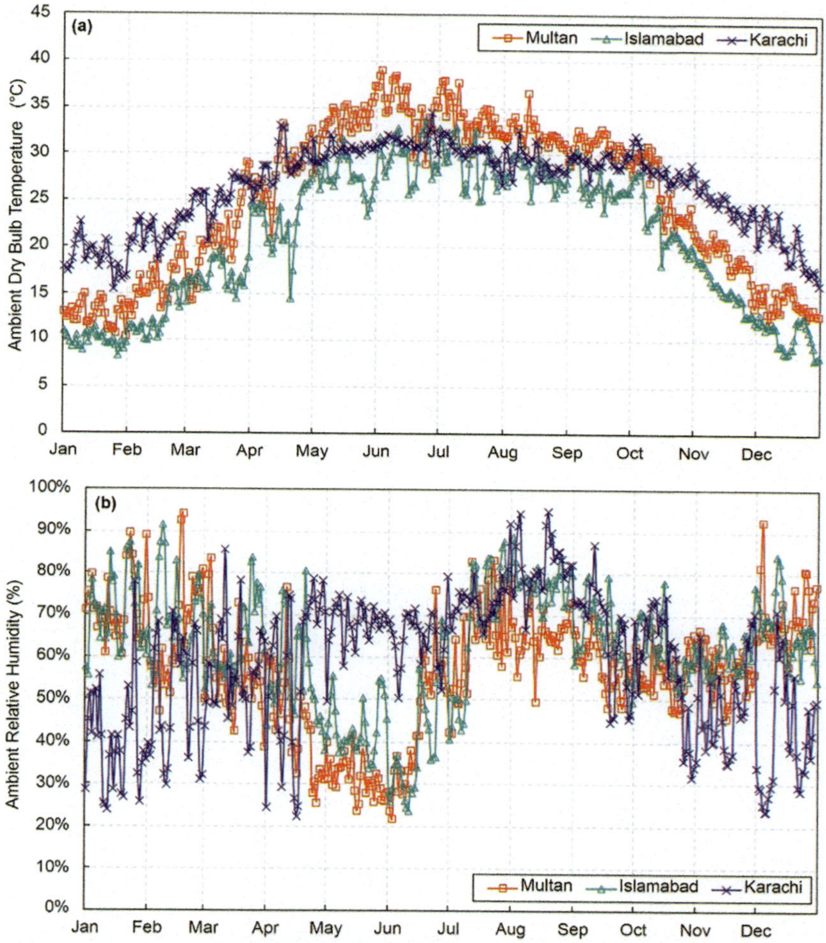

Fig. 10.4 Profile of (A) ambient dry bulb temperature and (B) ambient relative humidity of the selected study area locations.

AC is often required. Therefore, buildings across Pakistan use traditional AC systems for attaining indoor thermal comfort. However, these AC systems utilize higher amounts of energy in countries stricken with energy shortages. AC of buildings is a major contributor to a building's total energy consumption. Exhaust air from a conditioned space in summer and/or winter is a wasted form of recoverable energy using a membrane ERV. Therefore, across Pakistan, these ERV systems could be applied in building AC to attain the desired temperature and humidity conditions for human comfort (Fig. 10.5).

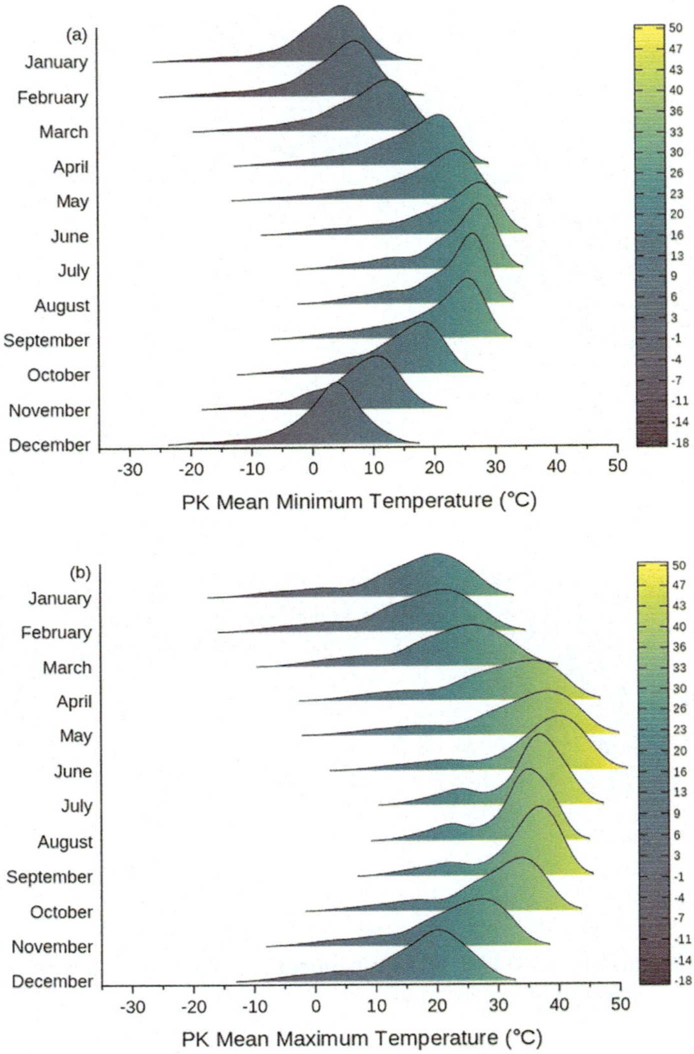

Fig. 10.5 Profile of mean (A) minimum and (B) maximum temperatures throughout a typical meteorological year across Pakistan.

10.3.2 Thermodynamic performance

Fig. 10.6 illustrates the results of the performance of the MEC-DX-ERV system coupled with a membrane-based energy recovery ventilator for the three selected cities across Pakistan. Fig. 10.6A shows that the MEC-DX -ERV system achieved an average temperature of 20°C throughout the year

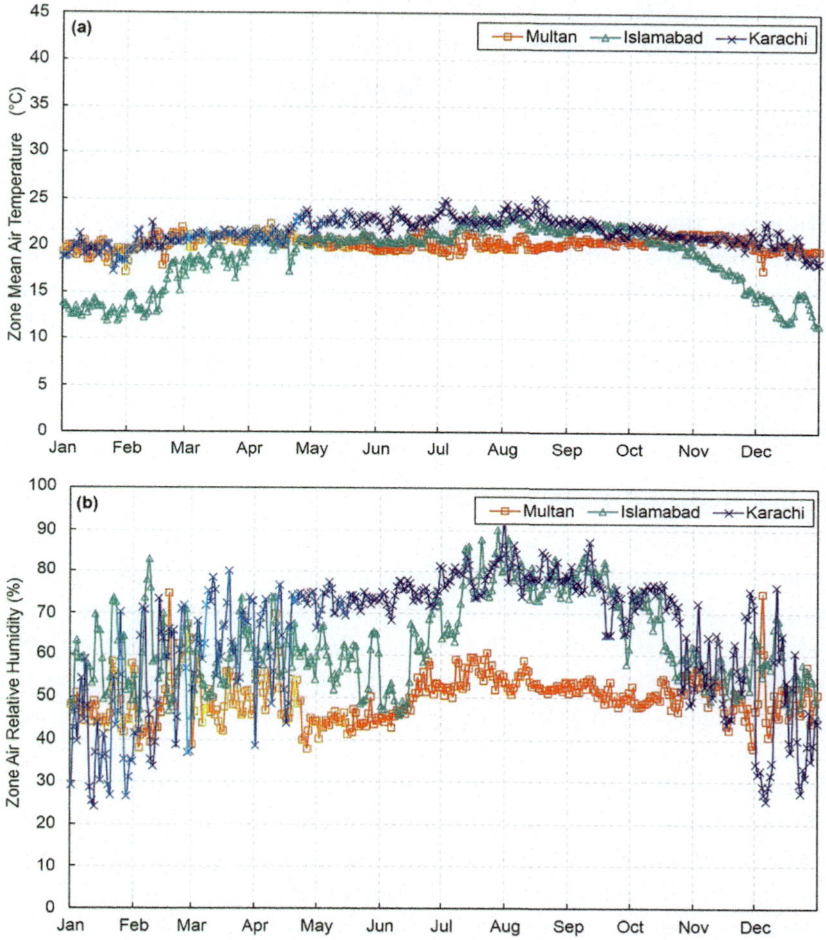

Fig. 10.6 Profile of (A) MEC-DX-ERV temperature and (B) MEC-DX-ERV relative humidity at the selected study area locations.

in Multan, whereas performance of the system slightly decreased in Karachi. On average, the system achieved ~22°C in Karachi, with peaks up to 25°C during the hot season. In the case of Islamabad, the system achieved an average temperature of ~19°C throughout the year, with depression peaking below 15°C in winter months of January, February, and December. In terms of relative humidity, Fig. 10.6B illustrates varied performance of the MEC-DX-ERV system across the year for the three selected cities. Accordingly, in Multan, the MEC-DX-ERV system achieved an average relative humidity of 50% with a stable trend throughout the year.

Meanwhile, in Karachi, the MEC-DX-ERV system achieved an average relative humidity of ~65% with a higher level of variation throughout the year. Whereas in the case of Islamabad, the MEC-DX-ERV system achieved an average relative humidity of ~64% with a medium level of variation throughout the year. From the results, it can be deduced that the variations in the performance of the system are climate dependent. In Islamabad, the relative humidity in ambient air is comparatively higher due to frequent rainfall, which leads to higher system relative humidity. Similarly, in Karachi, the city being closer to the Arabian Sea, relative humidity in ambient air is higher, which leads to higher system relative humidity. While in Multan, relative humidity in ambient air is comparatively lower due to infrequent rainfalls, which leads to lower system relative humidity.

10.3.3 Human thermal comfort

PMV is a measure used to quantify the thermal sensation of a group of humans, whereas PPD is a quantifiable measure of dissatisfaction with the supplied air temperature and humidity. MRT is the average temperature radiating from the surrounding bodies. When PMV approaches zero, it indicates a state of thermal neutrality and optimal comfort. Higher PMV values suggest thermal discomfort due to either feeling too warm or too cold. Similarly, elevated PPD values correspond to a greater number of dissatisfied individuals. Fig. 10.7 illustrates the PMV and PPD results obtained from the MEC-DX-ERV. According to Fig. 10.7A, the system achieved an average of −0.9 PMV in Multan throughout the TMY. It is worth mentioning that the MEC-DX-ERV achieved a stable PMV of −1 throughout the summer months up to the end of September. In the case of Karachi, the system achieved a relatively slightly warmer PMV, i.e., −0.7 on average throughout the year. It is worth mentioning that the system achieved variation in PMV (ranging from −0.2 to −1.0) throughout the duration of the summer months all the way to the end of September. In the case of Islamabad, the system achieved, on average, −1.2 PMV with stable PMV in summer months and −2.0 PMV in winter months.

MRT impacts comfort by influencing the heat exchange between the human body and its surroundings. Deviations from the desired MRT can lead to discomfort, as surfaces that are significantly cooler or warmer than the body can affect thermal balance negatively. Fig. 10.8 shows the profile of MRT of the MEC-DX-ERV system for the three selected cities across

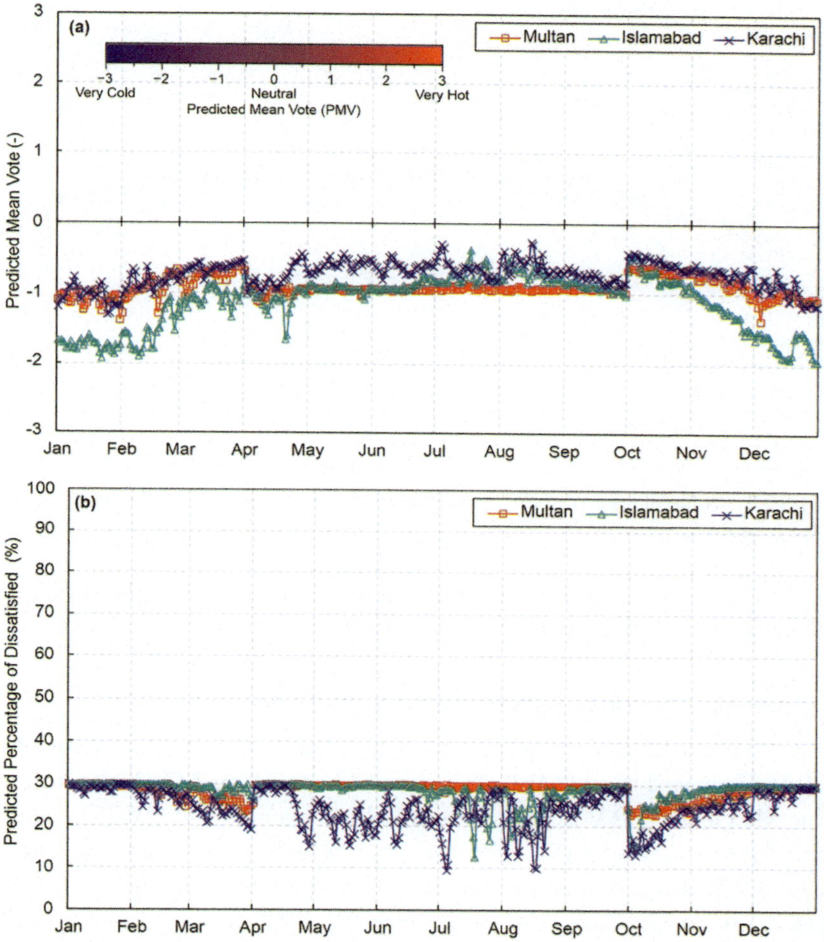

Fig. 10.7 Profile of (A) MEC-DX-ERV predicted mean vote and (B) MEC-DX-ERV predicted percentage of dissatisfied at the selected study area locations.

Pakistan throughout the year. According to Fig. 10.8, the MEC-DX-ERV system achieved an MRT of 25.5°C on average for the climatic conditions of Multan throughout the year, closely triumphed by an MRT of 26°C in Karachi. Whereas in Islamabad, the MEC-DX-ERV system achieved an MRT of 23°C on average throughout the year. In Islamabad, the system achieved lower peaks of 15°C in winter months and upper peaks of 30°C in summer months. Whereas in the case of Karachi and Multan, the system achieved lower peaks of 20°C in winter conditions and upper peaks of 30°C

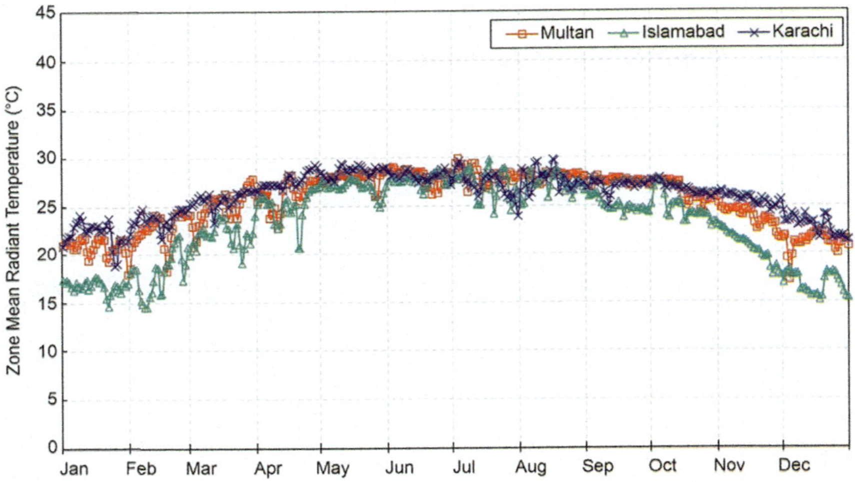

Fig. 10.8 Profile of mean radiant temperature of the MEC-DX-ERV at the selected study area locations.

in summer months. Essentially, in terms of thermal comfort, the MRT must be close to or equal to the air temperature surrounding the subject body.

10.3.4 Energy consumption and CO_2 emission

Fig. 10.9 illustrates the kWh units of electricity consumed by MEC-DX-ERV compared to the MEC-DX system for building AC. According to a, the MEC-DX-ERV system consumed on average 895 kWh in Multan compared to the MEC-DX system, which consumed on average 1390 kWh throughout the year. As per results obtained, in b, the MEC-DX-ERV system consumed on average 819 kWh in Karachi compared to the MEC-DX system, which consumed on average 1461 kWh throughout the year. As per results presented in c, the MEC-DX-ERV system consumed on average 612 kWh in Islamabad compared to the MEC-DX system, which consumed on average 922 kWh throughout the year.

Fig. 10.10 illustrates the CO_2 emissions from the building in cases of MEC-DX-ERV and MEC-DX only, respectively. According to Fig. 10, the MEC-DX system emitted on average 702 $kgCO_2/kWh$, whereas it was only 452 $kgCO_2/kWh$ using MEC-DX-ERV in Multan, Pakistan. In Islamabad, the MEC-DX system emitted, on average, 466 $kgCO_2/kWh$, whereas it was 309 $kgCO_2/kWh$ in the case of MEC-DX-ERV. In Karachi, the MEC-DX system emitted, on average, 738 $kgCO_2/kWh$, whereas it was 413 $kgCO_2/kWh$ in the case of MEC-DX-ERV.

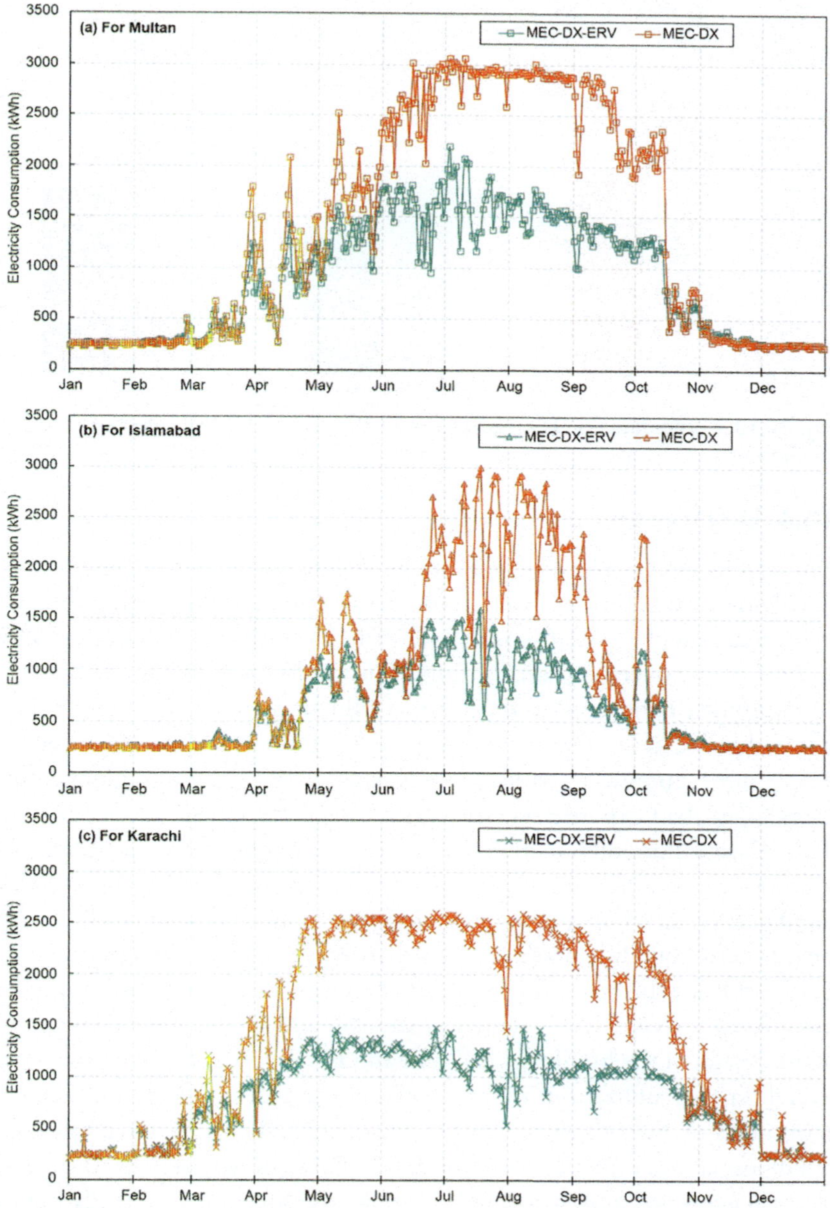

Fig. 10.9 Comparative electricity consumption at the facility for the selected study area locations.

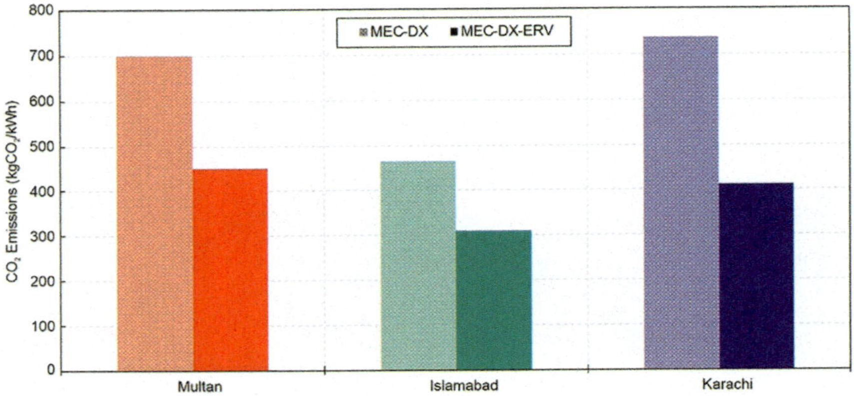

Fig. 10.10 Comparative CO_2 emissions from the facility for the selected study area locations.

10.4 Discussion

The results of the performance of the membrane-based energy recovery ventilator for building indoor comfort are discussed in this section to understand whether the system could potentially be utilized for indoor comfort and energy recovery at the same time. The results show that the temperature and humidity conditions in the cities of Multan, Islamabad, and Karachi are not within the desired human thermal comfort range, making AC necessary. The summer months in Multan and Karachi experience high temperatures, while Islamabad experiences the lowest temperatures throughout the year. The relative humidity in the cities varies throughout the year, with the monsoon season bringing increased relative humidity levels. These findings highlight the need for measures to improve indoor comfort in these cities, such as better building design and ventilation systems. In addition, the results obtained from this study indicate that the local climatic conditions must be considered for an efficient design of the AC and ERV system. The appropriate design of the buildings' AC unit and ERV system can help reduce the overall energy utilization, increase human thermal comfort, and keep the economics of AC large-scale buildings in check. The appropriate design of buildings, an uncommon phenomenon in Pakistan, could help advance the national economy and overall quality of life. Results from the study indicate that the ERV systems are climate-dependent, and their performance is also impacted by the outside air. For example, in Multan, the overall performance of sensible heat recovery of the ERV system was considerably higher compared to other cities. Whereas in the case of latent heat load recovery, the ERV system performed relatively better in Islamabad and Karachi. The findings from the study indicate that some mitigation strategies could be incorporated into everyday life and the building construction sector for efficient electricity use. The use of shading devices, orientation of the building, green roofs, and green walls could potentially lead to even lower energy requirements for building AC. In addition, awareness campaigns regarding the efficient design of buildings should be conducted for all the potential stakeholders. Further, policies regarding appropriate construction of buildings, their orientation, and construction material should be developed and enforced upon the stakeholders. The harsh climatic conditions in Pakistan result in a potentially lower standard of living for most people. In summers, across Pakistan, people suffer from heatstroke through severe heat waves. This could be mitigated using energy-efficient AC technologies, which can operate in different conditions. In common practice, the standard EC systems are being used in Pakistan as a low-cost alternative to the traditional AC systems. However, use of the EC systems is limited to only a few months of summer. As soon as the monsoon period starts, due to more frequent rainfall, the humidity in the ambient air increases to a level of discomfort. In such cases, EC systems fail to provide the desired temperature and humidity conditions for human comfort. Therefore, common folk turn to traditional AC systems, which lead to higher energy input and ultimately unbearable electricity costs. Therefore, the results from this study suggest that ERV technology could be adopted with traditional AC systems to avoid compromising the desired temperature and humidity level and electricity costs. Moreover, the design of buildings across Pakistan could be improved using adequate insulation materials, providing green spaces, and appropriate shading. These measures could help reduce the scorching load of the sun onto the buildings. A recent study in Multan, Pakistan (Ashraf et al., 2022), indicated that MEC-DX-ERV systems could prove to be beneficial mainly from the energy-saving point of view. The authors concluded that the mentioned system could potentially solve the problems regarding AC of buildings in Multan. However, the study failed to investigate the broader application of ERV technology across varying climatic conditions. Therefore, the present study investigated the already established and highlighted technology, i.e., MEC-DX-ERV, in Islamabad and Karachi and compared it with the results from Multan. In conclusion, the results of the present study demonstrate the potential benefits of using efficient AC systems in Pakistan and highlight the importance of considering such systems as part of a comprehensive approach to building design and energy management. By working to reduce energy consumption, improve indoor comfort, and contribute to sustainable development, such systems could play a critical role in shaping the future of the country. From an experimental point of view, the membrane-based energy recovery ventilation systems are under-explored in Pakistan. Therefore, the future research directions must include experimental setups, prototypes of the technology, and dissemination of the developed technology at a larger scale across the country.

10.5 Conclusions

This study aimed to explore spatial variation in the performance of MEC-DX-ERV for building AC. The system was modeled and simulated for three different major cities (i.e., Multan, Karachi, and Islamabad) of Pakistan. The thermodynamic performance of the system was analyzed in terms of mean zone temperature and relative humidity, PMV and percentage, PPD, electricity consumption, and CO_2 emissions. According to the results, the system coupled with a membrane energy recovery ventilator performed comparatively better than the system without the recovery option in terms of temperature, humidity, MRT, human thermal comfort, electricity consumption, and CO_2 emissions. However, the system performed better in a relatively colder climate (i.e., Islamabad) in the summer months compared to the other two cities. The performance of the system varies in colder climates by efficiently recovering heat and managing moisture, while in warmer climates it focuses on sensible and latent heat recovery and condensation control. In conclusion, the evaporative cooling assisted with a conventional vapor compression system coupled with this membrane energy recovery ventilator technology can help achieve the desired temperature and humidity conditions for human thermal comfort. Moreover, it can also help keep the electricity bills in check and CO_2 emissions from building AC under control.

Acknowledgment

The United Kingdom Government funds this research (and/or) pilot activity through the Project, APP47457, titled "Super–efficient Sustainable Cooling Solution for All Applications (S2Cool)," under the Ayrton Challenge Programme of the UK Research and Innovation (UKRI); however, the views expressed do not necessarily reflect the UK Government's official policies. The Project, APP47457 is implemented/led by Northumbria University, UK, together with partners to develop sustainable cooling solutions for several applications.

Funding statement

The United Kingdom Government funds this research (and/or) pilot activity through the Project, APP47457, titled "Super–efficient Sustainable Cooling Solution for All Applications (S2Cool)," under the Ayrton Challenge Programme of the UK Research and Innovation (UKRI); however, the views expressed do not necessarily reflect the UK Government's official policies. The Project, APP47457, is implemented/led by Northumbria University, UK, together with partners to develop sustainable cooling solutions for several applications.

CRediT authorship contribution statement

Hadeed Ashraf was involved in the conceptualization, methodology, software, formal analysis, investigation, writing of the original draft. **Muhammad Sultan, Muhammad Wakil Shahzad, and Faheem Hassan Akhtar** were involved in the conceptualization, methodology, validation, resources, visualization, supervision, project administration, writing (review and editing), funding acquisition. **Muhammad Asif, Nadia Riaz, Naveed Husnain, Uzair Sajjad, Muhammad Farooq, Muhammad Yasar Javaid, and Saliha Qamar** were involved in the conceptualization, methodology, software, formal analysis, investigation, writing (review and editing).

Conflicts of interest

The authors declare no conflict of interest.

Data availability statement

The datasets used and/or analyzed during the current study are available from the corresponding authors upon request.

Declaration of generative AI and AI-assisted technologies in the writing process

No generative AI/AI-assisted technology is used in this paper.

References

Alonso, M.J., Mathisen, H.M., Aarnes, S., Liu, P., 2017. Performance of a lab-scale membrane-based energy exchanger. Appl. Therm. Eng. 13594311. 111, 1244–1254. Available from: http://www.journals.elsevier.com/applied-thermal-engineering/. doi:10.1016/j.applthermaleng.2015.11.119.

American Society of Heating and Refrigerating and Air-conditioning Engineers, Inc., 2017. ASHRAE Handbook—Fundamentals (SI), American Society of Heating/Refrigerating and Air-conditioning Engineers, Inc, Atlanta, Atlanta, GA.

Ashraf, H., Sultan, M., Shamshiri, R.R., Abbas, F., Farooq, M., Sajjad, U., Md-Tahir, H., Mahmood, M.H., Ahmad, F., Taseer, Y.R., Shahzad, A., Niazi, B.M.K., 2021. Dynamic evaluation of desiccant dehumidification evaporative cooling options for greenhouse air-conditioning application in multan (Pakistan). Energies. 19961073. 14 (4). Available from: https://www.mdpi.com/1996-1073/14/4/1097/pdf. doi:10.3390/en14041097.

Purushothama, B., 2009. Air-conditioning units. In: Purushothama, B. (Ed.), Humidification and Ventilation Management in Textile Industry. Elsevier, pp. 174–193. doi: 10.1533/9780857092847.174.

Ashraf, H., Sultan, M., Farooq, M., Ibrahim, S.M., Sajjad, U., Shahzad, M.W., Khan, M.U., Jamil, M.A., 2022. Potential investigation of membrane energy recovery ventilators for the management of building air-conditioning loads. Energies. 19961073. 15 (6). Available from: https://www.mdpi.com/1996-1073/15/6/2139/pdf. doi:10.3390/en15062139.

Chancel, L., 2022. Global carbon inequality over 1990–2019. Nat. Sustain. 23989629. 5 (11), 931–938. Available from: https://www.nature.com/natsustain/. doi:10.1038/s41893-022-00955-z.

Çengel, Y.A., Ghajar, A.J., 2019. Heat and Mass Transfer: Fundamentals & Applications. McGraw-Hill Education. https://books.google.com.pk/books?id=RiXCvgEACAAJ.

Cho, H.J., Cheon, S.Y., Jeong, J.W., 2023. Experimental analysis on energy recovery ventilator with latent heat exchanger using hollow fiber membrane. Energy Convers. Manag. 01968904. 278. Available from: https://www.journals.elsevier.com/energy-conversion-and-management. doi:10.1016/j.enconman.2023.116706.

Engarnevis, A., Huizing, R., Green, S., Rogak, S., 2018. Heat and mass transfer modeling in enthalpy exchangers using asymmetric composite membranes. J. Membr. Sci. 18733123. 556, 248–262. Available from: www.elsevier.com/locate/memsci. doi:10.1016/j.memsci.2018.03.007.

Fehrm, M., Reiners, W., Ungemach, M., 2002. Exhaust air heat recovery in buildings. Int. J. Refrig. 25 (4), 439–449. doi:10.1016/S0140-7007(01)00035-4.

Huang, K., Jintian, X., Guan, J., Feng, G., Liu, X., 2023. Optimization of a collector-storage solar air heating system for building heat recovery ventilation preheating in the cold area. Energy Build. 03787788. 284, 112875. doi:10.1016/j.enbuild.2023.112875.

Jafarinejad, T., Shafii, M.B., Roshandel, R., 2019. Multistage recovering ventilated air heat through a heat recovery ventilator integrated with a condenser-side mixing box heat recovery system. J. Build. Eng. 23527102. 24, 100744. doi:10.1016/j.jobe.2019.100744.

Liu, P., Alonso, M.J., Mathisen, H.M., Halfvardsson, A., Simonson, C., 2023. Understanding the role of moisture recovery in indoor humidity: an analytical study for a Norwegian single-family house during heating season. Build. Environ. 03601323. 229, 109940. doi:10.1016/j.buildenv.2022.109940.

Noor, S., Ashraf, H., Sultan, M., Khan, Z.M., 2020a. Evaporative cooling options for building air-conditioning: a comprehensive study for climatic conditions of Multan (Pakistan). Energies. 19961073. 13 (12). Available from: https://www.mdpi.com/1996-1073/13/12/3061. doi:10.3390/en13123061.

Noor, S., Ashraf, H., Sultan, M., Miyazaki, T., Mahmood, M.H., Khan, Z.M., 2020b. Investigation of direct and indirect evaporative cooling options for greenhouse air conditioning in Multan (Pakistan). Proc. Int. Exch. Innov. Conf. Eng. Sci. 2434-1436. 6, 110–115. doi:10.5109/4102474.

Kubba, S., 2012. Impact of energy and atmosphere. In: Kubba, S. (Ed.), Handbook of Green Building Design and Construction. Elsevier, Amsterdam, The Netherlands, pp. 385–492. doi:10.1016/B978-0-12-385128-4.00009-3.

Noor, S., Ashraf, H., Hussain, G., Sultan, M., Miyazaki, T., Shakoor, A., Mahmood, M.H., Riaz, M., 2021. Spatiotemporal investigation of evaporative cooling options for

greenhouse air-conditioning application in Pakistan. Fresenius Environ. Bull. 30 (3), 2858–2870.
Shahzad, K., Sultan, M., Bilal, M., Ashraf, H., Farooq, M., Miyazaki, T., Sajjad, U., Ali, I., Hussain, M.I., 2021. Experiments on Energy-Efficient Evaporative Cooling Systems for Poultry Farm Application in Multan (Pakistan). Sustainability. 2071-1050. 13 (5), 2836. doi:10.3390/su13052836.
Sultan, M., Ashraf, H., Miyazaki, T., Shamshiri, R.R., Hameed, I.A., 2021. Temperature and Humidity Control for the Next Generation Greenhouses: Overview of Desiccant and Evaporative Cooling Systems. IntechOpen. doi:10.5772/intechopen.97273.
Ullah, H.S., Ashraf, H., Bilal, M., Sultan, M., Shabir, F., Ashraf, S., Mahmood, M.H., Niaz, Y., Khan, Z.M., 2020. Evaporative cooling systems for thermal comfort of foreign cattle breeds: THI evaluation and system feasibility. Proc. Int. Exch. Innov. Conf. Eng. Sci. 2434-1436. 6, 98–103. doi:10.5109/4102473.
U.S. DoE, 2009. EnergyPlus Engineering Reference: The Reference to EnergyPlus Calculations. Lawrence Berkeley National Laboratory.
Wang, L., Curcija, D., Breshears, J., 2015. The energy saving potentials of zone-level membrane-based enthalpy recovery ventilators for VAV systems in commercial buildings. Energy Build. 03787788. 109, 47–52. doi:10.1016/j.enbuild.2015.10.009.
Woods, J., 2014. Membrane processes for heating, ventilation, and air conditioning. Renew. Sustain. Energy Rev. 13640321. 33, 290–304. Available from: https://www.journals.elsevier.com/renewable-and-sustainable-energy-reviews. doi:10.1016/j.rser.2014.01.092.
Yüksel, A., Arıcı, M., Krajčík, M., Civan, M., Karabay, H., 2022. Energy consumption, thermal comfort, and indoor air quality in mosques: impact of Covid-19 measures. J. Clean. Prod. 09596526. 354, 131726. doi:10.1016/j.jclepro.2022.131726.
Zhang, L.Z., Niu, J.L., 2001. Energy requirements for conditioning fresh air and the long-term savings with a membrane-based energy recovery ventilator in Hong Kong. Energy. 03605442. 26 (2), 119–135. Available from: www.elsevier.com/inca/publications/store/4/8/3/. doi:10.1016/S0360-5442(00)00064-5.

CHAPTER 11

Artificial neural network-based analysis of dehumidification performance of desiccant air-conditioning system

Muhammad U.B.U. Butt[a], Muhammad Hamid Mahmood[a], Muhammad Sultan[a], Muhammad Nauman Ashraf[a], Muhammad Wakil Shahzad[b], Muhammad Ahmad Jamil[b], Haseeb Yaqoob[b], Uzair Sajjad[c], and Muhammad Farooq[d]

[a]Department of Agricultural Engineering, Bahauddin Zakariya University, Multan, Pakistan
[b]Department of Mechanical and Construction Engineering, Northumbria University, Newcastle Upon Tyne, United Kingdom
[c]Interdisciplinary Research Center for Sustainable Energy Systems (IRC-SES), King Fahd University of Petroleum and Minerals, Dhahran, Saudi Arabia
[d]Department of Mechanical Engineering, University of Engineering and Technology, Lahore, Pakistan

11.1 Introduction

Desiccant dehumidification and air conditioning can be used as an environmentally friendly and economical alternative option for several air-conditioning requirements, such as space/building cooling and agriculture product storage. However, performance prediction of such systems is very crucial before considering their potential use as an alternative to traditional air conditioning systems. Traditional approaches for the performance prediction of desiccant dehumidification include experimental, analytical, and numerical methods. The problem with numerical and analytical approaches is that they involve the use of complex equations and assumptions; on the other hand, experimental studies are laborious and costlier. Artificial intelligence systems are becoming useful computational tools for solving complicated problems over the last two decades (Jani et al., 2017).

Artificial neural networks (ANNs) became a popular and widely used tool for data prediction, clustering, classification and pattern recognition (etc.) in a variety of fields. The ANNs are one of the types of models of machine learning (ML). ANNs are incredibly feasible to traditional statistical

Sustainable Hybrid Cooling and Dehumidification Strategies for Controlled Environments
https://doi.org/10.1016/B978-0-443-48555-8.00015-3

models with respect to effectiveness (Dave and Dutta, 2014). The peak potential of the ANN that heightened its need of research in this domain is the high processing speed in parallel execution (Izeboudjen et al., 2014). Data analysis factors such as performance, processing speed, accuracy, latency, fault tolerance, stability, convergence, and volume are used to estimate full potential of the ANN (Mozaffari et al., 2019; He and Garcia, 2009). The ANNs work on the principle of human intelligence. It is in fact a mechanical version of human intelligence. The concept to develop an electronic brain first came under discussion with the invention of the first programmable computer in 1940, an intelligent machine based purely on mathematical reasoning (Buchanan, 2005). In ANN the managing elements are known as neurons, which are interlinked by links known as weights. The simplest ANN model comprises three layers: the output layer, input layer, and a single supposed hidden layer (at least one). The ANN can be developed for language processing, image recognition, and much more. Nowadays, ANNs are primarily utilized for universal function approximation in numerical paradigms due to their self-learning ability, adaptivity, tolerance for error, etc. (Wang et al., 2018). The required input data type and the structure of the network determine the selection of layers. A simple ANN comprises of links/synapses, summation unit, and squashing/activation function. The bias is applied to adjust the activation function net input. The execution of the network is established by the values of weight and bias of each neuron (Jani et al., 2017). To obtain the required output, the network is first trained with the available data sets (Rojas, 1996).

In this study, different approaches are used for the performance estimation of a desiccant dehumidification system, i.e., ANNs, adaptive neuro fuzzy inference system (ANFIS) and support vector machine (SVM). The results yielded from such techniques are further compared based on different statistical parameters such as R^2, RMSE, has to ensure MAE to ensure the correctness of the results.

11.2 Research methodology

For this study, different soft computing approaches such as ANNs, SVMs, and ANFISs are employed to develop models for prediction and analyzing the dehumidification of a solid desiccant (silica gel) employed in a DAC system. The network architecture is formed using the MATLAB neural network toolbox. The models developed for this purpose using various techniques are evaluated on the basics of different statistical parameters. Experimental data used in this study is obtained from Aleem et al. (2020).

11.2.1 Feed forward neural network (FFNN)

Feed forward neural network (FFNN) is a class of ANN. It is commonly employed as multilayer feed forward neural network (MLFFNN). A simple MLFNN is comprised of three layers; one or more hidden layer(s) are sandwiched between an input and an output layer. Each layer contains several neurons, and various layers contain nonlinear transfer functions that help the network to learn linear and nonlinear associations between input and output vectors. The MLFFNN is commonly trained by applying the backpropagation (BP) algorithm for training. The hierarchy of three said layers is such that the first layer is the input layer that contains input data, the second is the hidden layer where all the transfer functions occur, and the third is the output layer. Fundamental correlations in the input layer data are absorbed by the neural network and kept as corrected weights or strengths. The input data layer is further linked to neurons having a specific weight. Every neuron processes the data of the input layer and passes it to an activation function to generate outputs. The model is trained to adjust/develop synaptic weights that further generate predicted outputs. These modeled/predicted outputs are compared with the actual data. A variation would occur between actual and predicted outputs if the network model is not well trained. In this case, the model will adjust the weights to minimize such variations. However, after completing these preliminary weight adjustment processing, the network further starts a thorough run of each input-output data set for more precise weight adjustment. Thus, it reduces the margin of error and variations among actual and predicted outputs. The procedure is repeated till the required tolerance zone of the error function is achieved. Such excessive reiteration of learning and modification of the weights is known as the BP algorithm.

The general network configuration of the ANN model and architecture of the MLFNN containing a single hidden layer can be seen in the literature. In this regard, the mathematical interpretation of a typical ANN function can be expressed by Eq. (1) (Shafiullah and Abido, 2018).

$$x_{\text{out}} = \beta\left[\sum_{j=1}^{h} W_{kj}\alpha\left(\sum_{i=1}^{n} W_{ji}x_i + \theta_{jo}\right) + \theta_{ko}\right] \quad (1)$$

where, x_i is the ith input node external value, α refers to a nonlinear activation function of the hidden layer, β is the linear activation function of the output layer, "n" and "h" symbolize the number of nodes in the input and hidden layers, respectively, θ_{jo} and θ_{ko} represent input and output node bias units of jth hidden node and kth output, respectively. W_{kj} represents the weight of the function that correlates the jth hidden node and kth output, and W_{ji} denotes the strength of the connecting node that joins input node "i" and hidden node "j." For the training of a typical FFNN, the mean square error (MSE) is used as a function of performance.

11.2.2 Adaptive neuro-fuzzy inference system

Fuzzy logic is a type of ANN that is established on a first-order Takagi–Sugeno fuzzy inference system (TSFIS) (Abdul, 2021; Kiran and Rajput, 2011). The structures of a typical ANN and ANFIS models are quite similar. The major difference between them is the processing method. The ANN model governs the utilization of neurons in the hidden layer, while ANFIS applies fuzzy rules in the hidden layers (Mahdevari and Khodabakhshi, 2021). The ANFIS model uses a parameter optimization technique to minimize variations/errors in actual and predicted datasets. However, the development of information systems and intelligent controls is accomplished using the fuzzy logic technique. ANFIS has been widely used in different engineering applications since the last decade. The complex nonlinear and dynamic systems can be modeled using the TSFIS fuzzy model.

The framework of a standard ANFIS model is composed of five different section/layers, i.e., fuzzification, rule, normalization, defuzzification, and output layer, as shown in Fig. 11.1 (Priyadarshi et al., 2022). A similar concept can also be seen in another study (Gill and Singh, 2017). In this figure, layer 1 characterized the fuzzification layer, layer 2 characterized the rule layer, layer 3 characterized the normalization layer, layer 4 characterized the defuzzification, and layer 5 characterized the output layer. In the first (fuzzification) layer, a membership function (MF) is formed against each input, and each node in this layer is adaptive. In the TSFIS model, if α and β represent E_i and F_i, respectively, then the structure can be defined by Eq. (2) (Haznedar and Kalinli, 2018).

$$\gamma = a_i\alpha + b_i\beta + c_i \tag{2}$$

where, inputs are shown by α and β, sets of if-then fuzzy rules parameters by a_i, b_i, c_i, and output by γ.

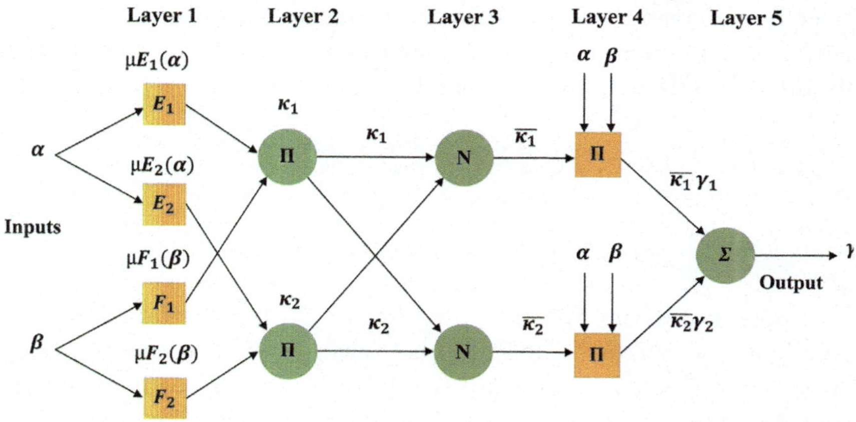

Fig. 11.1 Illustration of architecture and workflow of a simple ANFIS. *(From Priyadarshi, G., Baruah, D., Naik, B.K., 2022. Design and performance prediction of desiccant coated heat exchanger using ANFIS–AI tool and dynamic model. Appl. Therm. Eng. 119034. doi:10.1016/j.applthermaleng.2022.119034.)*

Moreover, every output node is connected to the MF and input values up to a certain degree known as membership degree. In this regard, Eqs. (3) and (4) (Haznedar and Kalinli, 2018) express the outputs of first layer nodes. The rules produced for each node by the FIS model are characterized in the second layer (rule layer). In the rule layer, cells are uniform, and cell count and rule count are equal. For every rule created in the fuzzification layer, the rule layer characterizes the weighted factor as given in Eq. (5) (Haznedar and Kalinli, 2018).

$$O_{1i} = \mu E_i(\alpha) \ldots\ldots i = 1, 2 \tag{3}$$

$$O_{1i} = \mu F_{i-2}(\alpha) \ldots\ldots i = 1, 2 \tag{4}$$

$$O_{2i} = k_i = \mu E_1(\alpha).\mu F_2(\beta), \quad i = 1, 2. \tag{5}$$

where O_{1i} shows fuzzification layer nodes and output while levels of membership among fuzzy parameters membership are given by μE_i and μF_i, O_{2i} is output for layer 2 (rule layer) and k_i is the weighing factor.

Similarly, the weighted factor of every input is gauged in the normalization layer. Normalization represents the ratio of the total sum of all the weighted factors to the ith rule weighted factor as given in Eq. (6) (Haznedar and Kalinli, 2018). Each node's output in the defuzzification layer is the creation of a normalized weighted factor and a first-order polynomial. Output obtained from the defuzzification layer can be assessed from Eq. (7) (Haznedar and Kalinli, 2018) as follows.

$$O_{3i} = \overline{k_i} = \frac{k_i}{k_1 + k_2} .. \quad i = 1, 2 \tag{6}$$

$$O_{4i} = \overline{k_i} . \ \gamma_i = \overline{k_i} . (a_i\alpha + b_i\beta + c_i) \ .. \ i = 1, 2 \tag{7}$$

where O_{3i} is the node output of layer 3/normalization layer, O_{4i} represents the output of nodes in the fourth layer, $\overline{k_i}$ is the normalized layer output, parameter set (a_i, b_i, c_i) in the fourth layer (conclusion parameters).

Finally, the output layer is used for the calculation of overall output summation from incoming signals. This layer is comprised of a single node known as the sum layer and indicated by Σ. The final output is calculated using Eq. (8) (Haznedar and Kalinli, 2018).

$$O_{5i} = \gamma = \sum \overline{k_i} . \ \gamma_i = \frac{\sum k_i . \gamma_i}{\sum k_i} \ .. \ i = 1, 2 \tag{8}$$

where O_{5i} is node outputs in layer 5

11.2.3 Support vector machine

The initial SVM presented by Vapnik (Cortes and Vapnik, 1995) is built on structural risk management (SRM) and statistic learning theory. It is categorized as a new algorithm of a neural network for forecasting (Dong et al., 2005). Therefore, its usage in research and industry is increasing because of its greater capacity to model and resolve nonlinear problems (Zhao and Magoulès, 2012). A typical SVM is composed of a collection of supervised ML models along with related kernel functions used for executing non liner classification and regression. SVM takes a low-dimensional input vector and plots it to a higher-dimensional feature space by generating an ideal separation hyperplane. It is to assure high generalization network capacity utilizing the dimension theory of Vapnik–Chervonenk as a base of operation demonstrated in Fig. 11.2 (Otchere et al., 2021).

The SRM (structural risk minimization) induction principle is used by the SVM as a statistical learning theory, and it shows a great deal of significance on the foundation of this algorithm. The SVM generates

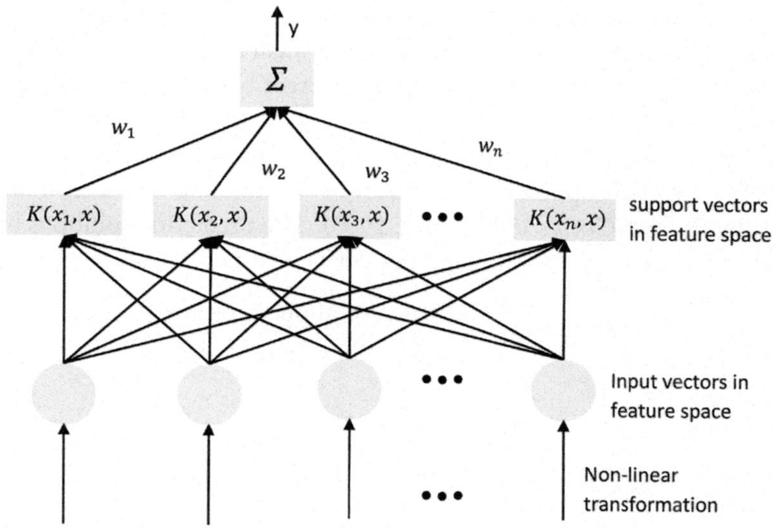

Fig. 11.2 General structure of the SVM in dimensional space. *(From: Otchere, D.A., Ganat, T.O.A., Gholami, R., Ridha, S., 2021. Application of supervised machine learning paradigms in the prediction of petroleum reservoir properties: comparative analysis of ANN and SVM models. J. Pet. Sci. Eng. 200, 108182. doi:10.1016/j.petrol.2020.108182.)*

light, more broadly generalized, and distinctive solutions with a straightforward geometric interpretation (Jones, 2018). The use of the SVM has an advantage that it can work with thin training data. Moreover, SVM is not much affected by overfitting. It is because of the margin maximization between the minimized error for the training data set and vector classes (Gharagheizi et al., 2017).

The other advantage of the SVM over traditional mathematical or statistical methods is that it can be easily used against vague or unclear data sets and shows high efficiency in nolinear relationship recognition between input and output data sets (Gholami and Ansari, 2017; Noi and Kappas, 2018). On the other hand, the performance of the SVM is centered/dependent on the regularization and kernel function parameters. This dependency on regularization and kernel function parameters is the major flaw in it (Anifowose et al., 2015). The regression practice of the SVM is established on minimization of architectural hazards as described by Vapnik (Cortes and Vapnik, 1995). It can be obtained by concurrently ignoring the data for training outside the model margin and governing the model complexity. However, this approach only essentially employs a tiny section of the training data for the output data set prediction. The ideal hyperplane acts as a maximal margin, using small data points for training to define a separation margin with minimum error between the two classes of vectors.

As the hyperplane is achieved, the supposed probability error value (i.e., ratio of expected support vector values to the training data constraints) is given by Eq. (9) (Cortes and Vapnik, 1995). The prediction solutions by the SVM are given by Eq. (10) (Tipping, 2001).

$$E[P_r(\text{error})] \leq \frac{E(\text{no. of support vectors})}{\text{no. of training vectors}} \tag{9}$$

$$y(x) = \sum_{n=1}^{N} w_n K(x, x_n) + w_o \tag{10}$$

where, w_n shows weights of the model and $K\ (x,\ x_n)$ represents the kernel function.

The fundamental notion of support vector cataloging is the employment of a learning method that establishes the ideal hyperplane separating two distinct courses in an n-dimensional space. It runs on the concept of learning the parallels between the groups rather than the differences and uses them as a support vector, unlike other algorithms, which use differences (Ccoicca, 2013). Margin is a term defined as the gap between the hyperplane dividing the different classes. The margin always needs to be maximized for better classification. An optimal SVM is found on the bases of a_i and b by minimization. The minimization is consequently optimized and expressed by Eq. (11) (Tang et al., 2020).

$$\max\left(\sum_{i=1}^{N} a_i - \frac{1}{2}\sum_{i=1}^{n} \times \sum_{j=1}^{n} y_i a_i k\left(\overrightarrow{x_i}, \overrightarrow{x_j}\right) y_i a_i\right) \text{such that}$$

$$\sum_{i=1}^{n} a_i y_i = 0 \text{ where } 0 \leq a_i \leq C,\ \ i = 1, \ldots, n \tag{11}$$

where, $k\left(\overrightarrow{x_i}, \overrightarrow{x_j}\right)$ shows kernel function and C represents regularization parameter

11.2.4 Model performance assessment parameters

Models generated using different techniques are checked to determine its usability/acceptability for the purpose. In this regard, different statistical parameters are used to check the goodness of the model and their predicted output (Betiku et al., 2021). The selection of an ideal statistical parameter for model performance evaluation is mainly driven by the model type utilized in the study and the type of outcomes it provides. However, it is suggested that the model performance evaluation also include a relative error measure, one absolute error measure, and measurement of fit (Legates and McCabe, 1999). Therefore, different statistical parameters are utilized in this study for the performance evaluation of the developed models. These parameters include, Coefficient of determination (R^2), mean absolute error (MAE), and RMSE. The R^2 is given by Eq. (12) (Hosoz et al., 2007).

$$R^2 = 1 - \left(\frac{\sum_{i=1}^{N} (x_i - y_i)^2}{\sum_{i=1}^{N} (y_i)^2} \right) \tag{12}$$

where, x_i and y_i represents actual output nodes and obtained/predicted outputs, respectively. R^2 ranges between 0 and 1. The best is 1 or near to 1 whereas the poorest is zero. The MAE is presented in Eq. (13) (Abedinpour et al., 2012). Lesser value of MAE represents the better performance of the models.

$$\text{MAE} = \frac{\sum_{i=1}^{N} \left(\theta_{\text{predicated}} - \theta_{\text{experimental}}\right)}{N} \tag{13}$$

where, N is the number of iterations, θ is the input signal for projected and experimental values.

The root mean square error (RMSE) is presented in Eq. (14) (Hosoz et al., 2007), where, x_i and y_i are number of nodes for actual output and predicted output (Hosoz et al., 2007). The variance between the projected value and regression line is determined by the RMSE. Lower value of RMSE reflects that model have minimum or small error.

$$\text{RMSE} = \sqrt{\frac{1}{N} \sum_{i=1}^{N} (x_i - y_i)^2} \tag{14}$$

11.3 Results

11.3.1 FFNN models

A feed forward NN model is developed for dehumidification performance estimation of the solid desiccant (silica gel) material used in the DAC system. Generally, the model is trained by using 70% of the available data, while the rest is used for testing the model. However, data such as 5%, 10%, and 15%, can be used for validating purposes. The input model is trained using different algorithms such as Levenberg–Marquardt, Scaled Conjugate Gradient, and Bayesian Regularization. The data for training is obtained from literature. The number of hidden layers in developed models is selected by the trial-and-error method.

The graphical diagram of the FFNN used in the study is demonstrated in Fig. 11.3. It contains three input layers, ten hidden layers, and two output layers. Moreover, "*w*" and "*b*" are representing weights and biases, respectively. In this study, different FFNN models with several training algorithms and varying number of hidden layers are generated to find out the most suitable model for the output data sets. It is determined that the FFNN model with tainlm algorithm with 10 hidden neurons yields the best output data set among the generated models. The prediction results of this model in comparison with the actual data for desiccant outlet humidity ratio (X_{out}) (g/kg of dry air) are shown in Fig. 11.4. Here the *y*-axis, in the figure represents the observed and network-modeled/predicted outputs. Whereas the *x*-axis corresponds to the number of observations. It can be seen that (Fig. 11.4.) the model predicted, and actual/observed humidity ratio values are in a close relation and both values show almost similar trend. Moreover, statistical parameters i.e., R^2, RMSE, and MAE, also show satisfactory values for X_{out} (g/kg of dry air) prediction. In this regard, R^2, RMSE, and MAE are calculated as 0.9901, 0.0282, and 0.0244, respectively.

The experimental output and FFNN-predicted output for desiccant outlet temperature (T_{out}) are shown in Fig. 11.5. The y-axis here denotes the observed and FFNN-predicted profiles of the (T_{out}) (°C), whereas the x-axis corresponds to the number of observations. It is evident that the predicted values (Fig. 11.5) represent flocculation as compared to actual data. It means that the model may over and/or underpredict the data at some observations. However, the overall prediction is close to the actual

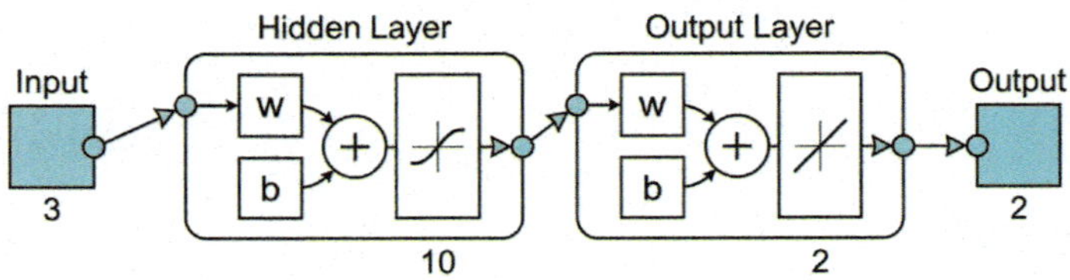

Fig. 11.3 Pictorial representation of structure of the developed feed-forward neural networking (FFNN) model in MATLAB.

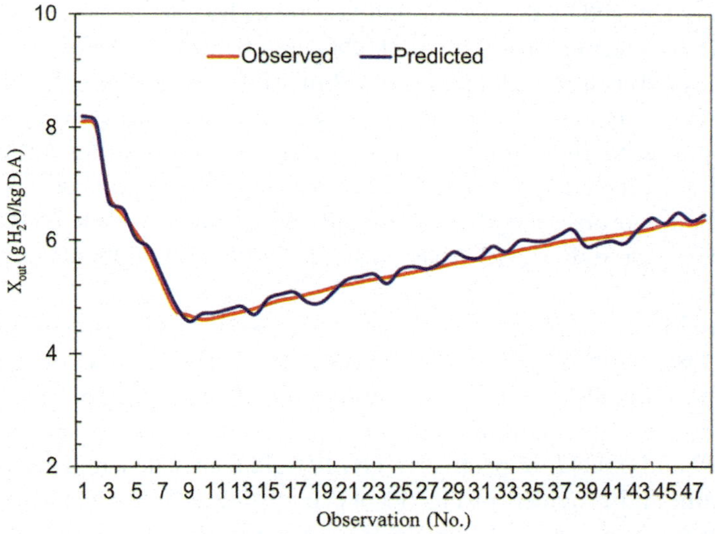

Fig. 11.4 Comparison of observed/experimental and FFNN model predicted humidity ratio (g/kg of dry air) profiles.

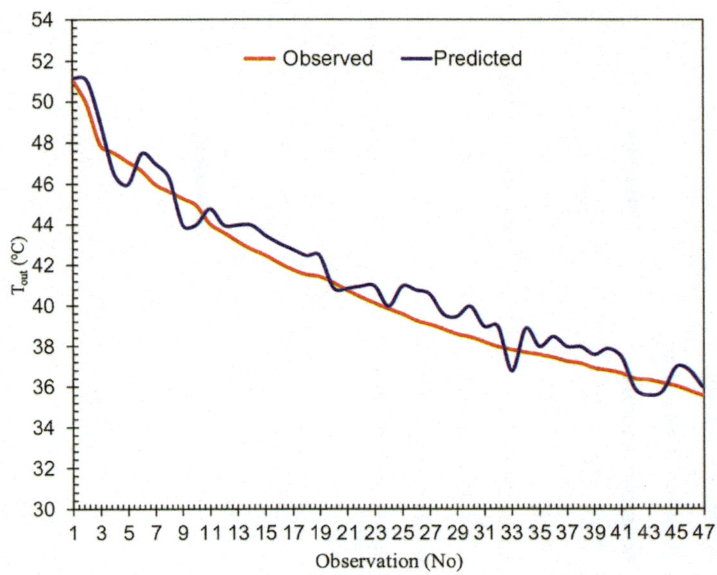

Fig. 11.5 Profiles of experimental/observed and FFNN-predicted temperatures.

values as depicted by the statistical indicators such as, R^2, RMSE, and MAE, which are determined as 0.9976, 0.0283, and 0.0242, respectively.

The RMSE against different algorithms of the FFNN model is shown in Fig. 11.6A. The y-axis denotes the RMSE values, whereas the x-axis corresponds to the algorithms used. The minimum value (0.0282) of RMSE is observed against the algorithm trainlm-10 (that with 10 number of hidden neurons). On the other hand, the maximum value of RMSE is calculated for trainscg-15 (that with 15 number of hidden neurons) as 1.6564.

The MAE values are also calculated against each algorithm, as shown in Fig. 11.6B. Like RMSE, the minimum value of MAE was observed for trainlm-10, while the maximum value was determined for trainscg-15. Accordingly, the maximum and minimum values of MAE are calculated as 1.399 and 0.0242, respectively. The R^2 values are calculated against each algorithm and provided in Fig. 11.6C. It is observed that the error minimum for trainlm-10 as calculated 0.9976. On the other hand, the maximum value of error is 0.5628 against trainscg-15.

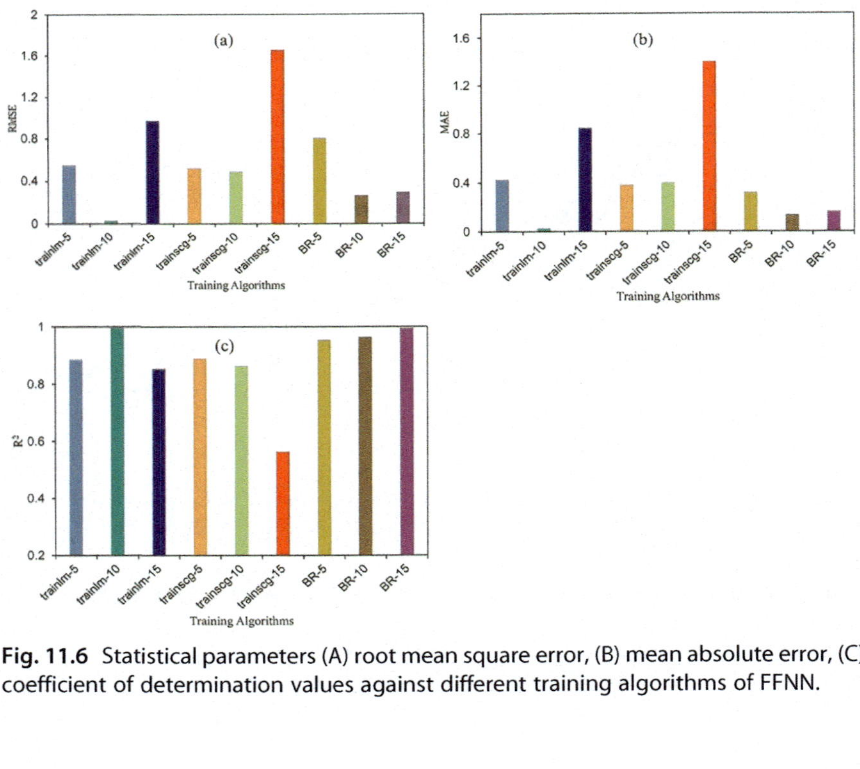

Fig. 11.6 Statistical parameters (A) root mean square error, (B) mean absolute error, (C) coefficient of determination values against different training algorithms of FFNN.

11.3.2 Adaptive neuro-fuzzy inference system

An ANFIS works by modifying the MFs and selecting the fuzzy rules that dictate the decision-making process.

However, there is no rule of thumb available when it comes to selecting the parameters such as MF, type of membership containing the total number of MF, fuzzy inference system (FIS), number of epochs, and error tolerance. Therefore, the trial-and-error method dictates these parameters. Various kinds of MFs used in ANFIS are shown in Fig. 11.7.

The figure (Fig. 11.7) shows a variety of MFs, i.e., triangular MF (trimf), trapezoidal MF (trapmf), Gaussian MF (gaussmf), piecewise linear MF (plmf), and generalized bell-shaped MF (gbellmf). In this study, four most widely used MFs are considered. These MFs vary from each other based on their shape. In a typical fuzzy system, TSFIS and Mamdani (M) inference systems are commonly utilized. Both systems operate with different procedures but the TSFIS gives more accurate results as compared to the others (Takagi and Sugeno, 1985). After selecting the MF and fuzzy system, the generated FIS is trained by choosing the minimum error tolerance and number of epochs. The general structure of the FIS with 14 rule layers is shown in Fig. 11.8.

A comparison of the observed/experimental and ANFIS-predicted humidity ratio (X_{out}) (g/kg of dry air) is shown in Fig. 11.9. The y-axis in the figure represents the observed/experimental and ANFIS-predicted humidity ratio values, whereas the x-axis corresponds to the number of observations. The figure represents that the model does not predict the data as quite well as the FFNN. Some values are overpredicted, while the others are underpredicted by the model. The statistical parameters, i.e., R^2, RMSE, and MAE, for the humidity ratio prediction are 0.7625, 0.5163, and 0.4528, respectively.

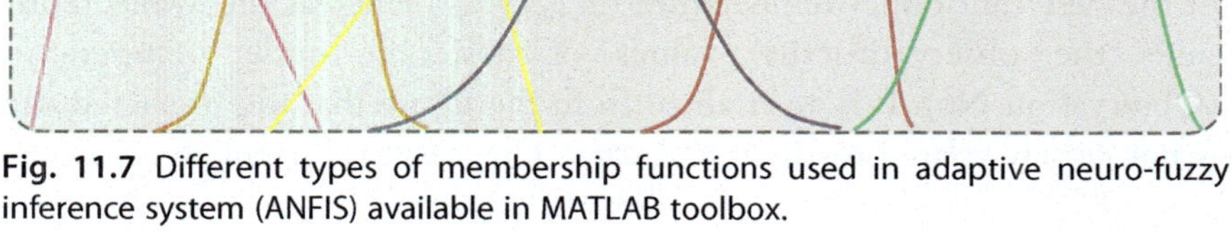

Fig. 11.7 Different types of membership functions used in adaptive neuro-fuzzy inference system (ANFIS) available in MATLAB toolbox.

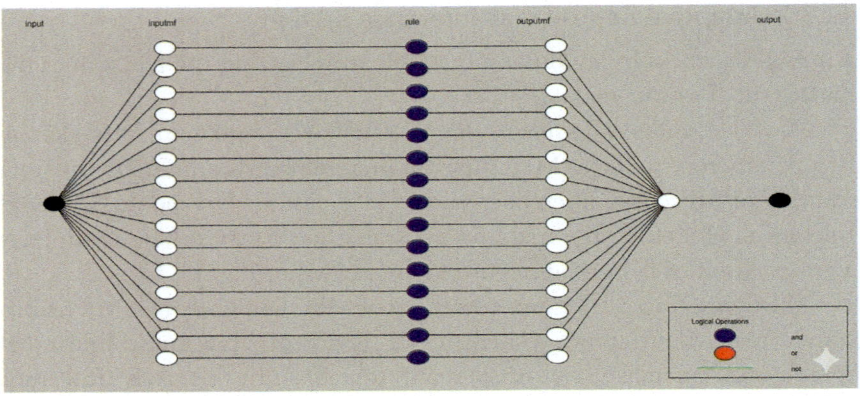

Fig. 11.8 FIS Structure with 14 rule layers developed in MATLAB.

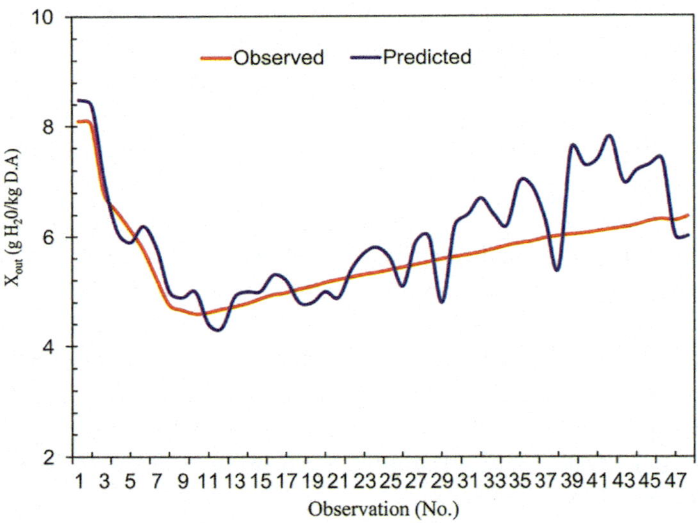

Fig. 11.9 The trends of observed/experimental and ANFIS predicted humidity ratio profiles.

Similarly, a comparison of observed/experimental and ANFIS-predicted dry bulb temperature at the desiccant outlet is shown in Fig. 11.10. The y-axis in the figure represents the network-modeled values of the desiccant outlet temperature (T_{out}) (°C), whereas the *x*-axis represents the observed/actual values of desiccant outlet temperature (Observation No). It is ascertained from the figure that the predicted data is not closely related to the actual data. The statistical parameters i.e., R^2,

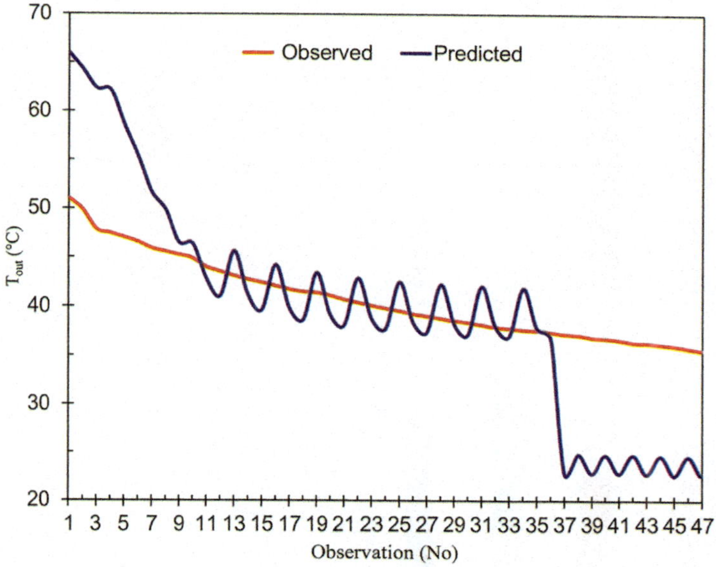

Fig. 11.10 The comparison of observed/experimental and ANFIS-predicted desiccant outlet temperature profiles.

RMSE, and MAE, are found to be 0.8512, 0.4631, and 0.4521, respectively.

Moreover, a comparison of RMSE, MAE, and R^2 values for different membership functions is also presented in this study. The RMSE values against different MFs are illustrated in Fig. 11.11A. The figure represents the RMSE value on the *y*-axis and the membership functions on the x-axis. It is clear from the figure that the RMSE value is minimum for gbellmf (Generalized Bell Membership Function). The RMSE value obtained using gbellmf is 0.4631, and it's found to be the best value. On the other hand, the worst value of RMSE is obtained in trapmf (Trapezoidal MF) and calculated as 6.4171. Similarly, a comparison of MAE values against different membership functions is displayed in Fig. 11.11B. It can be clearly examined from the figure that the error is minimum for gbellmf (Generalized Bell Membership Function). The MAE value obtained under gbellmf is 0.4521 and it's found to be the best value. On the other hand, the MAE value obtained using trapmf (Trapezoidal MF) is 9.6732 and found to be the worst. The R^2 against different membership functions is demonstrated in Fig. 11.11C. It can be clearly examined from the figure that the error is minimum for gbellmf (Generalized Bell MF) and calculated as

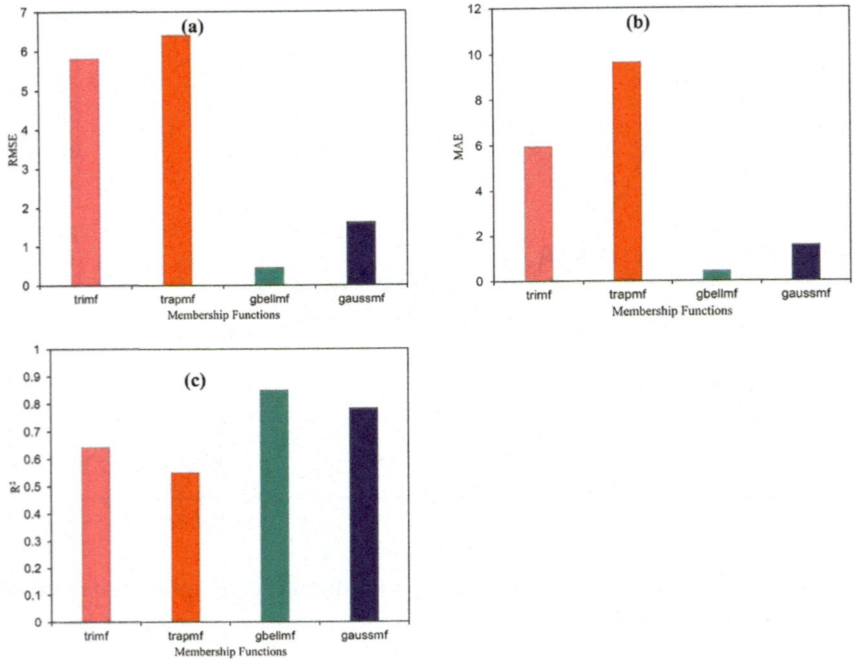

Fig. 11.11 Comparison of (A) root mean square error, (B) mean absolute error, and (C) coefficient of determination values among different membership functions of ANFIS.

0.8512. On the other hand, the MAE value obtained using trapmf (Trapezoidal MF) is 0.5514, and it is found to be the worst.

It can be easily deciphered from the graphs that ANFIS models yield near-acceptable values of statistical indicators but not satisfactory to fulfill the required criteria for the modeling. The ANFIS models can be used for modeling the desiccant air conditioning system, but as compared to FFNN models, they do not generate as likeable outputs. Another problem with the ANFIS models is that the processing time is greater than that of FFNN.

11.3.3 Support vector machine

SVM modeling works more effectively with a different input data division. Therefore, input data was divided such that 80% of input data was used for training and the remaining 20% for testing and validation, i.e., 10% for each. The main purpose of the training set is to understand and learn the model. While the validation is used to check the performance of the model. Due to its frequent evaluation, which is used to update the hyperparameters, it is said that the data sets for validation can indirectly affect the model. The training data set provides impartial evaluation of the final model. However, the testing data set does not impact the model as it is only used to estimate the performance. In this study, six distinct support vector machine models, i.e., linear, quadratic, cubic, fine, medium, and coarse Gaussian are used to predict subjective parameters. To compare the accuracy of outputs from each of these models, various sorts of validation are conducted. All support vector machine models are trained using the specification, i.e., 5 folds for cross-validation.

The graphical representation of the relationship between the true and predicted response for (A) linear SVM, (B) quadratic SVM, (C) cubic SVM, (D) fine Gaussian SVM, (E) medium Gaussian SVM, and (F) coarse Gaussian SVM models is shown in Fig. 11.12. The x-axis represents the true response, while the y-axis corresponds to the projected response. However, a best-fit line is also shown on the graph to represent the deviation of prediction from the observed/experimental data set. Fig. 11.12A shows that the linear SVM model has predicted the values quite effectively, which shows that the model is good for this data set. Moreover, the values of statistical indicators i.e., RMSE, MAE, R^2 are also satisfactory and calculated as 0.58265, 0.45106, and 0.96, respectively. Similarly, the relationship between true and predicted responses for quadratic SVM is shown in Fig. 11.12B. The figure shows that the model has predicted the values quite effectively and the model is good for this data set at five number of cross validations. The values of the statistical parameters RMSE, MAE, and R^2 used to check the performance of the model are 0.28231, 0.21904, and 0.97, respectively. Moreover, the relationship between true and predicted responses for the cubic SVM prediction model is shown in Fig. 11.12C. The figure shows that the model has predicted the values quite effectively, meaning the model is good for this data set at five numbers of cross-validations.

The values of the statistical parameters (RMSE, MAE, and R^2) used to check the performance of the model are 0.28139, 0.2013, and 0.99, respectively. Moreover, the relationship between true and predicted responses for the cubic SVM prediction model is shown in Fig. 11.12C. The figure shows that the model has predicted the values quite effectively meaning the model is good for this data set at five number of cross validations. The values of the statistical parameters (RMSE, MAE, R^2) used to check the performance of the model are 0.28139, 0.2013, and 0.99, respectively. The relationship between true and predicted responses calculated under fine Gaussian SVM model is shown in Fig. 11.12D. The figure shows that the model did not

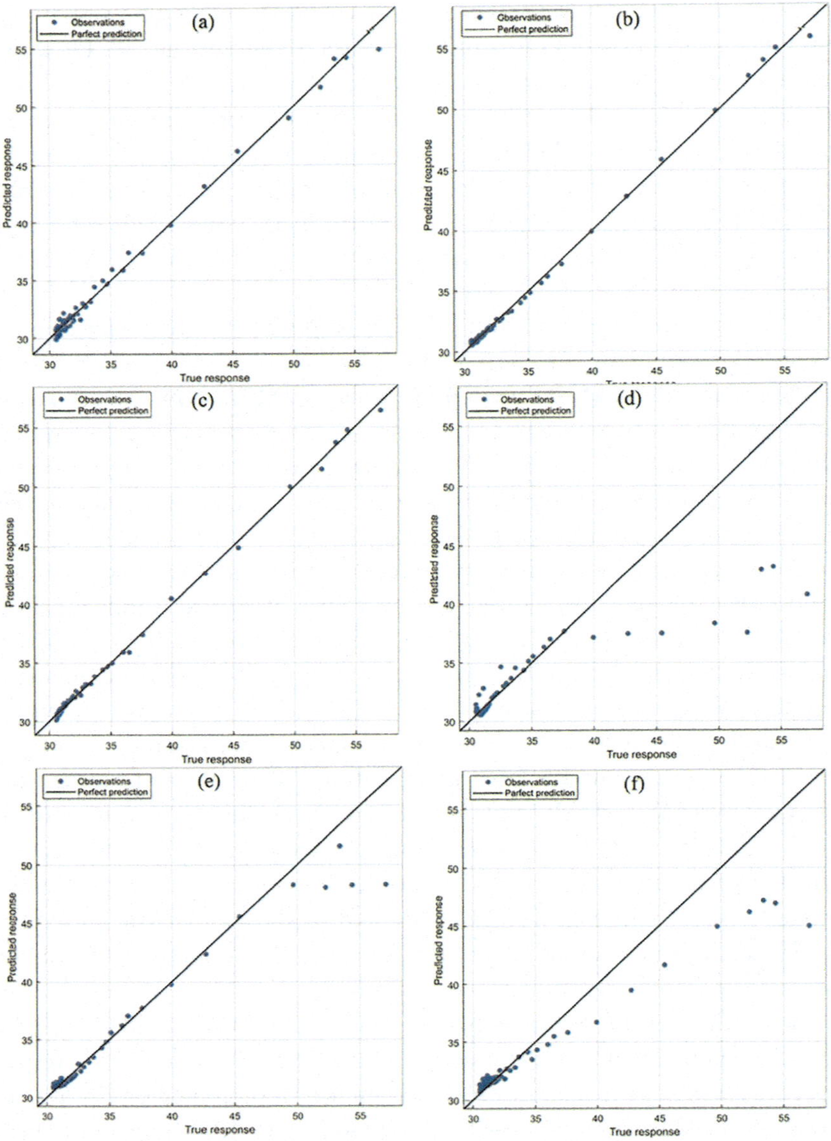

Fig. 11.12 Relationship between true and predicted responses under (A) linear SVM, (B) quadratic SVM, (C) cubic SVM, (D) fine Gaussian SVM, (E) medium Gaussian SVM, and (F) coarse Gaussian SVM models.

predict the values in the permissible range of modeling. The values of the statistical parameters (RMSE, MAE, R^2) used to check the performance of the model are 4.4841, 2.007, and 0.60, respectively. Medium Gaussian SVM model is shown in Fig. 11.12E. The figure shows that the model has predicted the values in the permissible range of R^2 but not for the other statistical parameters. The values of the statistical parameters (RMSE, MAE, R^2) used to check the performance of the model are 1.719, 0.69641, and 0.94, respectively. Moreover, the relationship between true and predicted responses calculated under coarse Gaussian SVM model is shown in Fig. 11.12F. The figure shows that the model has predicted the values in the permissible range of R^2 but not for the other statistical parameters. The values of the statistical parameters (RMSE, MAE, R^2) used to check the performance of the model are 2.7028, 1.3948, and 0.85, respectively.

The analysis based on the above results shows that the cubic SVM generates the best results for the performance prediction of solid desiccant. The comparison of RMSE values calculated under different SVM models is shown in Fig. 11.13A. The figure represents the RMSE value on the *y*-axis and the type of models used on the *x*-axis. It can be clearly examined from the figure that the error is minimum for cubic SVM and found to be 0.28139. On the other hand, the highest/worst value of RMSE is found to be 4.4841 in the fine Gaussian SVM model. Similarly, the comparison of MAE values calculated under different SVM models is shown in Fig. 11.13B. It can be clearly examined from the figure that the error is minimal for cubic SVM. The MAE value obtains by using Cubic SVM is 0.2013 and it's found to be the best value. On the other hand, the MAE value obtained by using fine Gaussian SVM is 2.007, and it's found to be the worst. Finally, the comparison of R^2 values calculated under different SVM models is shown in Fig. 11.13C. It can be clearly observed from the figure that the value is highest for cubic SVM and calculated as 0.99 which is found to be the best value among all the models. On the other hand, the R^2 value obtained using a fine Gaussian SVM is the worst among all and found to be 0.60.

In conclusion, it is ascertained that most of the SVM models have good prediction performances that are satisfactory to be used for prediction modeling. However, the prediction performance of FFNN models is superior to SVM and ANFIS models.

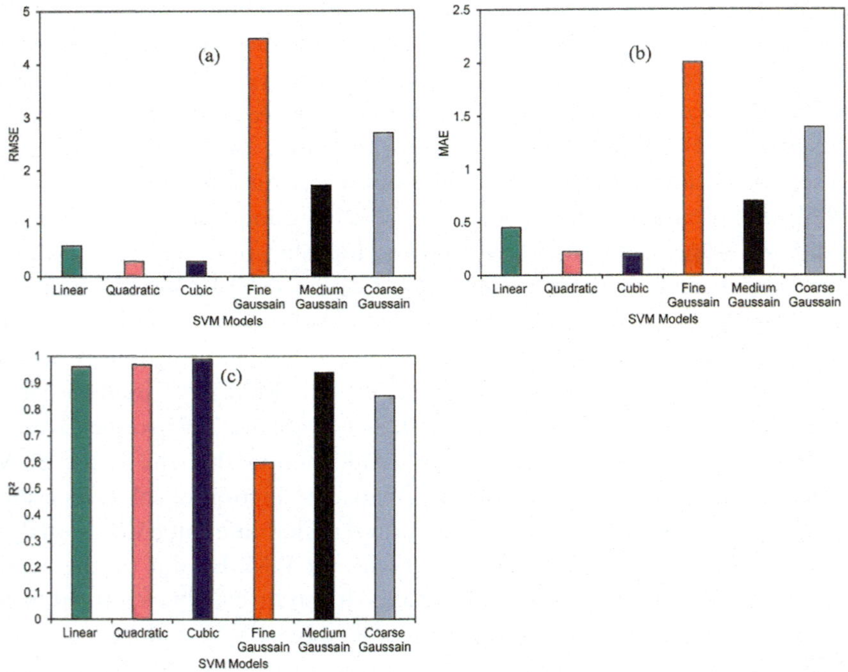

Fig. 11.13 Comparison of (A) root mean square error, (B) mean absolute error, and (C) coefficient of determination values among different SVM models.

11.4 Discussion

The performance prediction of a desiccant air conditioning system was analyzed using various ANN techniques, with experimental data from literature serving as the basis for training and testing. The results demonstrate that the ANFIS exhibited limited predictive accuracy, as indicated by relatively low statistical performance metrics ($R^2=0.8512$, RMSE = 0.4631, MAE = 0.4521), suggesting its inadequacy for modeling the nonlinear behavior of solid desiccant dehumidification processes. In contrast, the Support Vector Machine (SVM), particularly the cubic kernel variant, showed significantly improved performance ($R^2=0.99$, RMSE = 0.28139, MAE = 0.2013), highlighting its suitability for handling complex multidimensional data. However, the FFNN, optimized using the Levenberg–Marquardt (trainlm) algorithm with 10 hidden neurons, achieved the highest accuracy ($R^2=0.9976$, RMSE = 0.0282, MAE = 0.0242), thereby establishing itself as the most effective model among those studied. These findings underscore the potential of FFNN as a superior alternative to conventional methods for predicting the performance of solid desiccant systems. Although both FFNN and SVM demonstrated strong predictive capabilities, the choice between them may depend on specific application contexts. Furthermore, the integration of larger experimental datasets in future research could enhance model generalization and reliability, while the exploration of hybrid or ensemble learning techniques may further advance the accuracy of performance prediction in HVAC applications.

11.5 Conclusions

Performance of the desiccant air conditioning system is predicted and analyzed using different ANN techniques. The experimental data from literature were used for training and testing of different ANN models. The performance of these models was further compared based on different statistical performance indicators. ANFIS did not predict the data within the acceptable margin of error, yielding and R^2, RMSE, and MAE of 0.8512, 0.4631, and 0.4521, respectively. SVM is a better technique than ANFIS for the prediction of solid desiccant dehumidification, achieving an RMSE, MAE, and R^2 of 0.28139, 0.2013, and 0.99, respectively. FNN predicted the solid desiccant dehumidification performance effectively, particularly the trainlm (Levenberg-Marquardt) training algorithm with 10 hidden neurons, which yielded an R^2, RMSE, and MAE of 0.9976, 0.0282, and 0.0242, respectively. The ANN model and the cubic SVM yield the best overall result for the dehumidification performance of solid desiccants. However, the studied techniques have preferences over each other based on the statistical parameters. FFNN could be used for performance prediction of solid desiccant instead of conventional approaches. Larger experimental datasets may be deployed in the FFNN models for optimal training and prediction.

Acknowledgment

The United Kingdom Government funds this research (and/or) pilot activity through the Project, APP47457, titled "Super–efficient Sustainable Cooling Solution for All Applications (S2Cool)," under the Ayrton Challenge Programme of the UK Research and Innovation (UKRI); however, the views expressed do not necessarily reflect the UK Government's official policies. The Project, APP47457, is implemented/led by Northumbria University, UK, together with partners to develop sustainable cooling solution for several applications.

Funding Statement

The United Kingdom Government funds this research (and/or) pilot activity through the Project, APP47457, titled "Super–efficient Sustainable Cooling Solution for All Applications (S2Cool)," under the Ayrton Challenge Programme of the UK Research and Innovation (UKRI); however, the views expressed do not necessarily reflect the UK Government's official policies. The Project, APP47457, is implemented/led by Northumbria University, UK, together with partners to develop sustainable cooling solution for several applications.

Conflicts of interest

The authors declare no conflict of interest.

Declaration of generative AI and AI-assisted technologies in the writing process

No generative AI/AI-assisted technology is used in this paper.

Credit Authorship Contribution Statement

Muhammad U.B.U. Butt was involved in the conceptualization, methodology, software, formal analysis, investigation, writing of theoriginal draft. **Muhammad Hamid Mahmood**, **Muhammad Sultan**, and **Muhammad Wakil Shahzad** were involved in the conceptualization, methodology, validation, resources, visualization, supervision, project administration, writing (review and editing), funding acquisition. **Muhammad Nauman Ashraf**, **Muhammad Ahmad Jamil**, **Haseeb Yaqoob**, **Uzair Sajjad**, and **Muhammad Farooq** were involved in the conceptualization, methodology, software, formal analysis, investigation, writing (review and editing).

References

Abdul, H., 2021. Use of Soft Computing Approaches for Runoff Estimation: A Case Study of Tarbela Dam, Pakistan, Bahauddin Zakariya University.

Abedinpour, M., Sarangi, A., Rajput, T.B.S., Singh, M., Pathak, H., Ahmad, T., 2012. Performance evaluation of AquaCrop model for maize crop in a semi-arid environment. Agric. Water Manag. 110, 55–66. doi:10.1016/j.agwat.2012.04.001.

Aleem, M., et al., 2020. Experimental investigation of desiccant dehumidification cooling system for climatic conditions of Multan (Pakistan). Energies 13 (21). doi:10.3390/en13215530.

Anifowose, F., Labadin, J., Abdulraheem, A., 2015. Improving the prediction of petroleum reservoir characterization with a stacked generalization ensemble model of support vector machines. Appl. Soft Comput. 26, 483–496. doi:10.1016/j.asoc.2014.10.017.

Betiku, E., et al., 2021. Performance evaluation of adaptive neuro-fuzzy inference system, artificial neural network and response surface methodology in modeling biodiesel synthesis from palm kernel oil by transesterification. Biofuels 12 (3), 339–354. doi:10.1080/17597269.2018.1472980.

Buchanan, B.G., 2005. A (very) brief history of artificial intelligence. AI Mag. 26 (4), 53.

Ccoicca, Y.J., 2013. Applications of support vector machines in the exploratory phase of petroleum and natural gas: a survey. Int. J. Eng. Technol. 2 (2), 113. doi:10.14419/ijet.v2i2.834.

Cortes, C., Vapnik, V., 1995. Support-vector networks. Mach. Learn. 20 (3), 273–297. doi:10.1007/BF00994018.

Dave, V.S., Dutta, K., 2014. Neural network based models for software effort estimation: a review. Artif. Intell. Rev. 42 (2), 295–307.

Dong, B., Cao, C., Lee, S.E., 2005. Applying support vector machines to predict building energy consumption in tropical region. Energy Build. 37 (5), 545–553. doi:10.1016/j.enbuild.2004.09.009.

Gharagheizi, F., Mohammadi, A.H., Arabloo, M., Shokrollahi, A., 2017. Prediction of sand production onset in petroleum reservoirs using a reliable classification approach. Petroleum 3 (2), 280–285. doi:10.1016/j.petlm.2016.02.001.

Gholami, A., Ansari, H.R., 2017. Estimation of porosity from seismic attributes using a committee model with bat-inspired optimization algorithm. J. Pet. Sci. Eng. 152, 238–249. doi:10.1016/j.petrol.2017.03.013.

Gill, J., Singh, J., 2017. Performance analysis of vapor compression refrigeration system using an adaptive neuro-fuzzy inference system. Int. J. Refrig. 82, 436–446. doi:10.1016/j.ijrefrig.2017.06.019.

Haznedar, B., Kalinli, A., 2018. Training ANFIS structure using simulated annealing algorithm for dynamic systems identification. Neurocomputing 302, 66–74. doi:10.1016/j.neucom.2018.04.006.

He, H., Garcia, E.A., 2009. Learning from imbalanced data. IEEE Trans. Knowl. Data Eng. 21 (9), 1263–1284. doi:10.1109/TKDE.2008.239.

Hosoz, M., Ertunc, H.M., Bulgurcu, H., 2007. Performance prediction of a cooling tower using artificial neural network. Energy Convers. Manag. 48 (4), 1349–1359. doi:10.1016/j.enconman.2006.06.024.

Izeboudjen, N., Larbes, C., Farah, A., 2014. A new classification approach for neural networks hardware: from standards chips to embedded systems on chip. Artif. Intell. Rev. 41 (4), 491–534.

Jani, D.B., Mishra, M., Sahoo, P.K., 2017. Application of artificial neural network for predicting performance of solid desiccant cooling systems—a review. Renew. Sustain. Energy Rev. 80, 352–366. doi:10.1016/j.rser.2017.05.169.

Jones, H., 2018. Machine Learning: An Essential Guide to Machine Learning for Beginners Who Want to Understand Applications, Artificial Intelligence, Data Mining, Big Data and More. CreateSpace Independent Publishing Platform.

Kiran, T.R., Rajput, S.P.S., 2011. An effectiveness model for an indirect evaporative cooling (IEC) system: comparison of artificial neural networks (ANN), adaptive neuro-fuzzy inference system (ANFIS) and fuzzy inference system (FIS) approach. Appl. Soft Comput. 11 (4), 3525–3533. doi:10.1016/j.asoc.2011.01.025.

Legates, D.R., McCabe, G.J., 1999. Evaluating the use of 'goodness-of-fit' measures in hydrologic and hydroclimatic model validation. Water Resour. Res. 35 (1), 233–241. doi:10.1029/1998WR900018.

Mahdevari, S., Khodabakhshi, M.B., 2021. A hybrid PSO-ANFIS model for predicting unstable zones in underground roadways. Tunn. Undergr. Sp. Technol. 117, 104167. doi:10.1016/j.tust.2021.104167.

Mozaffari, A., Emami, M., Fathi, A., 2019. A comprehensive investigation into the performance, robustness, scalability and convergence of chaos-enhanced evolutionary algorithms with boundary constraints. Artif. Intell. Rev. 52 (4), 2319–2380.

Noi, P.T., Kappas, M., 2018. Comparison of random forest, k-nearest neighbor, and support vector machine classifiers for land cover classification using sentinel-2 imagery. Sensors 18 (1). doi:10.3390/s18010018.

Otchere, D.A., Ganat, T.O.A., Gholami, R., Ridha, S., 2021. Application of supervised machine learning paradigms in the prediction of petroleum reservoir properties: comparative analysis of ANN and SVM models. J. Pet. Sci. Eng. 200, 108182. doi:10.1016/j.petrol.2020.108182.

Priyadarshi, G., Baruah, D., Naik, B.K., 2022. Design and performance prediction of desiccant coated heat exchanger using ANFIS–AI tool and dynamic model. Appl. Therm. Eng., 119034. doi:10.1016/j.applthermaleng.2022.119034.

Rojas, R., 1996. Neural Networks: A Systematic Introduction, Springer.

Shafiullah, M., Abido, M.A., 2018. S-transform based FFNN approach for distribution grids fault detection and classification. IEEE Access 6, 8080–8088.

Takagi, T., Sugeno, M., 1985. Fuzzy identification of systems and its applications to modeling and control. IEEE Trans. Syst. Man Cybern. 15 (1), 116–132. doi:10.1109/TSMC.1985.6313399.

Tang, L., Zhang, M., Wen, L., 2020. Support vector machine classification of seismic events in the Tianshan Orogenic Belt. J. Geophys. Res. Solid Earth. 125 (1). doi:10.1029/2019JB018132.

Tipping, M.E., 2001. Sparse Bayesian learning and the relevance vector machine. J. Mach. Learn. Res. 1 (Jun), 211–244.

Wang, D., He, H., Liu, D., 2018. Intelligent optimal control with critic learning for a nonlinear overhead crane system. IEEE Trans. Ind. Informatics 14 (7), 2932–2940. doi:10.1109/TII.2017.2771256.

Zhao, H., Magoulès, F., 2012. A review on the prediction of building energy consumption. Renew. Sustain. Energy Rev. 16 (6), 3586–3592. doi:10.1016/j.rser.2012.02.049.

CHAPTER 12

Application of solar chimney assisted passive cooling in postharvest storage

Muhammad Irfan[a], Muhammad Hamid Mahmood[a], Muhammad Sultan[a], Muhammad Khalid[a], Muhammad Wakil Shahzad[b], Muhammad Ahmad Jamil[b], Haseeb Yaqoob[b], Uzair Sajjad[c], and Naveed Husnain[d]

[a]Department of Agricultural Engineering, Bahauddin Zakariya University, Multan, Pakistan
[b]Department of Mechanical and Construction Engineering, Northumbria University, Newcastle Upon Tyne, United Kingdom
[c]Interdisciplinary Research Center for Sustainable Energy Systems (IRC-SES), King Fahd University of Petroleum and Minerals, Dhahran, Saudi Arabia
[d]Department of Mechanical Engineering, Bahauddin Zakariya University, Multan, Pakistan

12.1 Introduction

The agricultural products are of a perishable nature and may rapidly spoil after harvest. It is because of high moisture content contained in most agricultural products. The moisture contents are reported to be about 60%–95% in the literature (Muhammad et al., 2016). Postharvest agricultural products constantly perform the respiration and transpiration processes (Mahmood et al., 2019). In ambient conditions the acid contents converted to soluble solids due to respiration (Muhammad et al., 2016). In the transpiration process, the water is continually removed from the agricultural product, and this process mainly depends upon the relative humidity. These processes may not be stopped but slow down by providing suitable temperature and relative humidity conditions. Most agricultural products need low temperature and high relative humidity (Mahmood et al., 2019). Mostly, the agricultural products contain the water content, which determines the freshness of agricultural products. Agricultural products shelf life becomes lower due to loss of moisture content, which results in bruising/degradation. The spoilage is also related to ambient temperature, relative humidity, improper postharvest management, and many other factors. The agriculture products contain important nutrients, minerals, vitamins, and fibers, which are necessary for the human body. Improper availability of

Sustainable Hybrid Cooling and Dehumidification Strategies for Controlled Environments
https://doi.org/10.1016/B978-0-443-48555-8.00009-8

these nutrients and vitamins for the human body causes many diseases in the human body (Islam and Morimoto, 2015a). The shortage of agricultural products caused 2.7 million deaths and 1.8% diseases in the world annually due to malnutrition (Lock et al., 2005). The lack of food causes the severe diseases/illness, as a balanced diet is needed for better health of humans. Agricultural products are the main way of consumable minerals and vitamins in the human body. In the developing countries, one-fifth of food processing and manufacturing industry units use agricultural products as a raw material. The agricultural products are available in short term of year, but the need of products remain whole the year. The proper storage fulfills the requirements for availability of products throughout the year (Ogbuagu et al., 2017).

The passive evaporative cooling systems can be used in buildings for the thermal comfort of humans and animals and for the storage of vegetable/fruit, etc. In the passive evaporative cooling system, the temperature of ambient air is reduced by the evaporation of water molecules. The latent heat of vaporization required during the evaporation process is gained from the dry air, which reduces the temperature of ambient air (Emdadi et al., 2016). The passive evaporative cooling chamber (PCC) used to decrease the temperature of the surroundings. Passive cooling methods are efficient and best for the environment because these systems cannot produce the hazardous gases that are harmful for human beings (Samuel et al., 2013). A natural ventilation system is the essential sustainable building strategy that was recognized in ancient times. The interest in natural ventilation systems is increasing gradually due to their potential benefits over the mechanical aeration system in terms of energy consumption and environmental effects. A huge amount of energy is required to run the mechanical aeration system. Mechanical aeration system is not suitable due to the emission of carbon dioxide, which has a harmful effect on the environment. A natural ventilation system provides a good quality of air, which is suitable for the thermal comfort of human beings (Khanal and Lei, 2011). The developing countries have limited sources of energy, especially the problems of electricity in the rural areas. The humid countries that have limited sources of energy to run the air conditioner system may depend on the natural ventilation system (Reddy et al., 2012). The main part of primary energy 30%–40%, is used in the buildings' aeration worldwide (Dodoo et al., 2011). A natural aeration system has the capability to reduce the consumption and cost of energy as compared to a mechanical ventilation system. The natural ventilation system consumes low or no energy; due to this, the system is being progressively proposed in terms of saving energy and improving the inside quality of buildings. Natural ventilation is used to save 10%–15% energy annually with suitable operational and climatic conditions. A passive ventilation system is an alternative to a mechanical ventilation system in terms of energy savings, operational cost, and low emission of gases. Moreover, passive ventilation with a solar chimney is an efficient way to increase the rate of natural aeration (Khanal and Lei, 2011). The control of temperature and humidity is important for the thermal comfort of human beings. In daily life air conditioning is required for thermal comfort in the sense of ventilation, cooling, heating, dehumidification, and humidification that consume a huge amount of energy (Sultan et al., 2015). Nowadays, air conditioning system is the basic requirement for buildings, offices, transports, and shopping malls. In this regard, the researchers have investigated the various technologies for the thermal comfort (Sultan et al., 2015). A passive cooling system is a building design to develop the internal thermal comfort for humans and nonhumans with low or no consumption of energy. In developing countries, there is a huge difference between demand and supply of electricity because the demanding rate of electricity is increasing at the same rate as supply. In developing countries there is hardly any investigation into the thermal comfort for the animals. Correspondingly, in the developing countries, they do not have proper storage buildings and systems for the postharvest storage of agricultural products. It is due to the higher initial cost of conventional air conditioning systems for product storage.

The passive cooling can be used for the storage of harvested products to increase shelf life. Nowadays, growers and sellers are wasting their products due to a lack of storage structure. In a PCC, the product is stored at the farm level, and by use of this chamber without spoilage, the growers sell out their products after some time of storage in the chamber (Basediya et al., 2013). The working process in a PCC is shown in Fig. 12.1. The outside wall is made by bricks, and for evaporative media, sand is used. Water is used for sand wetting and, thereby, evaporation. When ambient air contacts the outer wall, the inner surface of wall, which is in a wet condition, absorbs the ambient air and transfers it to evaporative media. Due to this process, the water content of the wet media releases this cool air in the surroundings of the chamber. Therefore, due to evaporation of water, cooling is produced, which decreases the temperature of the chamber and increases the relative humidity.

The brinjal (*Solanum melongena*) is a perishable vegetable that belongs to the family of Solanaceae. Brinjal is a valuable agricultural product in terms of

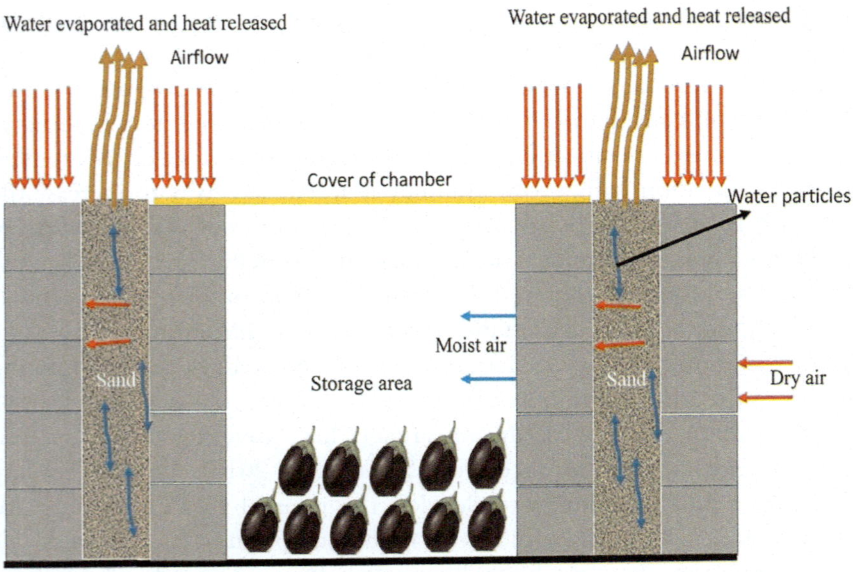

Fig. 12.1 Working principle of passive evaporative cooling chamber (PCC).

high nutritional content. The favorable temperature required for the growth of brinjal in daytime and nighttime is 27°C–32°C and 21°C–27°C, respectively (Lawande and Chavan, 1998). The temperature and relative humidity suitable for storaging brinjal are 10°C–12°C and >90%, respectively (Passam and Karapanos, 2008), and such storage conditions can be achieved by active cooling systems. Brinjal is the rich source of vitamin A, ascorbic acid, thiamine, polyphenols, and riboflavin. Moreover, brinjal is the antioxidant potential and help to avoid chronic and deteriorating disease. Brinjal has an extremely perishable nature due to this disclosure of interior cellular tissues caused by cracking, which increases the respiration rate, metabolism process, production of ethylene, loss of water content, and microbial adulteration in brinjal. Due to these processes, the storage life of brinjal decreases (Manca et al., 2017). The storage life of brinjal increases by controlling all these factors.

In this regard, a chimney-equipped PCC was built. It is a low-cost and environmentally friendly storage option for farmers and vendors. Thermodynamic evaluation of PCC was determined by a series of experiments. Furthermore, physicochemical analyses of stored brinjals were carried out by physiological weight loss (PWL), visual appearance (VA), total soluble solids (TSS), titratable acidity (TA) and pH.

12.2 Research methodology

The developed setup of PCC (Khalid, 2022) at the Department of Agricultural Engineering, Bahauddin Zakariya University, Multan, Pakistan (30°15′49″N, 71°30′35″E) was used for brinjal storage. The PCC was developed by Khalid (2022) using locally and easily accessible materials. The representative diagram of the experimental setup is shown in Fig. 12.2. The inner and outer wall of the zero-energy cool chamber were developed of bricks. The outer wall has dimensions (Length × Width × Height = 5 ft × 3.5 ft × 2 ft), and the inner wall has dimensions (Length × Width × Height = 3.64 ft × 1.60 ft × 2 ft). It was built in a rectangular shape to enhance the air contact area with the walls, increasing heat transfer between the system and ambient air. The higher heat transfer through walls encourages the evaporation rate from the media. The space between

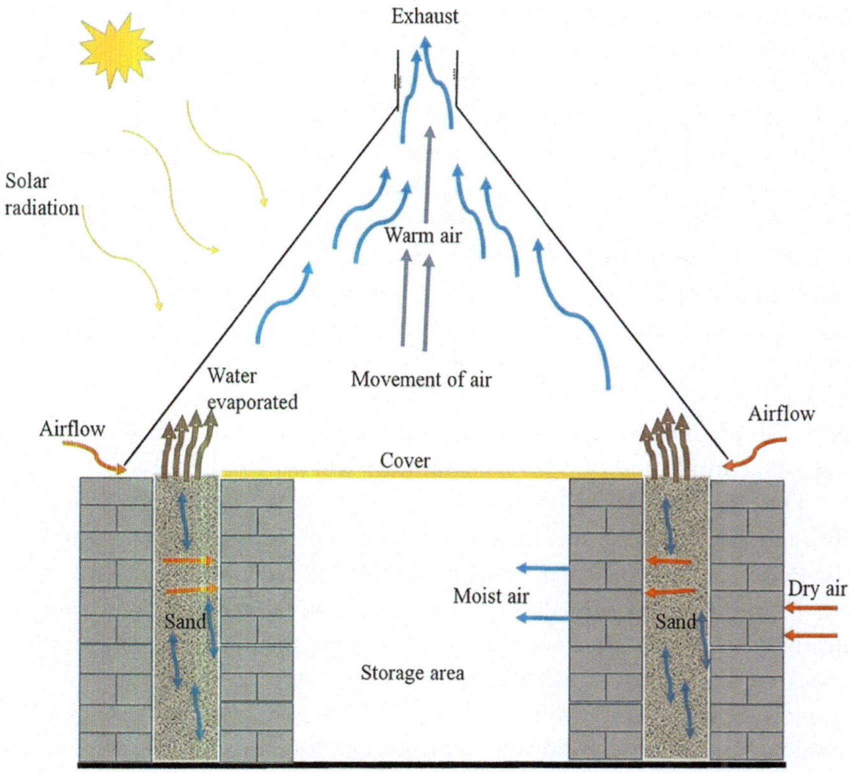

Fig. 12.2 Schematic diagram of chimney-equipped passive evaporative cooling chamber.

outer and inner wall is 0.33 ft, uniformly filled with the sand as an evaporative medium. Furthermore, the water was provided by a storage tank through PVC pipes to moisten the evaporative medium. The dimension of the nominated land is 7 ft × 5 ft, which is made by bricks.

In this study, a solar chimney was installed on developed PCC to enhance thermodynamic efficiency. Chimney are one of the most effective ways to increase passive evaporation to a create cooling effect for the storage application of agricultural products (Chakraborty and Fonseca, 2005). The chimney was made by a locally available material, a galvanized iron sheet. The prototype of the chimney-equipped PCC was developed with dimensions of Length = 68.58 cm, Width = 53.34 cm, and Height = 80 cm. The chimney has dimensions of Length = 66 cm, Width = 50.8 cm, and Height = 33 cm with a chimney hole of 3.3 cm to exit the air. The side/inclination angle of the chimney mainly depends on the latitude, and the best/ideal inclination angle lies between 40 and 60 degrees (Mathur et al., 2006). In this study, the chimney side was developed at an inclination angle of 45 degrees to receive maximum radiation. In another study, Monghasemi and Vadiee (2018) used a 45 degree inclination angle for maximum ventilation. The better top gap/hole to height ratio is 1:10 for proper balance of cycle maintenance in a chimney (Monghasemi and Vadiee, 2018). Passive evaporative cooling depends on the air movement through the chimney to balance the pressure level. The chimney's role is to create the pressure difference to increase the rate of evaporation. In other words, the amount of evaporation will rely on the pressure of air moving from the evaporative medium. The chimney creates a buoyancy force that draws air from the lower sides and leaves from the top of the chimney. The buoyancy force balances the constant air movement cycle (Chakraborty and Fonseca, 2005). The buoyancy/drawing force on air movement inside the chimney is based on the wind density difference inside and outside the chimney (Khanal and Lei, 2011).

The efficiency of chimney-assisted cooling chamber was calculated through measuring temperature and relative humidity inside and outside of the PCC. The performance efficiency of the cooling chamber is determined by storing matured brinjal. The ripe, fresh brinjal was collected from the nearby market of Multan and then graded for uniform size for storage. Plastic crates were used for brinjal storage in three conditions: chamber, ambient, and room condition. The relative humidity and temperature in PCC, ambient, and room condition was measured by relative humidity/temperature digital sensor HTC-1. The wet-bulb temperature was calculated using the inside and outside environmental condition given by Eq. 1) (Stull, 2011).

The cooling efficiency of PCC as compared to outside environmental condition was calculated by given (Eq. 2) using dry-bulb ambient temperature inside and wet-bulb temperature (Riffat and Gan, 2019).

$$
\begin{aligned}
\mathrm{T_w} = {} & \mathrm{Tatan}\left[0.151977(\mathrm{RH\%}+8.313659)^{\frac{1}{2}}\right] \\
& + a\tan\left(T+\mathrm{RH\%}\right) \\
& - a\tan\left(\mathrm{RH\%}1.676331\right) \\
& + 0.00391838\left((\mathrm{RH\%})^{3/2}\right)a\tan\left(0.023101\mathrm{RH\%}\right) \\
& - 4.686035
\end{aligned}
\tag{1}
$$

$$
\text{Cooling effectiveness} = \frac{T_1 - T_2}{T_1 - T_w} \times 100 \tag{2}
$$

where T_1 is the dry-bulb temperature, T_2 is inside temperature of chamber, and T_w is the wet bulb temperature.

In agricultural products, brinjals are stored at three different places in passive chamber (PCC), room (RMC), and ambient (AMC) condition. To check the quality of brinjal different physical (PWL and VA) and chemical (pH measurement, TSS, and TA) analyses were performed on a regular daily basis. PWL decreases the taste and quality of brinjal. Digital balance method was used to calculate the PWL in percentage reported by Tefera et al. (2007) and Pinheiro et al. (2009). W_1 and W_2 were initial and final daily weight for PWL (W_L) calculation.

$$
W_L(\%) = \frac{W_1 - W_2}{W_1} \times 100 \tag{3}
$$

The VA of brinjal is determined by visualizing color change. Three same-color samples of brinjal were placed in three different environmental conditions. The brinjal color changes from purple to black with the passage of storage duration. TSSs were measured by digital refractometer. The method of measuring soluble solids is reported in literature (Tefera et al., 2007; Ishaque et al., 2019). In this method, first extract the juice with easily available material such as a blinder then filter the extracted juice with the help of mesh. Clean the refractometer with distilled water before using and properly dry it. It was calibrated with distilled water; on this stage it gives a zero value. Take 2–3 drops of the sample and put them on the prism of refractometer. The eyepiece focuses on the digits' value, which is zoomed by a moveable lid. Note the reading of brix in %, which sits at the base of blue line. Measurements were taken by putting 2–3 drops of sample juice on the prism of the refractometer.

TA is the measure of total acid content in brinjal. It was calculated by the titration method reported by Olosunde et al. (2016). In this method, a 5 mL sample solution was titrated with 0.1 NaOH molar solution by putting 3–4 drops of phenolphthalein as an indicator. It is titrated until the sample solution turns to pink with pH values of 8.3. TA was calculated by putting values in the given equation (Eq. 4) (Olosunde et al., 2016). Moreover, pH value was also measured with a digital sensor (Nkolisa et al., 2018).

$$TA_{\%} = \frac{NaOH_v.NaOH_n.6.706 \times 10^{-3}}{V_s} \times 100 \tag{4}$$

where $NaOH_v$ the volume of sodium hydroxide is, $NaOH_n$ is the normality of sodium hydroxide, V_s is the volume of sample, 10^{-3} is the equivalent of malic acid in milligram.

12.3 Results and discussion

12.3.1 Cooling efficiency

Performance evaluation of solar chimney-equipped PCC is mainly examined by measuring its relative humidity and temperature in comparison with other conditions such as ambient and room conditions. The variation in temperature and relative humidity for the inside of PCC, ambient, and room conditions is shown in Fig. 12.3. The maximum and minimum temperature was measured in PCC and AMC condition as 25.14°C, 23.03°C and 33.63°C, 34.51°C, respectively. The average temperature in PCC, AMC, and RMC conditions was 23.82°C, 34.92°C, and 30.11°C, respectively. The result gained from this study that the temperature inside the PCC was 10°C–11.10°C lower than the AMC temperature. This condition is suitable for the storage of brinjal for a short period. Islam and Morimoto (2015a) reported that the temperature difference between cooling chamber and outside condition was measured as 12°C. In another study (Rayaguru, 2010), it is stated that the temperature reduced in evaporative cooling chamber 5°C–8°C as compared to ambient condition. Dadhich (2008) investigated that the temperature in the evaporative cool chamber is 10°C–15°C lower than the outside condition. The temperature dropped 16°C inside the evaporative cooler as compared to ambient conditions (Mogaji and Fapetu, 2011). Ogbuagu et al., (2017) reported that the

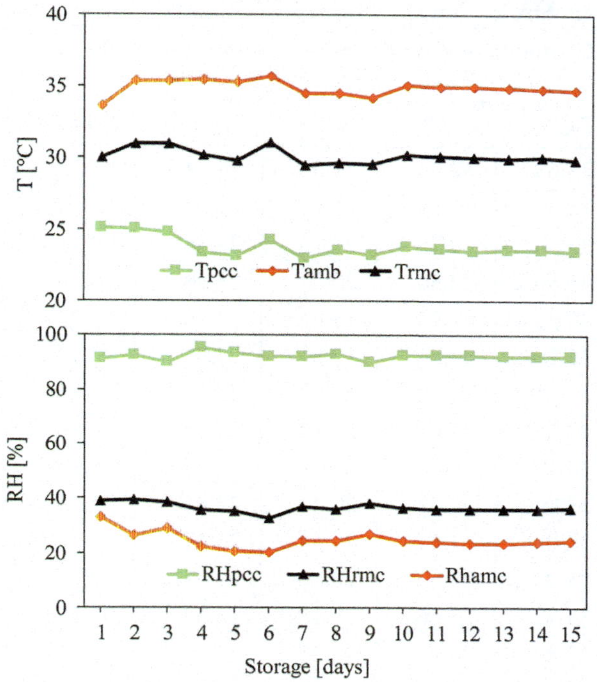

Fig. 12.3 The variation of temperature and relative humidity under different environmental conditions.

temperature of evaporative cooler reduced 11°C according to outside condition. Ronoh et al. (2018) reported 16.85°C maximum reduction in temperature inside the evaporative cooler.

The relative humidity was higher inside the PCC and lower in the ambient conditions during the process. Fig. 12.3 shows the changes in relative humidity under the PCC, RMC, and AMC conditions. The highest relative humidity measured in the PCC, RMC, and AMC was 95%, 39%, and 29%, respectively. In another similar study, Islam and Morimoto (2014) reported a maximum relative humidity of 90% inside the PCC compared to ambient conditions. In another study Islam and Morimoto (2015b) reported a relative humidity of 91.7% in a cool chamber compared to outside conditions. Furthermore, in this study the average difference in relative humidity between the PCC and AMC was 67%.

The cooling efficiency was measured from the lowering in temperature and rise in relative humidity inside of PCC. The inside and ambient temperature and relative humidity with cooling efficiency are given in

Table 12.1 Performance evaluation of passive evaporative cooling chamber.

Days	Environmental condition			PCC condition		Cooling efficiency
	T_{db}(°C)	RH (%)	T_{wt}(°C)	T_{PCC}(°C)	RH_{PCC}(%)	%
1	33.63	32.88	21.35	25.14	91.25	69
2	35.36	26.50	21.21	25.09	92.38	73
3	35.36	28.75	21.76	24.86	90.25	77
4	35.50	22.50	20.24	23.38	95.38	79
5	35.34	20.75	19.62	23.20	93.25	77
6	35.73	20.50	19.81	24.25	92.13	78
7	34.51	24.50	20.08	23.03	92.25	80
8	34.53	24.63	20.12	23.54	92.88	76
9	34.19	26.88	20.44	23.25	90.25	80
10	35.06	24.38	20.44	23.82	92.34	77
11	35.03	24.11	20.34	23.67	92.34	77
12	34.98	23.53	20.16	23.52	92.60	77
13	34.92	23.66	20.15	23.53	92.25	77
14	34.87	24.02	20.21	23.58	92.13	77
15	34.76	24.46	20.25	23.49	92.13	78
Average	34.92	24.80	20.41	23.82	92.25	77

Table 12.1. The cooling efficiency of PCC remains in the range of 69%–80% with the average efficiency being 77%. In another study, Zakari et al., (2016) reported 83% cooling efficiency of evaporative based cooling chamber. Another study (Nkolisa et al., 2018), reported a 67.17% average cooling efficiency of an evaporative cooling chamber.

12.3.2 Physicochemical analyses of stored product

Physicochemical analyses (PWL, VA, pH, TSS, and TA) are accomplished to check the quality of stored products (brinjal) with storage time. For the investigation of product quality, which is stored in the chamber, weight loss is the main parameter to find out the freshness of the product. Due to PWL, the amount of moisture content reduced day by day, which causes a decreasein the taste and quality of the product. PWL was determined at 1-day interval during the storage time (Islam and Morimoto, 2015b). With the passage of time, vegetables and fruits reduced in weight due to some biochemical and bacterial action. The same method was used for the calculation of the weight loss by Tefera et al. (2007), in which each day's calculated weight was measured with a digital balance and the quality of fruits was checked. The accumulative PWL was shown in percentage. For the investigation of weight loss of brinjal an equation is given below, which is given by Pinheiro et al. (2009). To measure the weight loss of stored product by using an electronic digital balance. The weight loss of brinjal is measured every day, and the percentage of weight loss is calculated. The freshness of brinjal is usually evaluated by the amount of water it contains. Brinjal will be fresh if there is a significant proportion of water present; otherwise, it may shrivel if exposed to the elements. Brinjal shriveling results in a loss of weight. PWL is the main factor in estimating the quality of stored brinjal. The PWL is calculated periodically with a 1-day interval. Three equal weights of brinjal were stored in PCC, ambient, and room conditions. The PWL variation for stored brinjal with the passage of time is given in Fig. 12.4. This figure shows that the weight loss in PCC is very low as compared to other environmental conditions.

The PWL for brinjal was calculated as 2.12%, 17.21%, and 24.95% on the third day stored in PCC, RMC, and AMC, respectively. The same trend shows on the sixth day of storage, PWL decreased to 3.8%, 30.01%, and 44.02% stored in PCC, RMC, and AMC, respectively. The PWL further decreased to 5.64%, 48.37%, and 65.92% for PCC, RMC, and AMC on the ninth day of storage, respectively. The same trend is shown

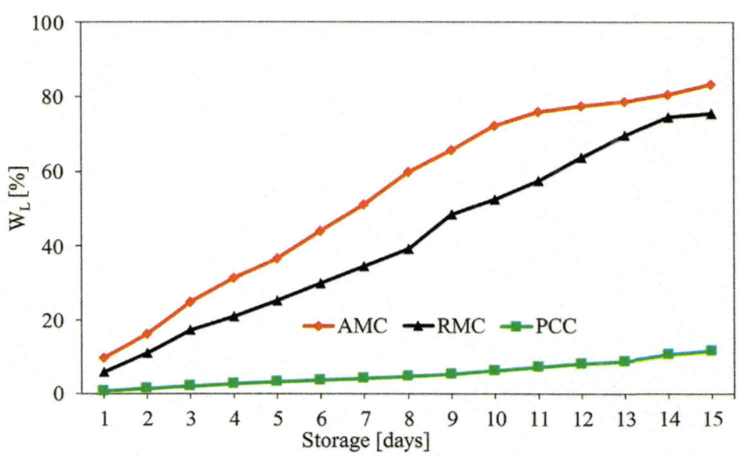

Fig. 12.4 Physiological weight loss of brinjal stored in room, ambient, and PCC conditions.

by Islam and Morimoto (2014): the PWL of brinjal was 5.01% on the ninth day in the cool chamber. The PWL at the 12th day of storage for stored brinjal was measured as 8.16%, 63.59%, and 77.91% in PCC, RMC, and AMC. The total weight loss in stored brinjal at the 15th was determined to be 11.65%, 75.48%, and 88.77%. About 5.9% loss of weight during storage time in brinjal was reported in the literature (Balogun and Ariahu, 2020).

The amount of TSSs in the sample is determined by refractometer. Its range was about 0–32°Brix (Tefera et al., 2007). The TSS values increased slowly with passage of storage period and loss of moisture content. In this process organic acids and sugar are consumed; due to this, concentration of TSS effects due to metabolism processes for long storage times. The ambient storage products show the high difference in TSS at an early stage (Ishaque et al., 2019). This method is suitable for juicy and ripe products. The citric acid mainly in brinjal transformed into fructose and sugar substances as it got full maturity. These brinjal substances represent the TSSs and are the indicator of good quality of the sample. Measurement of total solid contents (Brix index) is one of the important chemical parameters to estimate the quality of brinjal. The effect of storage conditions of PCC, RMC, and AMC on TSS is given in Fig. 12.5. A significant difference is measured in brinjal TSS under PCC, RMC, and AMC conditions with the passage of storage time. It is due to the conversion of acid into solid contents. On first day of storage the TSS of sample was 4.3°brix index for all conditions.

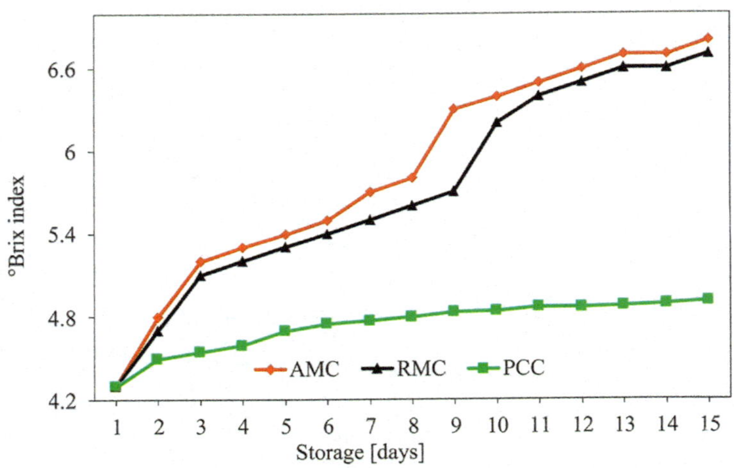

Fig. 12.5 Determination of soluble solids in stored brinjals.

With the passage of time it increased; on the third day its values were determined to be 4.55, 5.1, and 5.2 at PCC, RMC, and AMC conditions, respectively. On the sixth day of storage the brix index of stored brinjal was measured at 4.75, 5.4, and 5.5 at PCC, RMC, and AMC condition, correspondingly. With the passage of time the brix index increases; at on the ninth day of storage the brix index was examined 4.84, 5.7, and 6.3 at PCC, RMC, and AMC conditions, respectively. On the 12th day of storage the product the values of brix index were observed as 4.87, 6.5, and 6.6 in the PCC, RMC, and AMB conditions, respectively. At the end of storage day, the brix index was determined at the PCC, RMCM and AMC condition was measured at 4.91, 6.7, and 6.8, respectively. The same trend of TSS of brinjal was reported by Islam and Morimoto (2014): the TSS values of brinjal were 6.74 in ambient conditions and 6.85 in the evaporative cooling chamber after 15 days of storage. The range of TSS determined by Balogun and Ariahu (2020) was 3.6–4.75.

VA is important to note the freshness of brinjal by visual sense. Most of the people visualize the quality of brinjal through appearance for selling and as for eating purposes. The observation of the study concluded that the change in color of brinjal took place instantly in some days stored at three different conditions. The visual assessment of stored brinjal is shown in Fig. 12.6. It is due to rapid microorganism activity and greater production of ethylene from brinjal. On the first day of storage, three equal-weighted trials of uniform purple brinjal of almost equal size were taken, which were subjected to storage under three at different conditions. It is clearly seen that brinjal stored at ambient condition after 3 days, near to decay as they converted shrinkage of moisture. On the third day of storage, ambient and room-condition brinjal samples are completely decayed, and they are not eatable due to physical freshness deterioration. Furthermore, brinjal stored at PCC retain its quality. The product, which was stored at PCC, was eatable on the 15th day of storage. In PCC, the quality of brinjal was maintained for a couple of days due to lower temperature and higher relative humidity.

pH is the measure of acidity in brinjal, which is important for freshness. The variation of pH in stored brinjal is shown in Fig. 12.7. On the initial day of the storage, pH value measured as 4.2 for the brinjal sample. The pH value increased on the third day, from 4.2 to 4.25, 4.4, and 4.5 at PCC, RMC, and AMC, respectively. At sixth day, pH value measured as 4.4, 5, and 5.2 at PCC, RMC, and AMC, respectively. The pH level of stored product on the ninth day of storage was measured as 4.6, 5.5, and 5.6 at PCC, RMC, and AMC, correspondingly. pH of stored brinjal increases

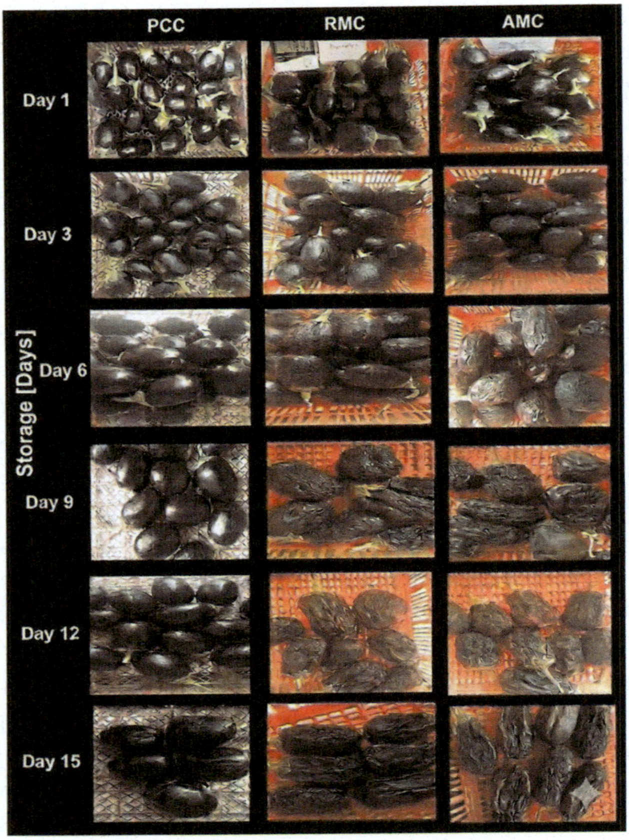

Fig. 12.6 Assessment of color change in stored brinjal under different conditions.

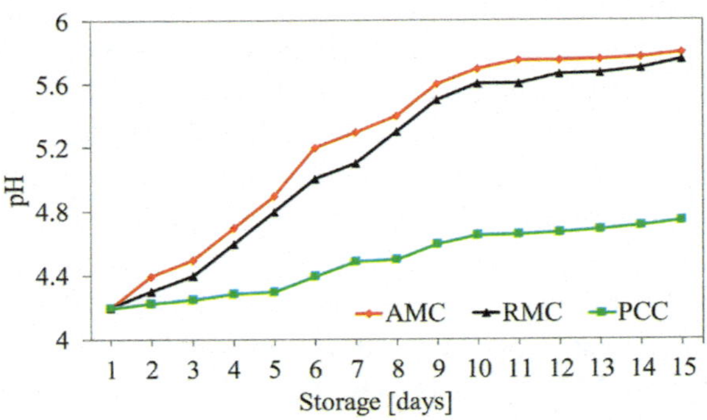

Fig. 12.7 pH level of brinjal sample stored under different conditions.

gradually on the 12th day of storage, measured as 4.67, 5.66, and 5.75 for brinjal stored in PCC, RMC, and AMC. The total pH level of brinjal sample stored in PCC, RMC, and AMC on 15th day, was 4.74, 5.75, and 5.8, respectively. The same trend of pH of brinjal given by Balogun and Ariahu (2020), which ranged was 3.89–4.14.

Different acids such as phenolic, ascorbic acid, and other are present in brinjal fruits. In which citric and phenolic acids are leading acid contents found in brinjal fruits. Acid concentration in phenolics decreases with storage duration due to conversion of acid into TSS as stated (Balogun and Ariahu, 2020). The TA of brinjal is measured for different storage conditions (PCC, RMC, and AMC) using Eq. (4). TA decreases in all storage conditions with the passage of storage days as shown in Fig. 12.8. The difference in TA decrement under storage conditions occurred mainly due to the variation in respiration rate and maturity stages (Islam and Morimoto, 2012). Thus, the conversion of acid into solid contents results in increasing TSS and pH and decreasing TA with storage time. The variation in TA shown in Fig. 12.8. On first day of storage: 0.87 for all storage conditions. On the third of storage, the TA of the product was measured 0.86, 0.8, and 0.79 at PCC, RMC, and AMC condition. The TA of the sample was examined on the sixth day of storage 0.83, 0.64, and 0.62

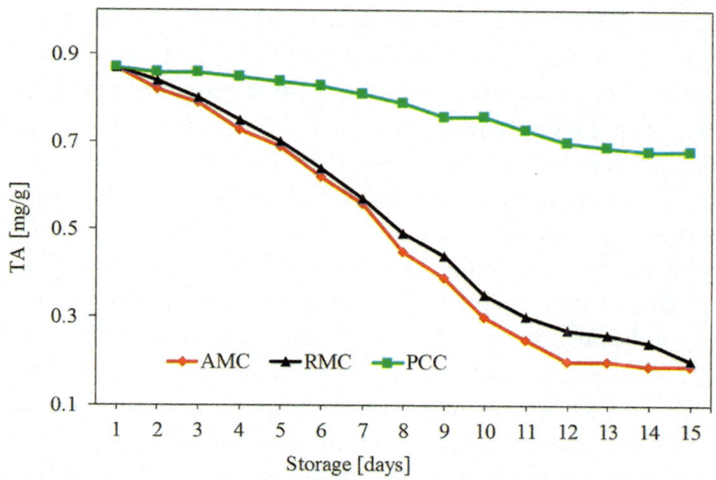

Fig. 12.8 Titratable acidity variation in brinjal stored in PCC, RMC and AMC condition. *(From Rodriguez-Jimenez, J.R., Amaya-Guerra, C.A., Baez-Gonzalez, J.G., Aguilera-Gonzalez, C., Urias-Orona, V., Nino-Medina, G., 2018. Physicochemical, functional, and nutraceutical properties of eggplant flours obtained by different drying methods. Molecules 23 (12). doi:10.3390/molecules23123210.)*

at PCC, RMC, and AMC conditions, respectively. On the ninth day of storage, the measure value of TA was 0.76, 0.44, and 0.39 in PCC, RMC, and AMC conditions, correspondingly. TA is decreased with increasing of maturity level of the products. On the 12th day of storage the TA of brinjal was examined 0.7, 0.27, and 0.2 at PCC, RMC, and AMC conditions, respectively. At the end of storage day, the TA was very low in AMC and RMC as compared to the PCC. On the 15th day of storage of brinjal, the TA was measured at 0.68, 0.2, and 0.19 at PCC, RMC, and AMC conditions, respectively. In another study, Rodriguez-Jimenez et al. (2018) reported TA values for brinjal ranging from 0.46–0.47. A similar trend was observed in another study by Balogun and Ariahu (2020), where the TA of eggplant measured during the experiment was 0.38–0.53.

12.4 Conclusions

Agricultural products are essential for maintaining human health. The agricultural country such as Pakistan produces an abundant amount of fruits and vegetables. However, huge amount of these products in developing countries is lost due to a lack of storage facilities. There is a need for storage facilities at the farm level and/or nearby to the farm. In this study, PCC is studied for the short term storage of brinjal at the farm level. The effect of a solar chimney for enhancing ventilation in the chamber is also investigated. The chimney is equipped over prototype PCC for the purpose. In PCC, the temperature drop was about 10°C and relative humidity remained greater than 90%. The cooling effectiveness of PCC was determined about 77%. However, the cooling effectiveness of the chamber can be enhanced by installing a solar chimney over the chamber. Trend of PWL of brinjal after 15 days of storage was measured as 11.65%, 75.48%, and 83.45% in PCC, room, and ambient conditions, respectively. TSSs were measured as 4.91° Brix, 6.7°Brix, and 6.8°Brix in passive evaporative cool chamber, room, and ambient condition, respectively. The pH value of stored brinjal after 15 days in passive evaporative cool chamber, room, and ambient condition were 4.74, 5.75, and 5.8, respectively. TA in brinjal stored in evaporative cool chamber, room, and ambient condition were measured as 0.68, 0.2, and 0.19 mg/g, respectively after 15 days of storage. The results clearly indicate that passive evaporative cool chamber is an effective alternative to conventional energy consuming storage systems for short-term agricultural product storage at farm level in harsh climatic conditions. The PCC maintain the quality of agricultural product during storage.

Acknowledgment

This research is conducted at the Department of Agricultural Engineering, Bahauddin Zakariya University, Multan, Pakistan with the financial support from the Higher Education Commission (HEC) of Pakistan under the research project NRPU-16962.

Funding statement

This research is financially supported by the Higher Education Commission (HEC) of Pakistan under the research project NRPU-16962.

CRediT authorship contribution statement

Muhammad Irfan was involved in the conceptualization, methodology, Software, formal analysis, investigation, writing of the original draft. **Muhammad Hamid Mahmood, Muhammad Sultan, and Muhammad Wakil Shahzad** were involved in the conceptualization, methodology, validation, resources, visualization, supervision, project administration, writing (review and editing), funding acquisition. **Muhammad Khalid, Muhammad Ahmad Jamil, Haseeb Yaqoob, Uzair Sajjad, and Naveed Husnain** were involved in the conceptualization, methodology, software, formal analysis, investigation, writing (review and editing).

Data availability statement

The datasets used and/or analyzed during the current study are available from the corresponding authors upon request.

Conflicts of interest

The authors declare no conflict of interest.

Declaration of generative AI and AI-assisted technologies in the writing process

No generative AI/AI-assisted technology is used in this paper.

References

Balogun, A.A., Ariahu, C.C., 2020. Quality evaluation of African eggplant stored in evaporative coolers. Asian Food Sci. J. 18 (1), 23–33. doi:10.9734/AFSJ/2020/v18i130207.

Basediya, A.L., Samuel, D.V.K., Beera, V., 2013. Evaporative cooling system for storage of fruits and vegetables—a review. J. Food Sci. Technol. 50 (3), 429–442. doi:10.1007/s13197-011-0311-6.

Chakraborty, J., Fonseca, E., 2005. Analysis and evaluation of a passive evaporative cool tower in conjunction with a Solar Chimney. Unchanged 1 (November), 485–490.

Dadhich, S.M., 2008. Comparative study on storage of fruits and vegetables in evaporative cool chamber and in ambient. Int. J. Food Eng. 4 (1). doi:10.2202/1556-3758.1147.

Dodoo, A., Gustavsson, L., Sathre, R., 2011. Building energy-efficiency standards in a life cycle primary energy perspective. Energy Build 43 (7), 1589–1597. doi:10.1016/j.enbuild.2011.03.002.

Emdadi, Z., Asim, N., Yarmo, M.A., Shamsudin, R., Mohammad, M., Sopian, K., 2016. Green material prospects for passive evaporative cooling systems: geopolymers. Energies 9 (8). doi:10.3390/en9080586.

Ishaque, F., Hossain, A., Sarker, A.R., Rahman, H., 2019. A study on low cost post harvest storage techniques to extend the shelf life of citrus fruits and vegetables. J. Eng. Res. Rep. 9 (1), 1–17. doi:10.9734/JERR/2019/v9i117009.

Islam, M.P., Morimoto, T., 2012. Zero energy cool chamber for extending the shelf-life of tomato and eggplant. Jpn. Agric. Res. Q. 46 (3), 257–267. doi:10.6090/jarq.46.257.

Islam, M.P., Morimoto, T., 2014. A new zero energy cool chamber with a solar-driven adsorption refrigerator. Renew. Energy 72, 367–376. doi:10.1016/J.RENENE.2014.07.038.

Islam, M.P., Morimoto, T., 2015a. Evaluation of a new heat transfer and evaporative design for a zero energy storage structure. Sol. Energy 118, 469–484. doi:10.1016/j.solener.2015.06.006.

Islam, M.P., Morimoto, T., 2015b. Progress and development in brick wall cooler storage system. Renew. Sustain. Energy Rev. 50, 277–303. doi:10.1016/j.rser.2015.05.019.

Khalid, M., 2022. Zero-energy cool chamber for tomato storage. In: Sultan, M., Miyazaki, T. (Eds.), Energy-Efficient Systems for Agricultural Applications. Springer. doi:10.1007/978-3-030-86394-4.

Khanal, R., Lei, C., 2011. Solar chimney—a passive strategy for natural ventilation. Energy Build 43 (8), 1811–1819. doi:10.1016/j.enbuild.2011.03.035.

Lawande, K.E., Chavan, J.K., 1998. Eggplant (brinjal). In: Salunkhe, D.K., Kadam, S.S. (Eds.), Handbook of Vegetable Science and Technology. CRC Press, pp. 243–262. doi:10.1201/9781482269871.

Lock, K., Pomerleau, J., Causer, L., Altmann, D.R., McKee, M., 2005. The global burden of disease attributable to low consumption of fruit and vegetables: implications for the global strategy on diet. Bull. World Health Organ 83 (2), 100–108. doi:10.1590/S0042-96862005000200010.

Mahmood, M.H., Sultan, M., Miyazaki, T., 2019. Significance of temperature and humidity control for agricultural products storage: overview of conventional and advanced options. Int. J. Food Eng. 1–21. doi:10.1515/ijfe-2019-0063.

Manca, C.S., et al., 2017. Acceptability and cold storage of fresh-cut eggplant (*Solanum Melongena* L .) using edible coatings and antioxidants. Nutr. Food Sci. Int. J. 2 (1), 1–7. doi:10.19080/NFSIJ.2017.02.555578.

Mathur, J., Mathur, S., Anupma, 2006. Summer-performance of inclined roof solar chimney for natural ventilation. Energy Build 38 (10), 1156–1163. doi:10.1016/j.enbuild.2006.01.006.

Mogaji, T.S., Fapetu, O.P., 2011. Development of an evaporative cooling system for the preservation of fresh vegetables. Afr. J. Food Sci 5 (4), 255–266.

Monghasemi, N., Vadiee, A., 2018. A review of solar chimney integrated systems for space heating and cooling application. Renew. Sustain. Energy Rev 81 (April), 2714–2730. doi:10.1016/j.rser.2017.06.078.

Muhammad, M.H., Sultan, M., Miyazaki, T., Koyama, S., 2016. Desiccant air-conditioning system for storage of fruits and vegetables : Pakistan preview. Evergreen 3 (1), 12–17. doi:10.5109/1657381.

Nkolisa, N., Samukelo, L., Seyoum, T., Chimphango, A., 2018. Evaluating evaporative cooling system as an energy- free and cost- e ff ective method for postharvest storage of tomatoes (*Solanum lycopersicum* L.) for smallholder farmers. Sci. Hortic. (Amsterdam) 241 (February), 131–143. doi:10.1016/j.scienta.2018.06.079.

Ogbuagu, N.J., Green, I., Anyanwu, C., Ume, J., 2017. Performance evaluation of a composite-padded evaporative cooling storage bin. Niger. J. Technol. 36 (1), 302–307.

Olosunde, W.A., Aremu, A.K., Onwude, D.I., 2016. Development of a solar powered evaporative cooling storage system for tropical fruits and vegetables. J. Food Process. Preserv. 40 (2), 279–290. doi:10.1111/jfpp.12605.

Passam, H.C., Karapanos, I.C., 2008. Eggplants, peppers and tomatoes : factors affecting the quality and storage life of fresh and fresh-cut (Minimally Processed) produce. Eur. J. Plant Sci. Biotechnol. 2 (1), 156–170. Available from: http://www.globalsciencebooks.info/Online/GSBOnline/images/0812/EJPSB_2(SI1)/EJPSB_2(SI1)156-170o.pdf.

Pinheiro, J., Alegria, C., Abreu, M., Gonçalves, E.M., Silva, C.L.M., 2009. Temperature effect on stored tomato (*Lycopersicum Esculentum* L.) quality parameters. Horticulture. 1, 10–13. Available from: http://hdl.handle.net/10400.14/5567.

Rayaguru, K., 2010. Water use optimization in zero energy cool chambers for short term storage of fruits and vegetables in coastal area. J. Food Sci. Technol. (August). doi:10.1007/s13197-010-0072-7.

Reddy, V.S., Premalatha, M., Ranjan, K.R., 2012. Experimental studies on solar chimney for enhanced ventilation. Int. J. Sustain. Energy 6451. doi:10.1080/1478646X.2010.534554.

Riffat, S., Gan, G., 2019. Experimental study of the potential of eucalyptus fibres for evaporative cooling. Renew. Energy 131, 250–260. doi:10.1016/j.renene.2018.07.005.

Rodriguez-Jimenez, J.R., Amaya-Guerra, C.A., Baez-Gonzalez, J.G., Aguilera-Gonzalez, C., Urias-Orona, V., Nino-Medina, G., 2018. Physicochemical, functional, and nutraceutical properties of eggplant flours obtained by different drying methods. Molecules 23 (12). doi:10.3390/molecules23123210.

Ronoh, E.K., Kanali, C.L., Ndirangu, S.N., Mang, S.M., John, A.W., 2018. Performance evaluation of an evaporative charcoal cooler and its effects on quality of leafy vegetables. J. Postharvest Technol. 06 (3), 60–69.

Samuel, D.G.L., Nagendra, S.M.S., Maiya, M.P., 2013. Passive alternatives to mechanical air conditioning of building: a review. Build. Environ 66, 54–64. doi:10.1016/j.buildenv.2013.04.016.

Stull, R., 2011. Wet-bulb temperature from relative humidity and air temperature. J. Appl. Meteorol. Climatol. 50 (11), 2267–2269. doi:10.1175/JAMC-D-11-0143.1.

Subr, A.K., Sawa, J., Parafiniuk, S., 2015. Practical deviation in sustainable pesticide application process. Agric. Agric. Sci. Procedia 7, 241–248. doi:10.1016/j.aaspro.2015.12.037.

Sultan, M., El-sharkawy, I.I., Miyazaki, T., Baran, B., Koyama, S., 2015. An overview of solid desiccant dehumidi fi cation and air conditioning systems. Renew. Sustain. Energy Rev. 46, 16–29. doi:10.1016/j.rser.2015.02.038.

Tefera, A., Seyoum, T., Woldetsadik, K., 2007. Effect of disinfection, packaging, and storage environment on the shelf life of Mango. Biosyst. Eng. 96 (2), 201–212. doi:10.1016/j.biosystemseng.2006.10.006.

Zakari, M.D., et al., 2016. Design and construction of an evaporative cooling system for the storage of fresh tomato. ARPN J. Eng. Appl. Sci. 11 (4), 2340–2348. doi:10.13140/RG.2.1.4748.4569.

CHAPTER 13

Sustainable greenhouse farming through integration of solar energy and desiccant dehumidification system using Design-Builder

Junaid Ramzan[a], Muhammad Sultan[a], Hafiz Muhammad Umar Raza[a], Muhammad Hamid Mahmood[a], Takahiko Miyazaki[b], Muhammad Wakil Shahzad[c], Muhammad Ahmad Jamil[c], and Uzair Sajjad[d]

[a]Department of Agricultural Engineering, Bahauddin Zakariya University, Multan, Pakistan
[b]Department of Advanced Environmental Science and Engineering, Faculty of Engineering Sciences, Kyushu University, Kasuga, Fukuoka, Japan
[c]Department of Mechanical and Construction Engineering, Northumbria University, Newcastle Upon Tyne, United Kingdom
[d]Interdisciplinary Research Center for Sustainable Energy Systems (IRC-SES), King Fahd University of Petroleum and Minerals, Dhahran, Saudi Arabia

13.1 Introduction

A controlled environment is necessary for greenhouse plant growth. To control the ambient air temperature and relative humidity to reach the thermal comfort range is known as air conditioning. Temperature and humidity control are important for high-yield food production in the greenhouse (Sultan et al., 2014). The most important climatic parameters for high-yield production in the greenhouse are temperature and relative humidity (Pasqualin et al., 2022). The relative humidity is the most complex climatic parameter to control within the greenhouse. Photosynthesis and evapotranspiration processes increase the relative humidity level inside the greenhouse (Sultan and Miyazaki, 2017). Photosynthesis is a natural process inside plants during the daytime. As a result, plant leaves generate water vapors within the greenhouse and impact the relative humidity of the surrounding air (Ashraf et al., 2021). High relative humidity inside the greenhouse leads to fungal diseases, leaf necrosis, plant calcium deficiency, and many other diseases that reduce product quality and may cause plant

Sustainable Hybrid Cooling and Dehumidification Strategies for Controlled Environments
https://doi.org/10.1016/B978-0-443-48555-8.00006-2

death. Excessive humidity causes plant diseases and deterioration of the crops in the greenhouse (Liu et al., 2019).

Furthermore, evapotranspiration is the process of evaporation (water vapors occur from the soil) and transpiration (water vapors occur from stomata) to add moisture inside the air of the greenhouse (Shamim et al., 2018). Photosynthesis and evapotranspiration are the basic phenomena in plant growth by which water vapors are released into the air, increasing the humidity in the greenhouse continuously (Çerçi et al., 2024). The psychrometric presentation shows the ideal growth of different crops in a greenhouse environment.

Temperature is the primary factor that affects plant growth and development. Plants require an ideal range of temperatures for properly growing in the greenhouse. The high air temperature inside the greenhouse slows down or even stops the growth of the plant; reduces the rate of photosynthesis, which reduces the growth of plants as well as crop yield; and increases the evapotranspiration rate, which increases the humidity inside the greenhouse (Iqbal et al., 2023). Due to the low temperature inside the greenhouse, the rate of photosynthesis and respiration processes is reduced, reducing plant growth inside the greenhouse and increasing the risk of plant diseases (Sultan et al., 2021).

Fig. 13.1 illustrates a solar-integrated desiccant dehumidification system is an economically effective air conditioning method, particularly in humid and hot regions (Hussain, 2023). This system is a possible alternative to conventional air-conditioning systems since it provides sensible and latent cooling loads independently (Noor et al., 2021). Several advantages have broadened the deployment of solar desiccant systems in recent years, including growing technology, no usage of chlorofluorocarbons, low energy consumption, standard human and nonhuman comfort, energy savings, and being environmentally friendly (Iqbal et al., 2023). The solar-integrated desiccant dehumidification system.

13.2 Research methodology

Fig. 13.2 illustrates the AutoCAD design of a greenhouse structure. (A) Isometric view. (B) Side view. (C) Front view. (D) Top view and selected study area: the greenhouse at Agricultural Engineering of Bahauddin Zakariya University, Multan, Punjab, Pakistan This study is designed the simulation to evaluate the energy efficiency and environmental management capabilities of a greenhouse that uses solar energy and

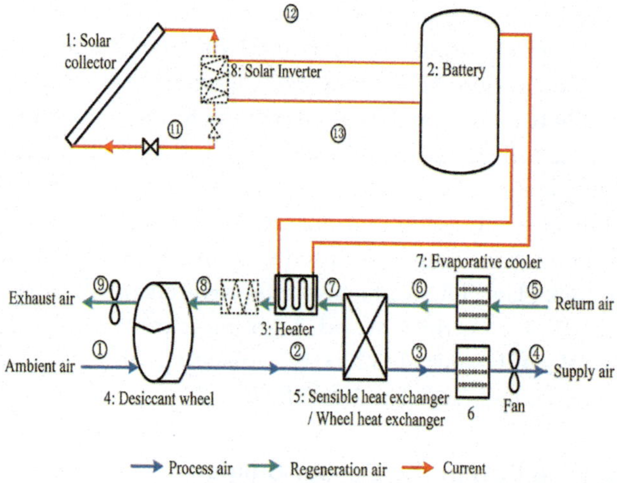

Fig. 13.1 Solar-integrated desiccant dehumidification system. *(Çerçi, K.N., Silva, I.R.O., Kaşka, Ö., Hooman, K., 2024. Summer period analysis of the rotary desiccant—hybrid cooling system combined with solid oxide fuel cells using human waste fuel. Clean. Eng. Technol. 26667908. 23. Available from: https://www.sciencedirect.com/science/journal/26667908. https://doi.org/10.1016/j.clet.2024.100818.)*

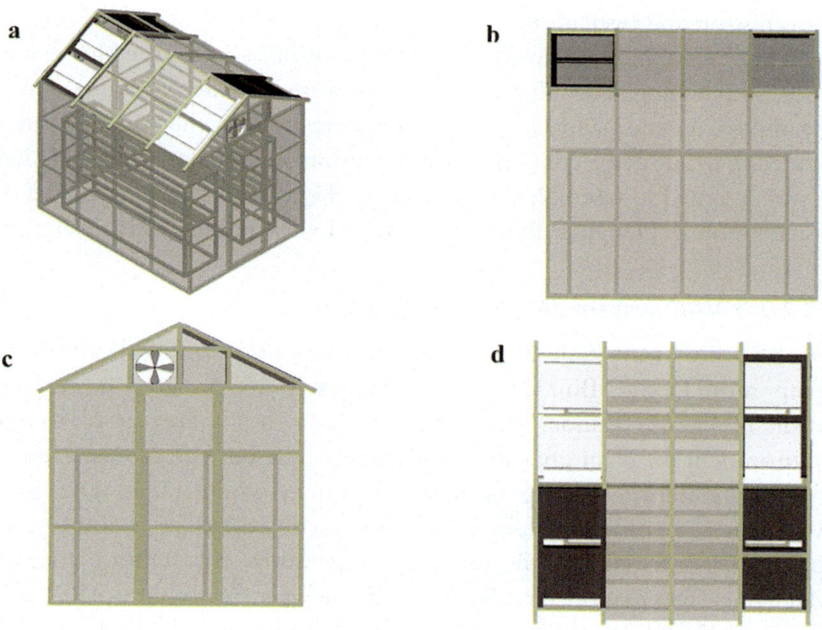

Fig. 13.2 Illustration of an AutoCAD design of a greenhouse structure. (A) Isometric view. (B) Side view (C) Front view. (D) Top view.

desiccant-based dehumidification. Design-Builder software is used for simulation, and the EnergyPlus engine is used for energy and environmental properties. A greenhouse building was occupying a floor of a 48 ft^2 area and a volume of 324 ft^3. The details of input parameters used for simulation are as follows: The design is based on ASHRAE 9.01 climate data for Multan, Pakistan. The greenhouse building occupied floor area of 48 ft^2 and volume of 324 ft^3. The greenhouse dimensions are 8 ft (length) by 6 ft (width) by 6 ft to 7.5 ft (height). Iron frames are used as construction material and glazed with 4 mm thick glass. The roof is also glass covered with an iron frame. The floor consists of 10 mm thick wood. Lighting is provided by 2 lights, each rated at 12 W. A heater of 3200 W capacity is used for heating. The ventilation rate is maintained at 480 CPM.

13.2.1 Model set up in the Design-Builder

A greenhouse model was developed using Design Builder version 7.0.2.006 to simulate environmental variables and energy efficiency. The greenhouse was 8 feet in length, 6 feet in width, and 6 feet in height of wall, with an additional 1.5 feet of sloping roof surface at the center. The greenhouse's 4 mm single glass panes are glazed with an iron frame for maximum light transmission and insulation.

The walls, which are depicted in the simulation as green; the blue external door; and the sloped roof for ideal ventilation are the external components of the model. Pitched and flat glass panels were used in the roofing system's modeling to allow for controlled heat retention and the flow of natural light. It was designed to resemble a wooden base. Fig. 13.3A illustrates model data visualization in Design-Builder and rendered view in Design-Builder.

13.2.2 HVAC system setup in Design-Builder

Fig. 13.4 shows the basic display panel for the simulation of HVAC system setup in the Design-Builder. The HVAC system for the greenhouse simulation in Design Builder is not just a system but a creator of an ideal environment for plant growth. It provides precise temperature and humidity management, resulting in an environment where plants can thrive. A variable air volume system with a water-cooled chiller is employed to maintain consistent humidity and temperature throughout the day. Mechanical ventilation circulates outside air at a steady rate of 0.1 m^3/s. The operational schedule was changed to "Office_OpenOff_Occ." This timeline defines when the ventilation system will be operational. The system was designed such that the zone equipment supplies full heating

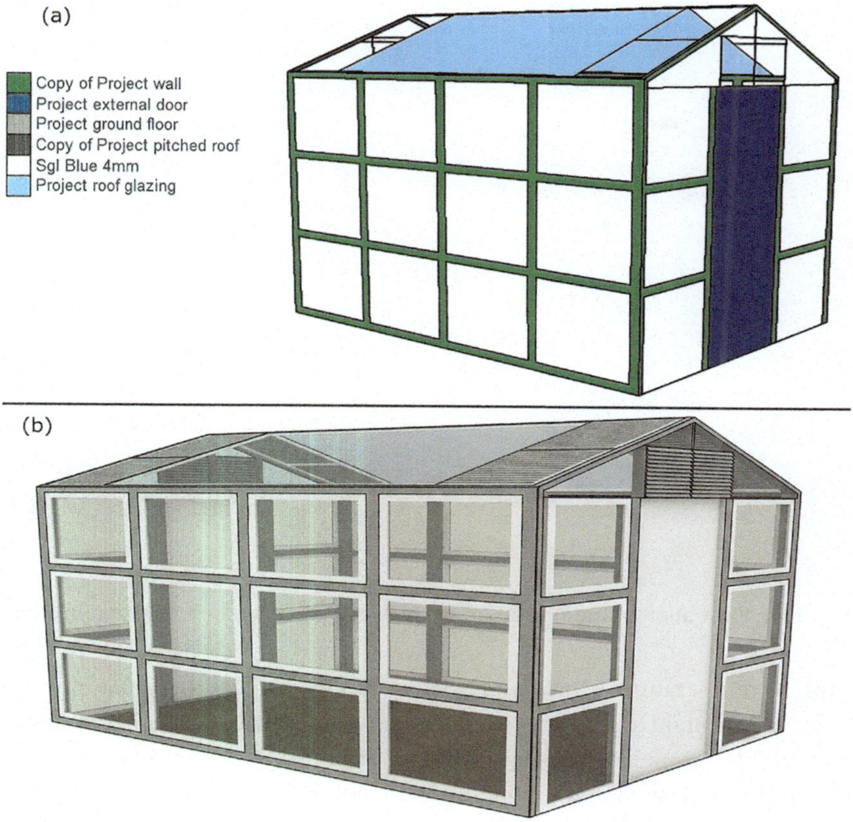

Fig. 13.3 Illustration of (A) model data visualization in Design-Builder and (B) rendered view in Design-Builder.

rather than recovering heat from ventilation, allowing you to have more control over the indoor temperature. The calibration of the heating and cooling system achieves thermal comfort. In this system, we select the maximum air temperature limit of 24 degrees Celsius and the 0.0156 g/g humidity ratio, which is limited for overdrying activity in the greenhouse system (Abdel-Salam et al., 2013). The cooling system, a reliable component of the HVAC system, maintains a stable temperature of 18°C and a humidity ratio of 0.0077 g/g in the greenhouse. Plant moisture levels remain constant at this temperature. The humidification and dehumidification systems were turned on to keep the humidity levels at their optimum. This allows the system to maintain ideal moisture levels regardless of external weather conditions, which is critical for greenhouse plant growth. In addition, natural ventilation is used to assist mechanical controls. When the

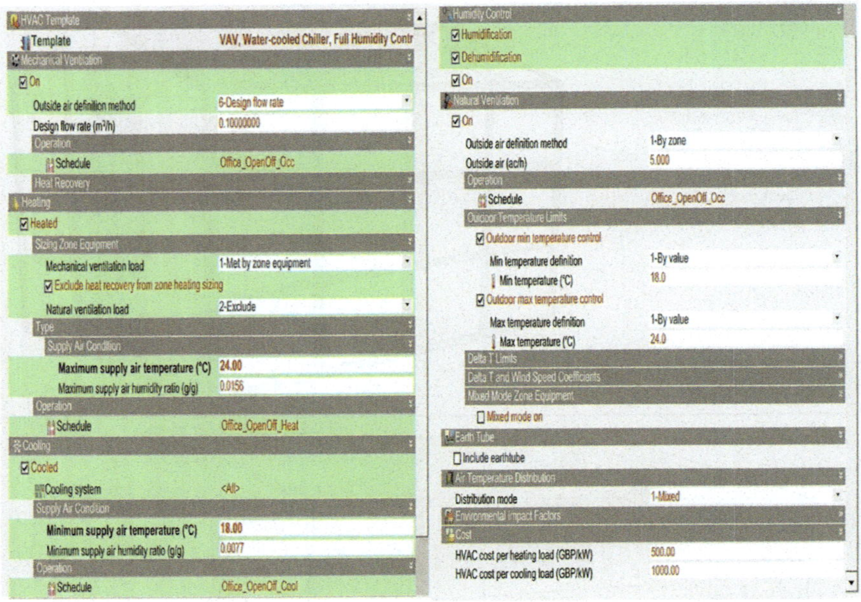

Fig. 13.4 Illustration of the simulation of HVAC system setup in the Design-Builder.

outside temperature is between 18°C and 24°C, natural ventilation takes over to maintain the indoor atmosphere as pleasant without depending excessively on mechanical equipment. The air exchange rate is set to 5 air changes per hour (ac/h), which ensures enough airflow when circumstances are good. Furthermore, the system operates in mixed mode, which means it shifts between mechanical and natural ventilation depending on outside conditions, therefore increasing energy economy while maintaining inside comfort. The HVAC system's heating cost is 500 GBP per kW, while cooling is 1000 GBP per kW, as shown in Fig. 13.4.

13.2.3 Solar system setup in Design-Builder

The solar energy system in this model generates power on-site using photovoltaic (PV) panels. The system is intended to increase the greenhouse's energy efficiency by assisting the desiccant dehumidification system.

Solar Collector Type: The selected collector type is "2-Photovoltaic," which is a popular option for power production. The installation fee is established at 1500 PKR per square meter (Jiaqiang et al., 2013).

The solar panels are affixed to the site's upper surface using bituminous felt as a mounting medium to provide enough support and integration.

A basic performance model is used, including a constant efficiency of 15% of photovoltaic panels, ensuring consistent performance in the conversion of solar energy to electricity. The heat transfer integration option is set to "Decoupled," which means that the thermal dynamics of the solar panels are modeled independently of the building's thermal mass.

Load Centers: The system comprises an electric load center that manages the distribution of produced electricity. The system calls for one electric load center, dubbed "Load Center 1," which contains a DC solar inverter. In this solar setup, the DC inverter is used to convert direct current into alternating current. This setup provides for the effective conversion of DC power from PV panels to AC for use in the greenhouse. By integrating this solar energy system with the desiccant dehumidification system, the model aims to optimize energy usage while reducing dependency on external power sources. The load generated at the site with solar panels is not enough to fulfill the greenhouse electricity requirements as shown in Fig. 13.5.

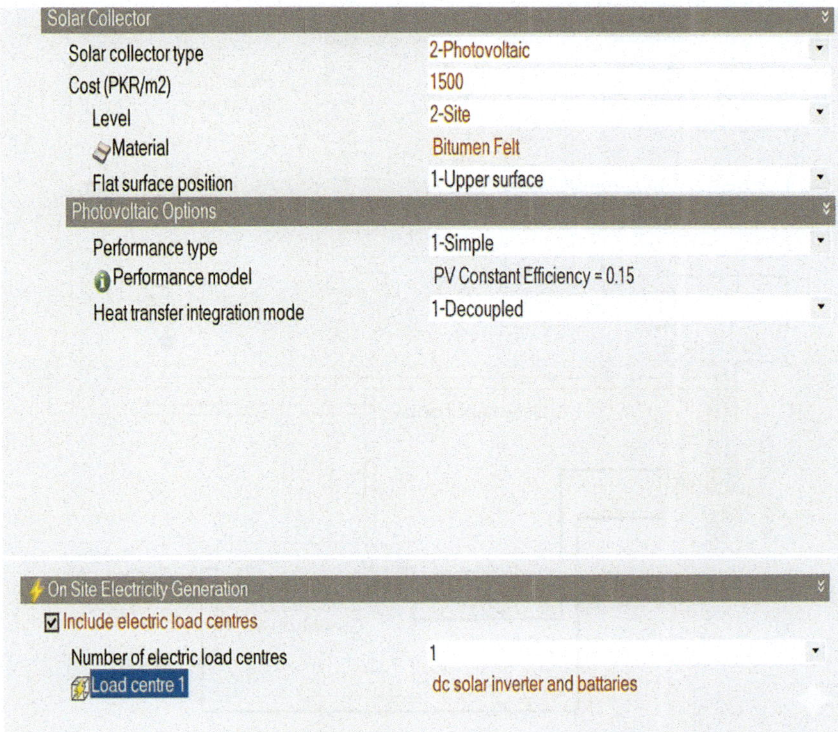

Fig. 13.5 Configuration of the solar energy system in Design-Builder.

13.2.4 Configuration of a desiccant dehumidification system

Fig. 13.6 illustrates the greenhouse's desiccant dehumidification system uses a desiccant wheel and an integrated ventilation system to regulate temperature and humidity. The desiccant loop draws air into the wheel, which collects moisture from the surrounding air. This dry air is subsequently cooled as it goes through a heat exchanger, therefore lowering both humidity and temperature. To create ideal growth conditions, the greenhouse circulates cooled, dehumidified air. Any extra heat is recycled by reactivating the desiccant wheel with a heating element, ensuring that it continues to collect moisture. The configuration also features an indirect evaporative cooling (IEC) system and a direct evaporative cooling (DEC) system to improve cooling and ventilation. The IEC cools the air before channeling it through wet pads for further cooling and humidity control. To maintain a balanced interior environment, the system also has ventilators for air exhaust and solar PV panels to increase energy efficiency. This complete

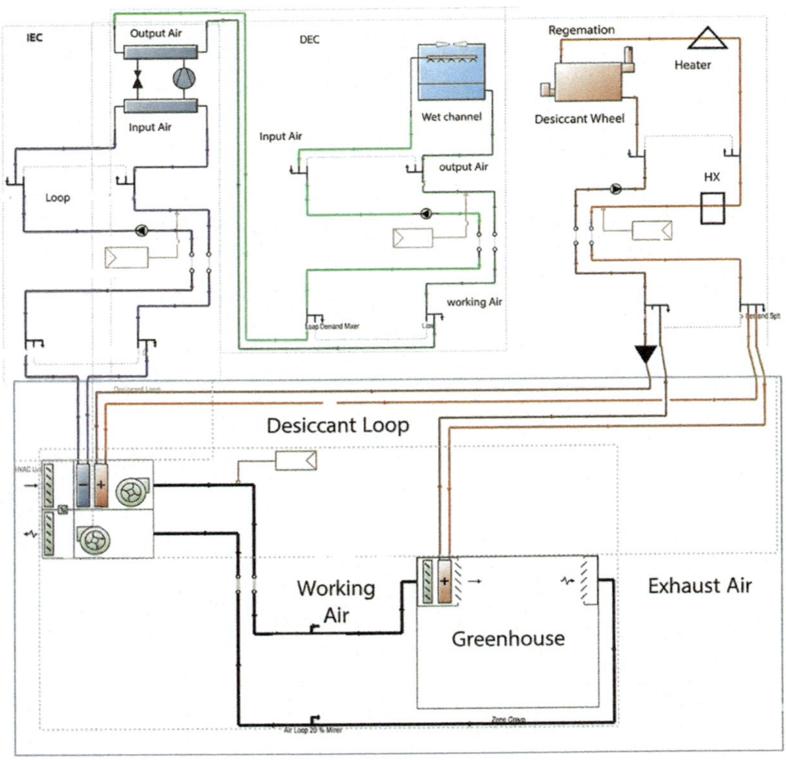

Fig. 13.6 Configuration of desiccant dehumidification system.

loop enables the greenhouse to maintain a regulated atmosphere, maintaining ideal development conditions via excellent regulation of air temperature, humidity, and ventilation.

13.3 Results and discussion

13.3.1 Thermal comfort analysis for cooling design simulation for March

Fig. 13.7 illustrates the results of the performance of environmental comfort levels in a greenhouse on March 15th. The gradual increase in internal air temperature, which reaches its peak between midday and 3:00 p.m. at 37°C–38°C, demonstrates significant heat retention within the greenhouse. The temperature starts at 20°C. In contrast, the outside dry-bulb temperature remains steady, peaking at about 25°C–26°C. The relative humidity, which had been at 55%, dropped swiftly to 40% by 8:00 a.m. as temperatures increased, then steadied for the rest of the day until late afternoon. Following a decrease to about 35% in the evening, there is a surge of >50%. Temperature fluctuations influence this pattern because warmer air holds

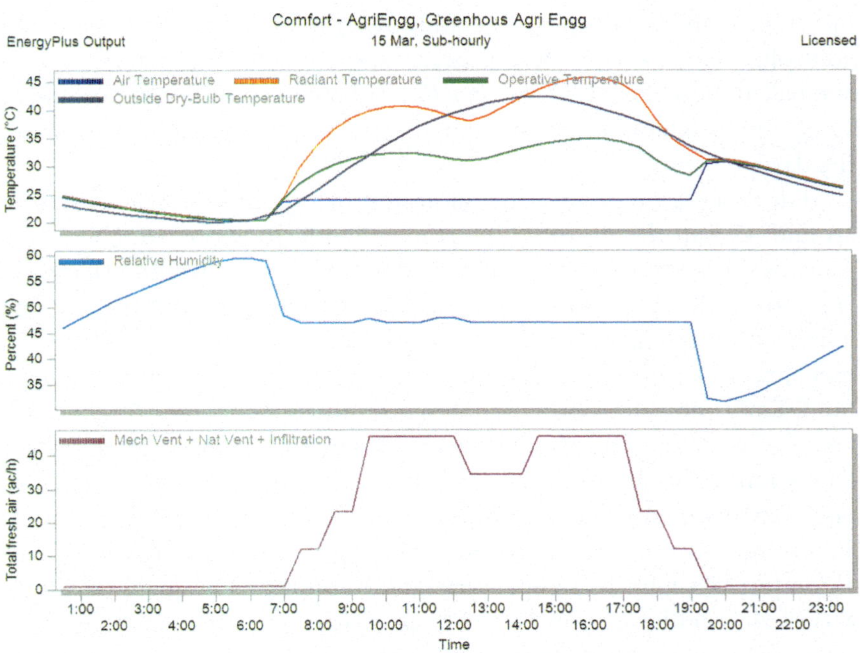

Fig. 13.7 Thermal comfort analysis for the greenhouse in March.

more moisture, lowering relative humidity during hot weather. Ventilation is essential for managing the indoor atmosphere. The entire fresh air exchange rate, which includes mechanical and natural ventilation and infiltration, is low at night and early in the morning before rapidly increasing by 7:00 a.m. During peak daytime hours (10:00 a.m. to 5:00 p.m.), the ventilation system generates up to 25 ac/h to reduce excessive heat and humidity. Ventilation drops in the late afternoon, hitting a low of 7:00 p.m. The greenhouse effectively uses ventilation to provide an ideal indoor climate for plant growth, with temperature, humidity, and ventilation all running on different daily cycles determined by solar gain and environmental control systems.

13.3.2 Thermal comfort analysis for cooling design simulation for April

This graph shows that temperature, relative humidity, and ventilation levels throughout the day depict comfort metrics within the greenhouse environment on April 15. The temperature screen displays the ambient temperature (blue line), radiant temperature (red line), and operating temperature. The maximum radiant temperature indicates that the greenhouse's surfaces heat up more than the air. The operating temperature suggests a warm but controlled environment during peak sunshine, which includes both air and radiant heat. The dark blue line displays the outdoor dry bulb temperature, which is always lower than the indoor temperature throughout the day.

The relative humidity panel indicates that the humidity level starts at around 45% in the morning, increases slightly to about 48%, and then remains quite stable until around 5:00 p.m., when it declines sharply to approximately 35%. Then, it gradually rises once again into the evening. The ventilation panel displays the air exchange rate by combining mechanical ventilation, natural ventilation, and infiltration (purple line). Early morning ventilation is weak (nearly nonexistent); nevertheless, it begins to climb dramatically at about 7:00 a.m. and reaches a maximum rate of around 40 ac/h by 10:00 a.m. This high ventilation rate is maintained until around 5:00 p.m., when it gradually declines to almost nothing. During peak heat, the high ventilation rate maintains a comfortable interior temperature and humidity. The graph illustrates a typical greenhouse interior climate that is well-regulated, with air temperature, humidity, and ventilation maintained at a consistent level

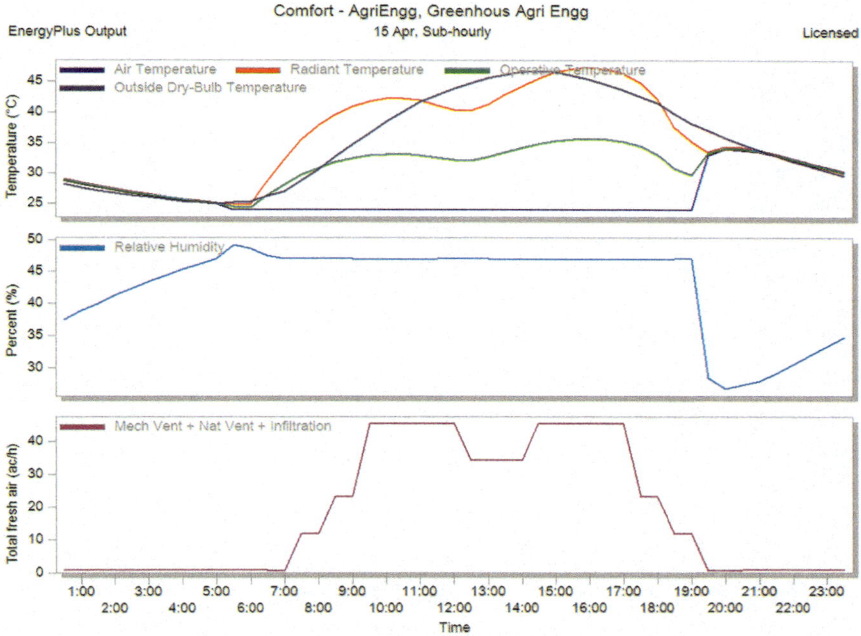

Fig. 13.8 Thermal comfort analysis for the greenhouse in April.

throughout the day to optimize plant growth and occupant comfort, as shown in Fig. 13.8.

13.3.3 Thermal comfort analysis for cooling design simulation for May

Fig. 13.9 illustrates the graph showing the temperature, relative humidity, and daily ventilation needed for a comfortable greenhouse atmosphere on May 15. Temperature Starting at 25°C, the air temperature quickly rises to 43°C–45°C between 12:00 p.m. and 3:00 p.m. before falling in the evening. Rising above the air temperature and peaking at >45°C, the dazzling temperature indicates considerable heat accumulation radiated from greenhouse surfaces. Show the total felt temperature; the operational temperature, which includes both air and radiant temperatures, follows a similar trend but at much lower levels. To show the interior temperature regulation of the greenhouse, the dark blue line, the exterior dry-bulb temperature steadily drops to 35°C. The relative humidity panel indicates that the humidity starts at 45%, rises slightly to almost 50% in the early morning, and stays constant throughout the day until roughly 7:00 p.m., when it falls

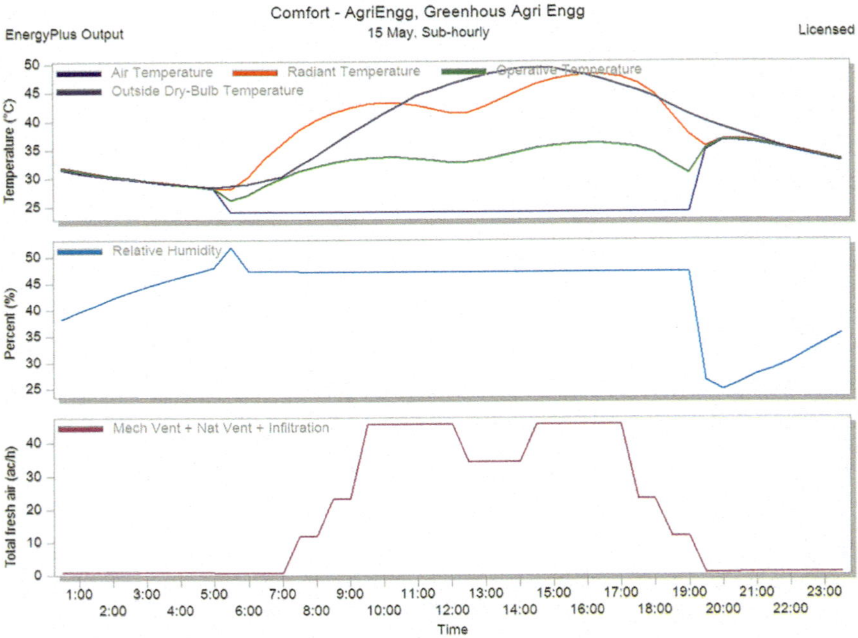

Fig. 13.9 Thermal comfort analysis for the greenhouse in May.

sharply to about 30%. This drop matches an internal temperature rise, as hotter air may retain more moisture, hence reducing relative humidity. The purple line of the ventilation panel shows the combined effects of mechanical and natural ventilation as well as infiltration. Early morning ventilation is low; it then rises at 7:00 a.m., reaching 9 ac/h. Peak temperatures help to explain this fast speed, which helps to manage humidity and cool. Ventilation first declines, then finally recovers to minimal levels around 8:00 p.m. The graph shows the adaptive comfort approach of the greenhouse, which combines temperature control, humidity management, and ventilation to provide ideal growing circumstances during times of strong heat and solar gain.

13.3.4 Thermal comfort analysis for cooling design simulation for June

In Fig. 13.10, on June 15, the graph shows the comfort factors of a greenhouse, including temperature variations, relative humidity, and daily ventilation rates. The temperature screen displays outdoor dry-bulb temperature (dark blue line), operational temperature (green line), radiant

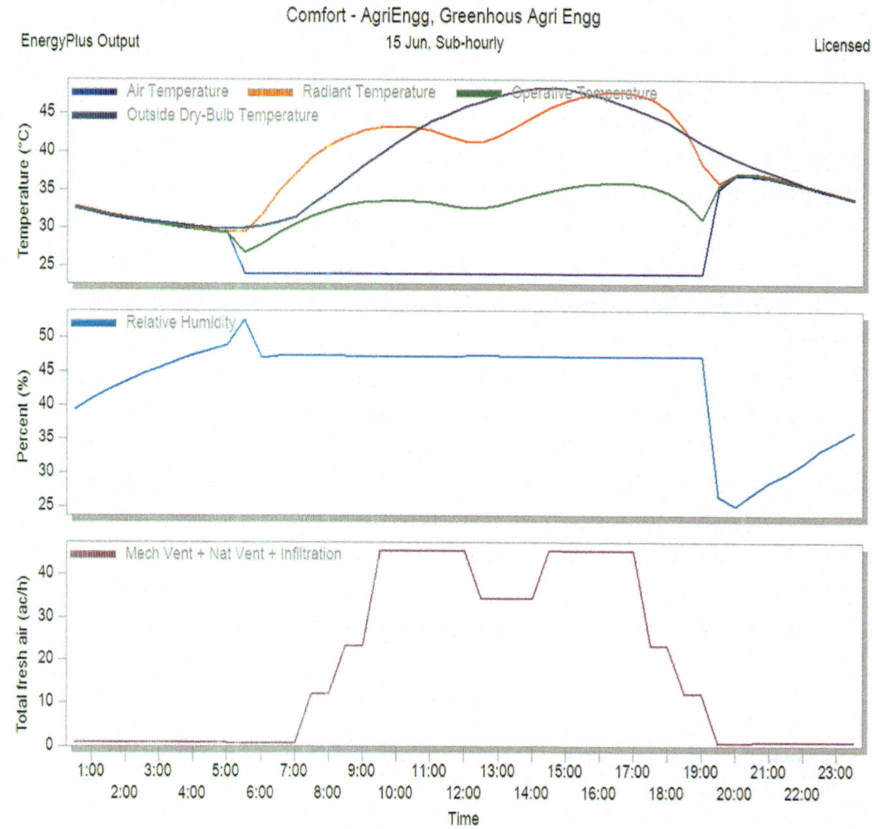

Fig. 13.10 Thermal comfort analysis for the greenhouse in June.

temperature (red line), and air temperature (blue line). <Early in the day, the temperature ranges from 25°C to 30°C; air and radiant temperatures rise rapidly at 7:00 a.m. and peak at 40°C–45°C about midday. The great temperature suggests heat collection from surfaces within the greenhouse as it surpasses the ambient temperature.

The operational temperature, which is a combination of ambient air and radiant heat, also follows a regular pattern, suggesting complete thermal comfort. The low outside dry-bulb temperature continues to be low throughout, reaching 35°C, at which point the inner greenhouse environment seems to be warmer. The relative humidity panel shows that humidity levels start at 45% and gradually climb to about 50% in the early morning. As the temperature rises, relative humidity stabilizes before rapidly falling to roughly 35% by late afternoon, corresponding with peak internal temperatures. The drop-in humidity corresponds to an increase in temperature since heated air retains more moisture and lowers relative humidity. As temperatures cool in the evening, relative humidity rises again.

The graph demonstrates a successful combination of temperature adjustment, humidity management, and ventilation that generates a pleasant and optimum environment for plant development within the greenhouse, especially during periods of maximum solar radiation and heat.

13.3.5 Thermal comfort analysis for cooling design simulation for July

Fig. 13.11 shows how comfortable a greenhouse is on July 15 by showing how the temperature, relative humidity, and air flow change throughout the day. The blue line on the weather panel shows the air temperature. It starts at 28°C–30°C in the morning and rises quickly after sunrise, hitting 40°C–45°C around noon. It then starts to drop in the evening. The radiated temperature (red line), which shows the heat that surfaces give off, follows a similar pattern but has slightly higher peaks; the dark blue line shows that the outdoor dry-bulb temperature stays lower throughout the day. This

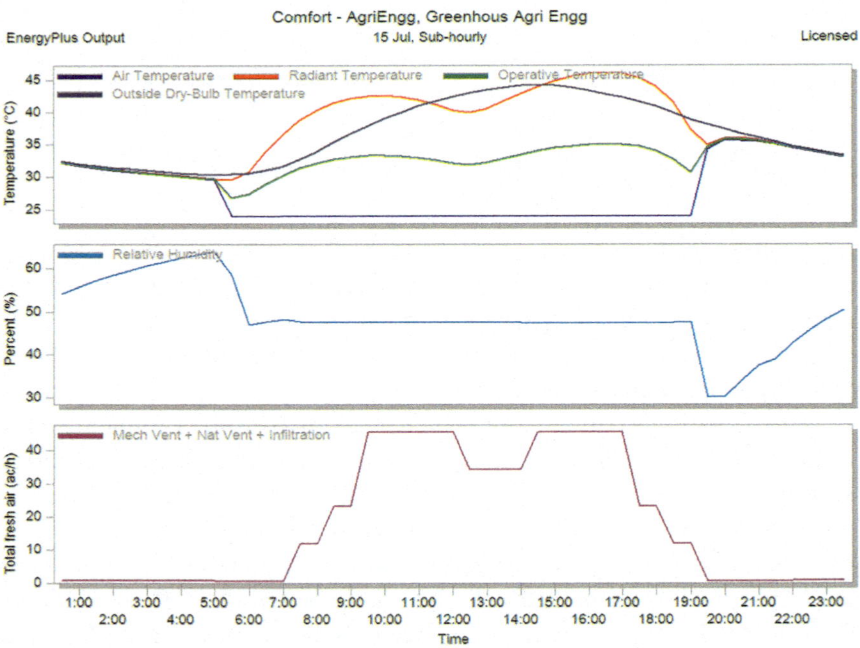

Fig. 13.11 Thermal comfort analysis for the greenhouse in July.

means that there is a temperature difference between the inside and outside. The relative humidity screen shows that the humidity is between 55% and 60% in the early morning. As the temperature inside rises, the humidity slowly drops. It stays between 40% and 45% during the hottest parts of the day and drops to about 35% by late afternoon. The ventilation screen (purple line) shows how much fresh air is coming in from natural and artificial ventilation, as well as entry. Early in the morning, ventilation is slow, but it picks up quickly around 7:00 a.m., and by 9:00, it has reached 35–40 ac/h. This high rate of airflow stays in place all day to keep the temperature and humidity inside under control during the hottest parts of the day. After 5:00 p.m., the airflow rate slowly slows down until it reaches its lowest point after dusk. The graph demonstrates how maintaining the greenhouse's temperature enhances the health of its occupants. Air circulation is good for plant growth because it prevents the temperature and humidity from becoming too high.

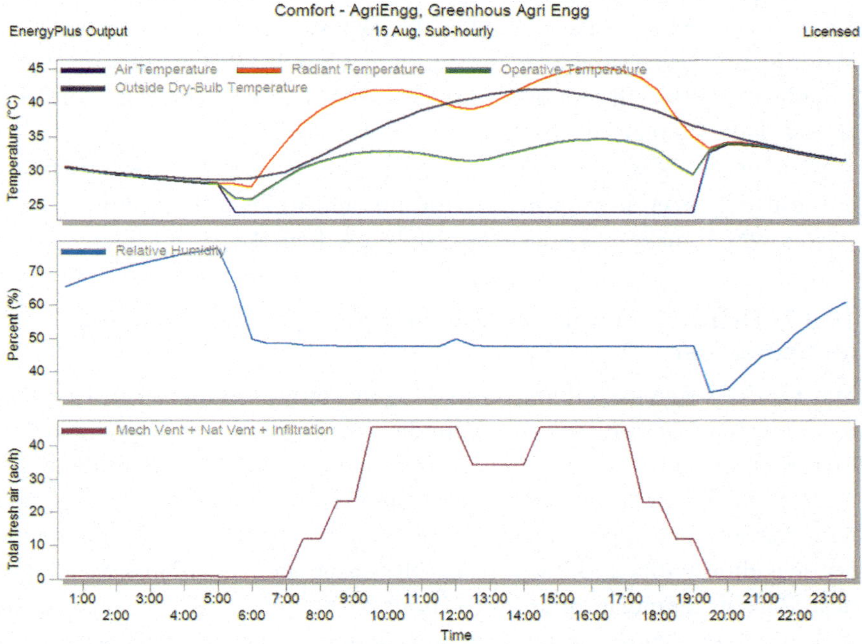

Fig. 13.12 Thermal comfort analysis for the greenhouse in August.

13.3.6 Thermal comfort analysis for cooling design simulation for August

Fig. 13.12 shows the day's changes in temperature, relative humidity, and air flow in a yard. On August 15, it shows the comfort limits. Blue lines on the screen that show the atmospheric temperature show that temperatures can be between 27°C and 30°C in the early morning. It quickly gets warmer after breakfast; by noon, the temperature is between 40°C and 45°C. In the evening, it cools down. The temperature of the red line, which shows that heat is being moved from inside the greenhouse, peaks a little higher than the temperature outside. The greenhouse's surfaces got hot very quickly. The green line shows the working temperature, which is found by adding up the rises in air and sun temperatures above noon. The dark blue line shows that the temperature of the dry bulb is a little cooler on the outside than on the inside. The greenhouse tells us why: it keeps heat in.

The relative humidity screen shows that the humidity is strongest in the early morning, at about 70%. At first, the number goes down, but it stays between 45% and 50% as the temperature rises during the hottest part of the day. It came back in the evening, when the temperature dropped to about 40%.

When fresh air gets into the greenhouse, the ventilation meter shows a purple line. This happens whether the air flow is natural or run-through. Venting isn't very strong in the morning, but by 9 a.m., it's up a lot, reaching 35–40 ac/h. After 5 p.m., ventilation slowly decreases until it's almost dark. By then, it's back to a somewhat low level.

The graph shows that the greenhouse's temperature is constantly being controlled to keep it from getting too hot and to keep the humidity level steady, which makes it perfect for plant growth all day.

13.3.7 Thermal comfort analysis for cooling design simulation for September

Fig. 13.13 displays daily fluctuations in many comfort indicators in a garden, including temperature, relative humidity, and airflow, on September 15. On the temperature display, blue is air temperature, red is radiant temperature, green is operating temperature, and dark blue is outside dry-bulb temperature. The top temperature of the radiating heat inside the greenhouse is almost 40°C, so it's clear that the inside surfaces are making a lot of heat. The air temperature drops to about 35°C–40°C, but the trend of the working temperature, which is the sum of the solar and air temperatures, is

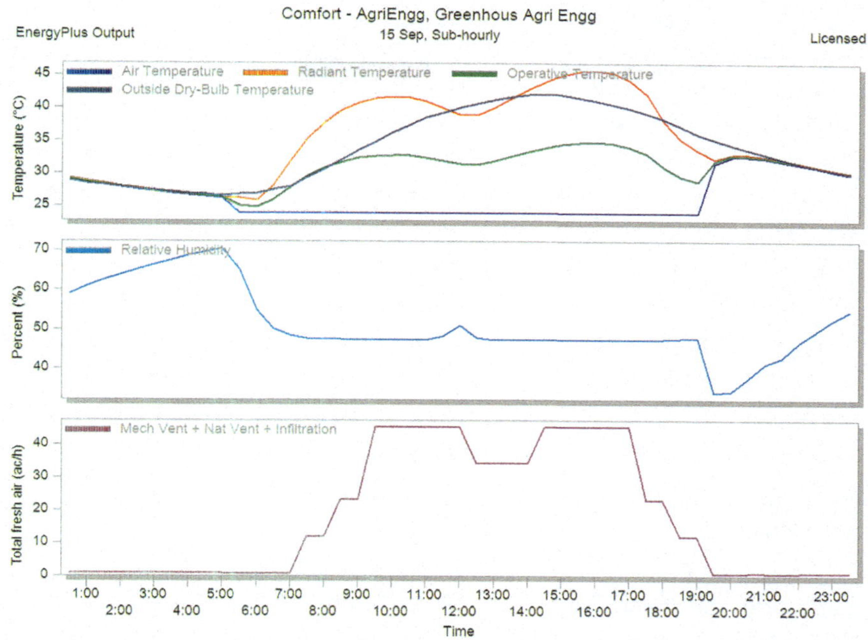

Fig. 13.13 Thermal comfort analysis for the greenhouse in September.

the same. The temperature of a dry bulb outside hits its highest point at 35°C and stays quite low after that.

The relative humidity screen says that the humidity is about 65%–70% in the morning, but it drops a lot as the temperature inside rises. At noon, the relative humidity settles out at 45%–50%, showing that higher temperatures make the humidity lower. The humidity goes up again in the evening when the sun goes down and the temperature drops.

On the ventilation panel, the purple line shows how much fresh air is coming in from entry, natural ventilation, and artificial ventilation as a whole. Ventilation rates are low in the early morning, but they quickly rise after 7 a.m. and reach their highest point of 35–40 ac/h by 9 a.m. By night, it's back to almost nothing. This graph shows how the greenhouse's temperature and humidity management system works to ensure optimum plant growth for better yield production.

13.3.7.1 Energy Consumption

Fig. 13.14 illustrates the January 2022 to December 2023 electricity consumption of a greenhouse, displaying blue for the consumption of the solar energy

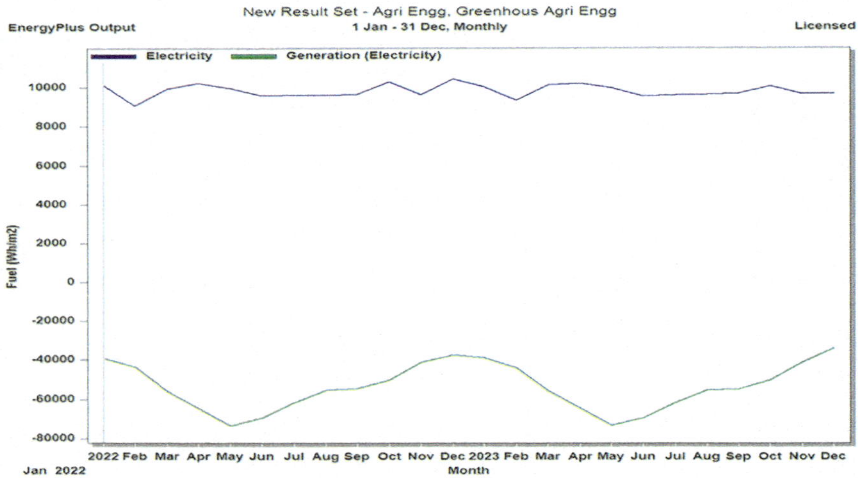

Fig. 13.14 Comparative electricity consumption at the facility for the selected study area locations.

generation and green for the greenhouse's use. With an average monthly electricity use of around 10 kWh/m^2, the blue line shows quite little variation. Some might find it shocking that demand rises in the warmer months of June and July because of more need for cooling. On the other hand, a drop of the same kind occurs in the winter. On the other hand, the trend of solar energy based on solar power showcases more clear seasonal fluctuations. Measuring solar energy from 6 to 8 kWh/m^2, the summer months of April through September topped the readings. They have reduced solar output results from shorter days and less sunlight throughout the winter, October through February. The lowest counts fell in January and December. Particularly in the summer, the solar energy production of the greenhouse meets the somewhat higher demand. The present solar power cannot satisfy the energy consumption. If the temperature and humidity of the greenhouse are maintained constant, it is more probable that extra energy sources or storage devices will be required to meet annual demand. The graph showing the January 2022 to December 2023 electricity consumption of a greenhouse displays blue for the consumption of the solar energy generation and green for the greenhouse's use. With an average monthly power usage of around 10 kWh/m^2, the blue line shows quite little variation.

13.4 Conclusions

This research investigated the potential for a solar energy system and a desiccant dehumidification system to collaborate in a greenhouse setting using the Design Builder program in conjunction with the EnergyPlus engine. The primary objective was "Simulation of small-scale greenhouse buildings by integration of solar energy and desiccant dehumidification systems using a Design-Builder" under real-world conditions. The software findings demonstrate the viability of combining solar energy with desiccant dehumidification for long-term usage in greenhouse farming. The following are the key findings of research results. To power the desiccant dehumidifier, lights, and heating, the greenhouse successfully incorporated a solar energy system, which supplied around 6–8 kWh/m^2 during peak months. The greenhouse relied heavily on renewable energy sources to power its operations, with monthly electricity use averaging about 10 kWh/m^2. Relative humidity was between 40% and 50%; summertime highs hit 45°C. Mixing mechanical and natural ventilation allowed us to achieve 35–40 ac/h, therefore greatly enhancing our power to control temperature and humidity during crop growth. With a maximum of 8 kWh/m^2, the summer months produced the most solar energy, therefore improving the sustainability of the system. In winter, the 2 kWh/m^2 alternative source or energy storage is needed. Combining solar power with desiccant dehumidification technology helped the greenhouse have a lesser carbon impact than conventional agricultural processes. This system is ecologically friendly for controlling temperature and humidity in greenhouses, and the results show that it performs best in regions with high sunshine and humidity. This approach is especially helpful in places with limited energy supply and high humidity, as it allows farmers to raise agricultural yields while reducing running costs and environmental repercussions. By means of quantitative solar energy collection, the greenhouse may meet up to 80% of its energy needs during peak months, therefore retaining perfect circumstances for plant development. Using this technology offers modern greenhouse farming a reasonably affordable and ecologically friendly alternative.

Acknowledgment

This research work is completed with the financial support from the Higher Education Commission (HEC) of Pakistan through the research Project NRPU-17029 awarded to Bahauddin Zakariya University. The United Kingdom Government funds this research (and/or) pilot activity through the Project: APP47457, titled "Super-efficient Sustainable Cooling Solution for All Applications (S2Cool)" under the Ayrton Challenge Programme of the UK Research and Innovation (UKRI); however, the views expressed do not necessarily reflect the UK Government's official policies. The Project, APP47457, is implemented/led by Northumbria University, UK, together with partners to develop sustainable cooling solution for several applications.

Funding statement

This research work is completed with the financial support from the Higher Education Commission (HEC) of Pakistan through the research Project NRPU-17029 awarded to Bahauddin Zakariya University. The United Kingdom Government funds this research (and/or) pilot activity through the Project: APP47457, titled "Super-efficient Sustainable Cooling Solution for All Applications (S2Cool)" under the Ayrton Challenge Programme of the UK Research and Innovation (UKRI); however, the views expressed do not necessarily reflect the UK Government's official policies. The Project, APP47457, is implemented/led by Northumbria University, UK, together with partners to develop sustainable cooling solution for several applications.

CRediT authorship contribution statement

Junaid Ramzan was involved in the conceptualization, methodology, software, formal analysis, investigation, writing of the original draft. **Muhammad Sultan**, **Takahiko Miyazaki**, **and Muhammad Wakil Shahzad** were involved in the conceptualization, methodology, validation, resources, visualization, supervision, project administration, writing (review and editing), funding acquisition. **Hafiz Muhammad Umar Raza**, **Muhammad Hamid Mahmood**, **Muhammad Ahmad Jamil**, **and Uzair Sajjad** were involved in the conceptualization, methodology, software, formal analysis, investigation, writing (review and editing).

Data Availability Statement

The datasets used and/or analyzed during the current study are available from the corresponding authors upon request.

Conflicts of Interest

The authors declare no conflict of interest.

Declaration of generative AI and AI-assisted technologies in the writing process

No generative AI/AI-assisted technology is used in this paper.

References

Abdel-Salam, A.H., Ge, G., Simonson, C.J., 2013. Performance analysis of a membrane liquid desiccant air-conditioning system. Energy Build. 03787788. 62, 559–569. doi:10.1016/j.enbuild.2013.03.028.

Ashraf, H., Sultan, M., Shamshiri, R.R., Abbas, F., Farooq, M., Sajjad, U., Md-Tahir, H., Mahmood, M.H., Ahmad, F., Taseer, Y.R., Shahzad, A., Niazi, B.M.K., 2021. Dynamic evaluation of desiccant dehumidification evaporative cooling options for greenhouse air-conditioning application in multan (Pakistan). Energies. 19961073. 14 (4). Available from: https://www.mdpi.com/1996-1073/14/4/1097/pdf. doi:10.3390/en14041097.

Çerçi, K.N., Silva, I.R.O., Kaşka, Ö., Hooman, K., 2024. Summer period analysis of the rotary desiccant—hybrid cooling system combined with solid oxide fuel cells using human waste fuel. Clean. Eng. Technol. 26667908. 23. Available from: https://www.sciencedirect.com/science/journal/26667908. doi:10.1016/j.clet.2024.100818.

Hussain, T., 2023. Optimization and comparative performance analysis of conventional and desiccant air conditioning systems regenerated by two different modes for hot and humid climates: experimental investigation. Energy Built Environ. 26661233. 4 (3), 281–296. Available from: www.keaipublishing.com/en/journals/energy-and-built-environment/. doi:10.1016/j.enbenv.2022.01.003.

Iqbal, W., Mahmood, M., Iqbal, S., Ali, M., Iqbal, M.H., Hussien, A.G., Kamel, S., 2023. Techno-economic and emission analysis of solar assisted desiccant dehumidification: an experimental and numerical study. Energy Rep. 23524847. 10, 2640–2654. Available from: http://www.journals.elsevier.com/energy-reports/. doi:10.1016/j.egyr.2023.08.077.

Jiaqiang, Y., Yulong, J., Jian, G., 2013. An intelligent greenhouse control system. TELKOMNIKA Indones. J. Electr. Eng 11 (8), 4627–4632. doi:10.11591/telkomnika.v11i8.3088.

Liu, J., Liu, X., Zhang, T., 2019. Performance investigation of a heat pump driven, vacuum liquid desiccant regeneration system. Energy Procedia 18766102. 158, 2435–2440. http://www.sciencedirect.com/science/journal/18766102. doi:10.1016/j.egypro.2019.01.307.

Noor, S., Ashraf, H., Sultan, M., 2021. Spatiotemporal investigation of evaporative cooling options for greenhouse air-conditioning application in PakistanFresenius Environ. Bull. 30 (3), 2858–2870. Available from: https://www.researchgate.net/publication/347496135.

Pasqualin, P., Lefers, R., Mahmoud, S., Davies, P.A., 2022. Comparative review of membrane-based desalination technologies for energy-efficient regeneration in liquid desiccant air conditioning of greenhouses. Renew. Sustain. Energy Rev. 13640321. 154, 111815. doi:10.1016/j.rser.2021.111815.

Shamim, J.A., Hsu, W.L., Kitaoka, K., Paul, S., Daiguji, H., 2018. Design and performance evaluation of a multilayer fixed-bed binder-free desiccant dehumidifier for hybrid air-conditioning systems: part I—experimental. Int. J. Heat Mass Transf. 00179310. 116, 1361–1369. Available from: http://www.journals.elsevier.com/international-journal-of-heat-and-mass-transfer/. doi:10.1016/j.ijheatmasstransfer.2017.09.051.

Sultan, M., Ashraf, H., Miyazaki, T., Shamshiri, R.R., Hameed, I.A., 2021. Temperature and Humidity Control for the Next Generation Greenhouses: Overview of Desiccant and Evaporative Cooling Systems. IntechOpen. doi:10.5772/intechopen.97273.

Sultan, M., Miyazaki, T., 2017. Energy-Efficient Air-Conditioning Systems for Nonhuman Applications. IntechOpen. doi:10.5772/intechopen.68865.

Sultan, M., Miyazaki, T., Koyama, S., Saha, B.B., 2014. Utilization of desiccant air-conditioning system for improvement in greenhouse productivity: a neglected area of research in Pakistan. Int. J. Environ. 4 (1), 1–10. Available fromhttp://www.benjapan.org/IJEARTICLEHISTORY.

CHAPTER 14

Integration of solar thermal collectors with desiccant cooling systems for controlled environments

Muhammad Waheed Azam[a], Ghulam Qadir Chaudhary[b], Giovani Sempirini[c], and Uzair Sajjad[d]
[a]Interdisciplinary Research Center for Industrial Nuclear Energy, King Fahd University of Petroleum & Minerals, Dhahran, Saudi Arabia
[b]Mechanical Engineering Department, MUST, Mirpur, Pakistan
[c]Department of Industrial Engineering (DIN), Alma Mater Studiorum - University of Bologna, Bologna, Italy
[d]Interdisciplinary Research Center for Sustainable Energy Systems (IRC-SES), King Fahd University of Petroleum and Minerals, Dhahran, Saudi Arabia

14.1 Introduction

In recent decades, it has been observed that there is a significant need for air conditioning (lying within the temperature range of 20°C–26°C) in the building sector due to climate change, economic growth, urbanization, cultural shifts, and increasing standards of living. As a result, energy demand for air conditioning is increasing steadily, which significantly contributes to global energy consumption (Ibrahim et al., 2026). The International Energy Agency reported that the building sector is considered for global energy consumption and carbon emissions, which contributed approximately 30% and 27%, respectively (United Nations Environment Programme, 2021). These resulting carbon emissions are caused by various building operations that require energy, including water heating, space heating, lighting, air conditioning, and cooking. Moreover, in residential and commercial buildings, these systems consume approximately 40% and 36%, due to continuously maintaining indoor temperature. According to researchers, the demand for air conditioning systems is predicted to triple by 2050 compared to 2019, meaning that 5.6 billion air conditioning units will be sold over the next 30 years due to market growth (International Energy Agency, n.d.).

Most commonly, the demand for air conditioning in both residential and commercial buildings is fulfilled through vapor-compression refrigerant-based systems. These systems consist of four major components (compressor,

Sustainable Hybrid Cooling and Dehumidification Strategies for Controlled Environments
https://doi.org/10.1016/B978-0-443-48555-8.10014-3

condenser, expansion valve, and evaporator). Briefly, refrigerants absorb heat in the evaporator and release it to the environment through the condenser. Moreover, among these four components, the compressor is most important because it compresses the refrigerant vapor and makes it possible for the refrigerant to continuously circulate throughout the system. However, these systems consume a significant amount of energy, have a high initial cost, and contribute to global warming (greenhouse effect, acid rain, and other generalized environmental pollution) and ozone depletion due to leakage of refrigerants. As a result of these concerns, researchers and scientists face many challenges to investigating and developing sustainable, environmentally friendly alternatives to conventional vapor-compression air conditioning systems. Desiccant cooling systems are an alternative solution to vapor-compression refrigerant-based systems that minimize energy consumption and environmental hazards (Duong, 2023). In this system, a desiccant wheel is used to dehumidify the air and then cooled in a sensible heat exchanger, and this air is further cooled by an evaporative cooling technique or the Maisotsenko cycle (MC) and finally directed into the required space. Furthermore, these systems not only maintain the indoor temperature but also maintain indoor humidity, which is an important parameter for human comfort. However, to restore the desiccant system to retain its drying ability, a consistent heat source (a temperature of approximately 60°C–95°C) is needed (Cuce et al., 2025). This stage is crucial, where the integration of the desiccant system with a solar thermal collector becomes highly critical. To restore the desiccant wheel by utilizing the solar heat instead of electricity or fossil fuels transforms a renewable and free resource into an innovative cooling impact, producing a climate control solution that is effectively sustainable (Shehram et al., 2025).

The following areas will be covered in this chapter:

- Comprehensive introduction to desiccant cooling systems and various thermal collectors.
- Integration strategies with thermal collectors.
- Application in a controlled environment.
- Challenges and future research directions.

14.2 Desiccant cooling system

The traditional vapor compression and absorption cooling systems can be replaced by desiccant cooling systems, which are heat-driven cooling devices. It is the best solution for addressing future energy demand challenges

Table 14.1 Summary of desiccant cooling system components.

Component	Function
Desiccant wheel (Shamim et al., 2025)	• It helps to remove the moisture from the air through adsorption or absorption
Regeneration heater (Verma and Yadav, 2025)	• Heats the regeneration air to the required temperature (typically 60°C–90°C) • Removes absorbed moisture from the desiccant
Heat exchanger (Zhao et al., 2021)	• It recovers heat between air streams • Preheats the regeneration air • Maintains thermal separation
Evaporative cooler (Zeoli et al., 2025)	• Cools the dehumidified (dry and warm) air • Uses water evaporation to lower the air temperature
Fans/blowers	• Circulate the air • Overcome pressure drops
Filters	• Cleans the air • Prevent fouling
Solar collector (Ge et al., 2014)	• Provide renewable heat • Supports energy-efficient and sustainable cooling solutions • Enables environmentally friendly operation • Maintains continuous desiccant operation in solar-assisted systems

while maintaining human comfort due to its capacity to adjust both temperature and humidity (Sultan et al., 2018a). The main components of the desiccant cooling system are the desiccant wheel, regeneration heater, heat exchanger, evaporative cooler, fans, filter, and solar thermal collectors. The main components and their functions in the desiccant cooling system are shown in Table 14.1.

These systems provide improved air quality, less energy consumption, and no refrigerants such as chlorofluorocarbons, hydrofluorocarbons, or hydrochlorofluorocarbons. However, these systems have a high initial cost, complex system design, and performance that depends on the environmental conditions. The system can be divided into two categories based on the condition of the materials used: liquid and solid desiccants. Each type has unique characteristics, working methods, and uses in air dehumidification and humidity control systems, explained below.

14.2.1 Solid desiccant cooling system

An energy-efficient air conditioning system known as a solid desiccant cooling system cools and dehumidifies air using a solid desiccant material, such as lithium chloride, zeolite, or silica gel. Silica gel is the most commonly used desiccant material in these systems that are currently on the market. It is stable, solid, insoluble, nontoxic, environmentally friendly, and also does not interact with most chemicals. The basic principle behind solid desiccant cooling is that as process air flows through a rotating dehumidifier, excessive water vapor is evaporated as a result. After that, the dehumidifier is regenerated by using hot air from a heater to absorb the moisture that has been accumulated, and a sensible heat recovery wheel and cooling coils are used to further cool the dry process air to a suitable room temperature (Lee et al., 2021).

To improve process efficiency and promote solar system utilization, a standard solar system may be integrated with solar collectors (which include water as a fundamental fluid and a separate water storage system). However, to connect a suitable air cooling system with the solar heating system, these integrated systems require a further air-to-water heat exchanger. Due to the high ambient humidity that precludes direct evaporative cooling, conventional desiccant cooling methods are inefficient in extremely humid places, such as the Mediterranean coast. This issue can be resolved by using more complex designs, such as desiccant air handling systems, which use enthalpy wheels and air coolers as extra inputs to supply cold water (Hanif et al., 2022).

The following are the primary benefits of using solid desiccant systems (Baniyounes et al., 2013):

- Materials used for desiccants can handle a range of high and low temperature variations.
- Between 60°C and 90°C, many solid desiccants can be regenerated, making them appropriate for waste heat or solar energy (Dincer and Rosen, 2007).
- It is suitable for both large-scale (commercial, industrial) and small-scale domestic applications.
- Compared to liquid materials, solid desiccants are more affordable and easily available
- Efficiently handles latent heat, reducing load on conventional cooling systems.
- Desiccant wheels that rotate 15–20 times per hour are used in solid desiccant systems (Xu and Chen, 2024).

- Any building cooling system that is currently in operation can be retrofitted.
- Due to less electricity consumption, there are reduced CO_2 emissions ultimately.

However, these systems have a number of disadvantages, including the need for a higher regeneration temperature, large size, and a comparatively greater cost as compared to liquid desiccant machines.

14.2.2 Liquid desiccant systems

A Liquid Desiccant Cooling System (LDCS) (Rekhade et al., 2025) is an advanced air-conditioning technology that cools and dehumidifies the air using a hygroscopic liquid, a material that draws moisture from the air. The most popular liquid desiccant materials on the market are tri-ethylene glycol, calcium chloride, lithium chloride, and lithium bromide. These materials are cheap, nontoxic, nonflammable, and odorless. The working principle of a LDCS is based on moisture absorption and regeneration (Liang and Li, 2025). The liquid desiccant in the conditioner absorbs moisture from the humid air and uses latent heat removal to cool it. After being heated and diluted, the desiccant is delivered to the regenerator, where it is heated (often with waste heat or solar energy) in order to release the absorbed moisture and restore its concentration so that it can be used again in the cycle. This technique is a very promising choice for increasing the use of solar-thermal air conditioning systems (Deng et al., 2011).

The following are the primary benefits of using liquid desiccant systems:

- When integrated into liquid-liquid heat exchangers, it demonstrates a high efficiency of heat transfer.
- Due to their limited pressure drop, liquid desiccant systems require a low regeneration temperature.
- The ability to regenerate the liquid desiccants at temperatures lower than 80°C makes the liquid desiccant system attractive.
- Liquid desiccant systems don't need to be continuously regenerated.
- Their ability to pump the liquid allows them to be found in small and compact sizes.

However, there are several drawbacks to liquid desiccant systems, such as the requirement for large amounts of desiccant and high-power pumps, the need for an additional heat exchanger for cooling, and a fan with a higher power because of the high pressure drop. The design of liquid desiccant systems is simple because the desiccant is typically sprayed directly over the incoming air to dehumidify it. There is no need for complex dehumidifier geometries because these systems may be regenerated utilizing inexpensive open-flow collectors or storage tanks. The water removal process in liquid desiccant systems operates in almost isothermal conditions, which contributes to steady thermal performance and improved energy efficiency overall. On the other hand, adiabatic operation of solid desiccant devices results in larger temperature fluctuations throughout the moisture removal process. Moreover, due to cooling tower water or chilled water providing internal cooling, liquid desiccants can provide dehumidification that is comparable to solid desiccant systems with a lower regeneration temperature (Liang and Li, 2025).

14.3 Solar-assisted desiccant air conditioning system

Solar thermal energy integration is a sustainable, eco-friendly technique to improve energy efficiency and reduce the carbon footprint of cooling operations in desiccant air-conditioning (DAC) systems (Alam et al., 2025). These systems are an eco-friendly alternative to traditional systems (conventional vapor compression systems), decoupling latent and sensible heat loads for improved energy efficiency. They can also reduce energy consumption, especially in humid climates, by using solar thermal energy for desiccant regeneration (Abbas et al., 2025). The various solar thermal technologies and integration strategies with desiccant systems are explained below.

14.3.1 Solar energy technologies

The production of a solar energy system depends on the unpredictable patterns of daily solar radiation, making it one of the most intermittent renewable energy sources, and it is widely recognized. As technology develops, solar is going to become gradually less expensive and, by 2030, the primary energy source for producing electricity in the majority of the world because the peak solar potential coincides with the building's peak cooling demands. Solar energy for achieving thermal comfort in summer is an attractive alternative to conventional air conditioning (Boutagount et al., 2025). Solar thermal collectors are devices that convert solar radiation into heat energy and drive the absorption and adsorption systems. The absorption cooling system is an effective method for both commercial and residential buildings, efficiently meeting cooling requirements by using solar thermal energy to operate a chemical cycle that produces cooling. Solar

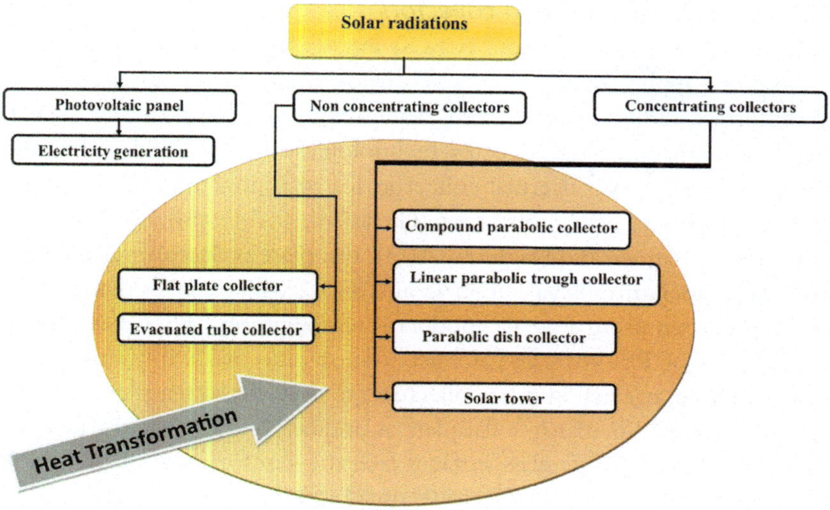

Fig. 14.1 Classification of solar energy technologies.

thermal systems are classified into several types based on geometry, working principle, and temperature range. The classification of solar thermal technologies is given in Fig. 14.1. The most common applications of solar thermal collectors are domestic water heating, space heating, direct and indirect power generation, desalination, and cooling processes. Residential applications use low-temperature collectors, whereas industrial and power-generation applications use medium- and high-temperature solar thermal collectors. Solar energy can be directly converted into electricity by using solar photovoltaic (PV) panels (Azam et al., 2025). Since solar PV panels don't require direct sunlight to operate, they can produce electricity even on cloudy days. These panels are more effective in relatively foggy conditions, since sunlight reflections can increase heat loss. The generated electrical energy can be supplied to conventional vapor-compression systems to produce the air-conditioning effect. However, the efficiency of PV panels is very low, i.e., about 15%–20% (Essa et al., 2025). A significant problem in warmer climates is that PV panels frequently perform less effectively in hot weather. Furthermore, solar energy is intermittent and therefore requires a storage system, but electric energy storage is more expensive than thermal energy storage. Hence, as an alternative, it is possible to use solar thermal collectors to harness solar thermal energy for heat-driven sorption cycles due to their higher efficiency compared to PV panels, i.e., up to 80% (Kadam et al., 2025).

14.3.1.1 Nonconcentrating collectors

Concentrating solar collectors (CSCs) are widely used in residential and commercial applications, are suitable for low-temperature applications, and absorb direct, diffuse, and reflected solar energy over a large surface area. The major nonconcentrating collector technologies are described in Fig. 14.1. Commonly used solar thermal collectors for solar air-conditioning applications include flat-plate collectors (Ge et al., 2014) and evacuated-tube collectors (Ensafjoee and Daghigh, 2023). Flat plate collectors perform effectively at temperatures between 35°C and 85°C, which makes them ideal for applications requiring low to moderate temperatures. They have an efficiency range of 30%–40% and a concentration ratio of ≤1. In terms of design and technique, evacuated tube collectors are superior to flat plate collectors. The sun is perpendicular to a flat plate at midday, but because the evacuated tube is tubular, all of its glass surfaces face the sun for most of the day (Ali et al., 2025). The flat plate collector is usually applicable for low-temperature applications. However, the evacuated tube collector can produce heat energy at elevated temperatures, i.e., up to 200°C.

14.3.1.2 Concentrating collectors

To achieve high temperatures, a CSC directs sunlight onto a small receiver area using mirrors or lenses. The major concentrating collector technologies are described in Fig. 14.1. In comparison to nonconcentrating collectors, this particular type of solar collector mostly operates at a higher temperature. A thermally operated cooling system is subsequently powered by this heat, including:

- Desiccant cooling systems
- Adsorption cooling systems
- Absorption cooling systems

The integration of solar-powered absorption cooling cycles and high-efficiency solar concentrating collectors is one of the most promising and environmentally friendly ways to meet modern-day cooling demands. This strategy improves system performance and overall energy utilization while lowering reliance on conventional sources of energy. Trombe and Foex conducted one of the first attempts in 1957 to achieve the desired result from an aqua-ammonia refrigerator driven by solar energy with respect to solar absorption cooling (Trombe and Foex, n.d.). It can be divided into three categories: single-effect [temperature range 80°C–100°C and coefficient of performance (COP) 0.7], double-effect (temperature range 120°C–150°C and COP 1.4), and multi-effect absorption cycles (temperature range 180°C–240°C and COP 1.8). This demonstrates that increasing the amount of absorption effects requires a higher driving temperature, which results in an increase in the COP. There are several types of concentrating collectors, including central receivers, linear Fresnel collectors, parabolic trough collectors, compound parabolic collectors, and parabolic dish collectors (Alsagri et al., 2020).

A desiccant wheel, which must be continuously regenerated with heat at temperatures normally between 60°C and 90°C, is used in desiccant systems to dehumidify air (Yang et al., 2021). Therefore, the high operating temperature of a parabolic trough collector has contributed to it being an effective thermal collector for a desiccant cooling system. The parabolic trough collector has two primary benefits: Firstly, it can collect sunlight in the early morning and late evening hours because it does not require a secondary reflector, resulting in high optical efficiency. Another type of concentrating collector used for air conditioning is a point-focus collector, also called a parabolic dish collector, which focuses sunlight onto a receiver positioned at the dish's focal point. Moreover, compound parabolic concentrators (CPCs) are usually preferred due to the suitability of the temperature range required for solar air conditioning applications. In conclusion, this technology significantly reduces the electrical energy used by a typical air conditioner for dehumidification by directly employing solar thermal energy for regeneration. This shifts the load from electricity to abundant, clean solar heat. As to result, concentrated solar-powered desiccant cooling is a highly effective method for controlling latent loads in humid conditions. It also lowers operating costs, reduces grid dependency, and greatly enhances indoor air quality (Patel and Modi, 2025).

14.3.2 Application of different thermal collectors

The solar thermal collectors are summarized with respect to efficiency trends and application range in Fig. 14.2. The primary vertical axis shows the collector efficiency, and the horizontal axis shows the ratio of the temperature difference (the difference between the average collector temperature and the ambient temperature) to the incident radiation. The efficiency curves correspond to test data for different collectors relative to an average ambient temperature of 25°C and an incident solar radiation of 1000 W/m^2.The choice of solar thermal collectors significantly impacts the performance and efficiency of desiccant cooling and absorption systems. Flat plate collectors with selective coatings, air collectors with selective coatings,

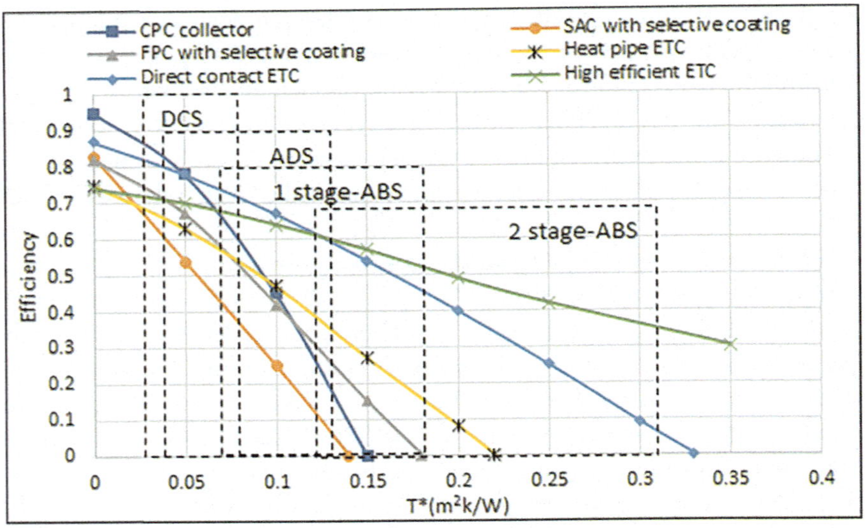

Fig. 14.2 Types of solar thermal collectors with their potential application. *(Reproduced with permission from: Henning, H.M., 2007. Solar assisted air conditioning of buildings - an overview. Appl. Therm. Eng 27 (10), 1734–1749. doi:10.1016/j.applthermaleng.2006.07.021.)*

and CPCs are particularly well-suited for desiccant cooling applications because they reliably operate around 70°C. This moderate operating temperature aligns well with the thermal requirements of desiccant cooling processes, ensuring effective desiccant regeneration while maintaining high energy efficiency (Buker and Riffat, 2015).

These same technologies are versatile and also find applications in adsorption and single-stage absorption cooling systems. The moderate temperature range they provide supports the regeneration of adsorbents and absorbents used in these systems, promoting steady performance and cost-effectiveness. However, an important limitation of flat plate collectors, air collectors, and CPCs is their efficiency sensitivity to temperature gradients. As the difference between the ambient temperature and the collector's average temperature increases, efficiency tends to decline more rapidly. This is due to larger thermal losses from the collector surface, which reduce the net useful heat gain. In contrast, heat-pipe evacuated-tube collectors demonstrate superior performance at larger temperature differences (Bozorgi et al., 2024a). Their design minimizes heat losses and maintains a more stable thermal efficiency when the temperature gradient increases, making them particularly favorable for applications where ambient conditions fluctuate or higher operating temperatures are required. For higher-temperature applications, such as two-stage absorption cooling systems operating around 140°C, high-efficiency evacuated tube collectors and direct-contact evacuated tube collectors are suitable options. These collectors can safely and efficiently achieve and sustain the elevated temperatures required by two-stage absorption cycles, boosting system performance and reliability while enhancing the overall solar thermal conversion efficiency. Their advanced heat transfer mechanisms and vacuum insulation enable them to handle higher thermal loads with less efficiency loss compared to conventional collectors (Bozorgi et al., 2024b).

Overall, selecting the appropriate collector type depends on the cooling system's temperature demands, ambient conditions, and efficiency requirements. Integrating collectors such as heat-pipe evacuated tubes and high-efficiency evacuated tubes can optimize system operation, especially for high-temperature absorption systems, while simpler flat-plate or CPC collectors remain efficient choices for moderate-temperature cooling applications.

14.3.3 Solar-assisted desiccant cooling system configurations

Air conditioning devices play a crucial role in providing a healthy, comfortable indoor environment by actively regulating temperature and humidity, which are fundamental to thermal comfort. This climate control is essential not only for occupant health and overall well-being, but also for boosting human productivity in various conditions (including homes, offices, factories, and other workspaces) where a stable and comfortable atmosphere helps individuals perform at their best. Thermal comfort itself is a complex concept that goes beyond simple temperature control; it is achieved through a balanced combination of temperature and moisture management, which advanced sorption techniques can effectively provide. Sorption methods, such as desiccant cooling systems, remove moisture from the air to maintain a desirable humidity level while also allowing specific management of sensible loads (temperature) and latent loads (humidity). This is achieved by integrating desiccant technology with an appropriate sensible cooling device, offering greater flexibility and efficiency across different indoor environments. There are various possible configurations of a solar-assisted desiccant cooling system (Hua et al., 2018). However, configuration selection depends on climate conditions, the type of application, and available space. Some commonly used configurations are discussed below.

14.3.3.1 Solar thermal autonomous desiccant cooling system

Solar autonomous desiccant cooling systems are suitable when air-conditioning is required only during the sunshine period. Furthermore, these systems are only ideal for loads corresponding to the direct solar heat gain, such as in the seminar rooms. There are two types of solar-autonomous desiccant cooling systems: (1) integrated design and (2) ambient air design. The schematic representation of integrated design is given in Fig. 14.3. In this configuration, the regeneration side air leaving the heat wheel is supplied to the air collector for heating. The air absorbs thermal energy and regenerates the desiccant wheel. This configuration is more efficient because less heat is required from the air stream to meet the desired regeneration demand (Robbins and Garimella, 2020).

Furthermore, the ambient air design configuration is shown in Fig. 14.4. In this configuration, the ambient air directly enters the thermal collector, where it absorbs thermal energy. Afterward, the heat collected by these solar thermal collectors is used to regenerate the desiccant wheel. The ambient air arrangement is especially efficient because it utilizes the use of solar energy, as we know that it is widely accessible, reducing reliance on electrical or fossil-fuel-driven heating systems. In addition, the direct utilization of ambient air in the heating process lowers system complexity, resulting in easier installation and maintenance needs. However, the efficiency of this arrangement is dependent on the amount of solar radiation that is available as well as the design of the thermal collector, which needs to be optimized for the local climate in order to make sure reliable regeneration performance.

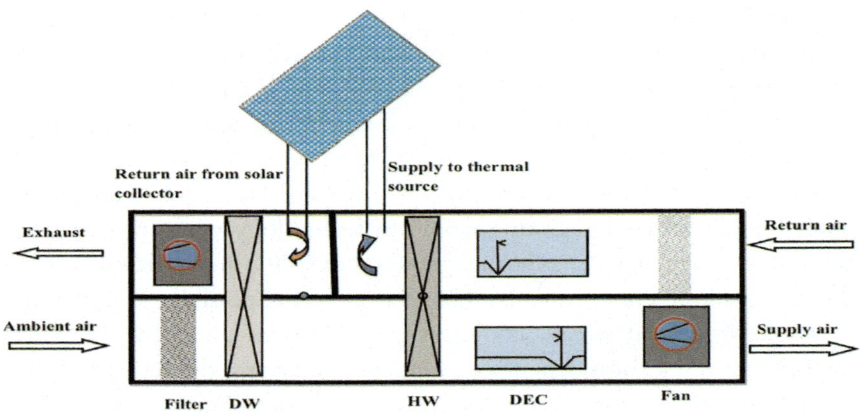

Fig. 14.3 Autonomous desiccant cooling system with integrated design: Author's work.

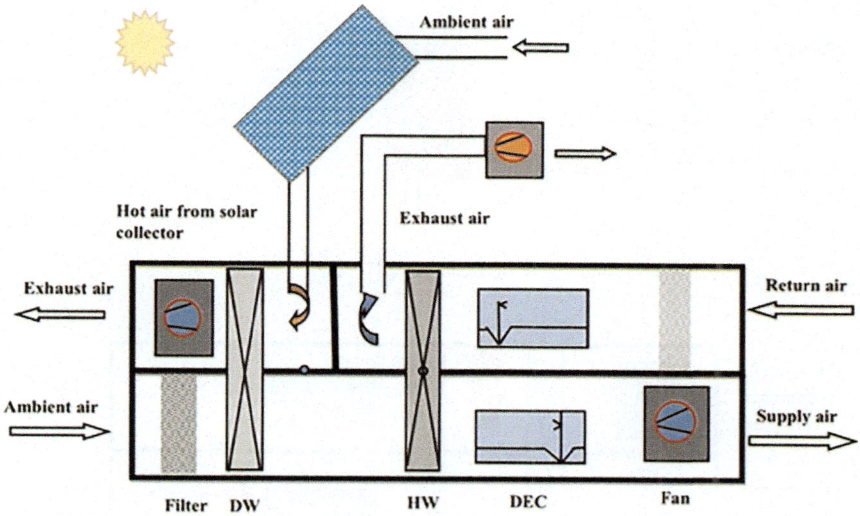

Fig. 14.4 Solar autonomous with ambient air conditions: Author's work.

14.3.3.2 Solar-assisted desiccant cooling system with auxiliary heater

An interesting and attractive alternative for building air conditioning applications is solar-assisted desiccant cooling technology. Air collectors are the main source of solar energy for the above configurations (such as the solar thermal autonomous desiccant cooling system); however, these designs have significant drawbacks when it comes to storing thermal energy for usage later on. In comparison to liquids, air has a lower heat capacity, which means it cannot effectively hold onto heat for long periods of time. Due to this, when solar energy is low, like during cloudy weather or at night, these devices lose some of their effectiveness. Therefore, to overcome these limitations, integrate thermal storage with solar-assisted desiccant cooling systems because it provides consistent and continuous functioning even when there is no direct sunlight available. This advanced system uses storage tanks to hold heated water, which acts as a reservoir for thermal energy. During sunny days, the collector loop warms the water in the storage tanks by absorbing solar energy. The desiccant material is subsequently regenerated by using this heated water to provide the thermal energy required for desorption, which is the process of extracting the absorbed moisture out of the desiccant wheel. Using water as the storage medium provides excellent thermal retention, enabling the system to handle cooling loads efficiently when solar radiation is insufficient (Zhang et al., 2025a).

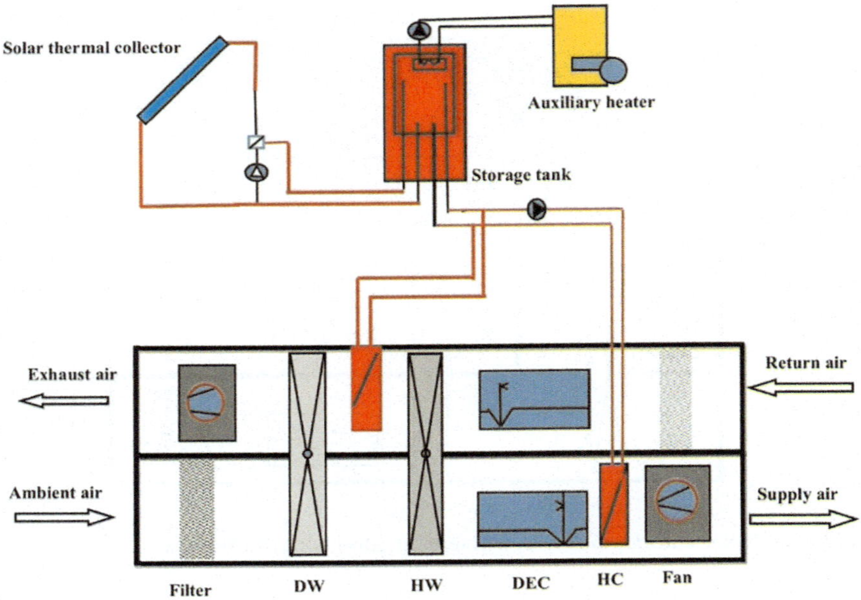

Fig. 14.5 Solar-assisted desiccant cooling system with auxiliary heater.

In addition, an auxiliary or backup heat source, such as an electric heater, gas burner, or other thermal unit, is integrated into the system to ensure the desired regeneration temperature is always reached, even under unfavorable weather conditions. This backup component enhances system reliability and makes it suitable for a wide range of climatic regions. The schematic in Fig. 14.5 illustrates how the solar collectors, storage tanks, and auxiliary heating unit are connected within the collector loop to provide a stable, continuous energy supply for the regeneration process. This hybrid configuration effectively combines renewable solar energy with thermal storage, offering greater flexibility, improved energy efficiency, and uninterrupted cooling performance across varying environmental conditions (Chaudhary et al., 2024).

14.3.3.3 Solar-assisted hybrid desiccant cooling system

Even though desiccant cooling systems offer a promising, energy-efficient alternative to traditional vapor-compression air conditioning systems, they do face several performance limitations. These systems commonly use both direct and indirect evaporative coolers to handle sensible cooling loads, but the effectiveness of each type depends mainly on the surrounding climatic conditions and the specific application requirements. In a direct evaporative cooling system, the process involves lowering the air temperature by evaporating water directly into the air stream. This method works efficiently when both indoor and outdoor humidity levels are low, as it adds moisture to the air while reducing its temperature (Hussain et al., 2022). However, when such systems operate in regions with high humidity or under conditions involving a high latent load (moisture load), their performance tends to decline. In these scenarios, the systems require higher regeneration temperatures to remove accumulated moisture from the desiccant material, thereby lowering the overall COP and increasing energy consumption (Khalid et al., 2009).

To address these drawbacks, indirect evaporative coolers are often employed. Unlike direct systems, they provide sensible cooling without adding moisture to the process air, making them suitable for more humid environments. The cooling effect is achieved through heat exchange between two separate air streams, one for the process air and another for evaporation, keeping the dehumidified air dry while still reducing its temperature. However, despite these advantages, indirect evaporative systems generally have a limited cooling capacity due to their dependence on wet-bulb temperature differences and system size.

A more effective solution is to integrate a hybrid configuration, as shown in Fig. 14.6. This hybrid system combines a solid desiccant dehumidifier, which efficiently handles latent loads (moisture removal), with a conventional vapor-compression unit, which provides sensible cooling (temperature reduction). By combining these two technologies, the system can achieve better control over both temperature and humidity, enhance energy efficiency, and maintain comfort across a broader range of climatic conditions. This hybrid approach effectively compensates for the limitations of individual cooling methods while optimizing system performance and operational reliability (Usman, 2025).

14.3.4 Design approach of efficient configurations

The design and development of solar air conditioning depend on numerous factors. The major factors include climate conditions, building type and loads, selection of solar thermal technology, and type of cooling system. Air-conditioning loads are generally composed of sensible and latent loads. Currently, various stand-alone work and heat-driven HVAC systems serve this purpose, but they exhibit certain performance limitations during cumulative load management, and their COP is also affected when subjected to handling both types of loads simultaneously. Therefore, integrating the two

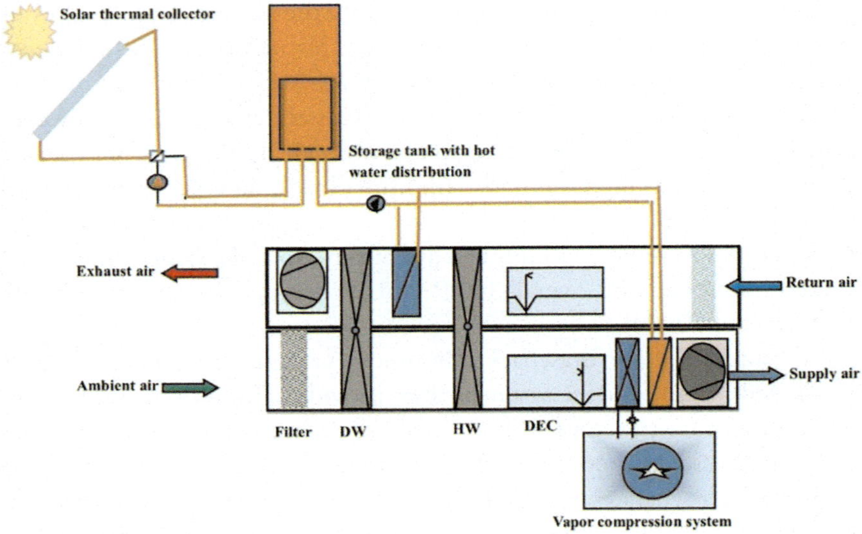

Fig. 14.6 Hybrid desiccant cooling system.

systems would be a more energy-efficient solution for handling separate loads (Lee et al., 2025). However, the innovative design approach should simultaneously achieve the desired comfort and improve efficiency. With the above-mentioned design strategies and requirements in mind, various efficient configurations for integrated solar air conditioning systems can be developed. These configurations can handle sensible and latent cooling loads separately. The detailed description of these configurations is given in the preceding sections.

14.3.4.1 Conventional DAC system

Human thermal comfort depends on the proper control of temperature and humidity. Therefore, desiccants can be employed in applications where moisture removal is the primary goal, given their capacity for water adsorption. These techniques use evaporative cooling (direct/indirect or Maisotsenko cooling) to control temperature and desiccant dehumidification to control humidity (Sultan et al., 2015). Fig. 14.7 shows a basic working principle of the DAC system. After passing through the desiccant wheel and entering the DAC at stage 1, the ambient air exits at stage 2. The air becomes dehumidified and warmer as a result of this process. Eventually, the process continues through a heat wheel for temperature decrease and heat recovery in stages 2–3. Moreover, the dehumidified air passes through

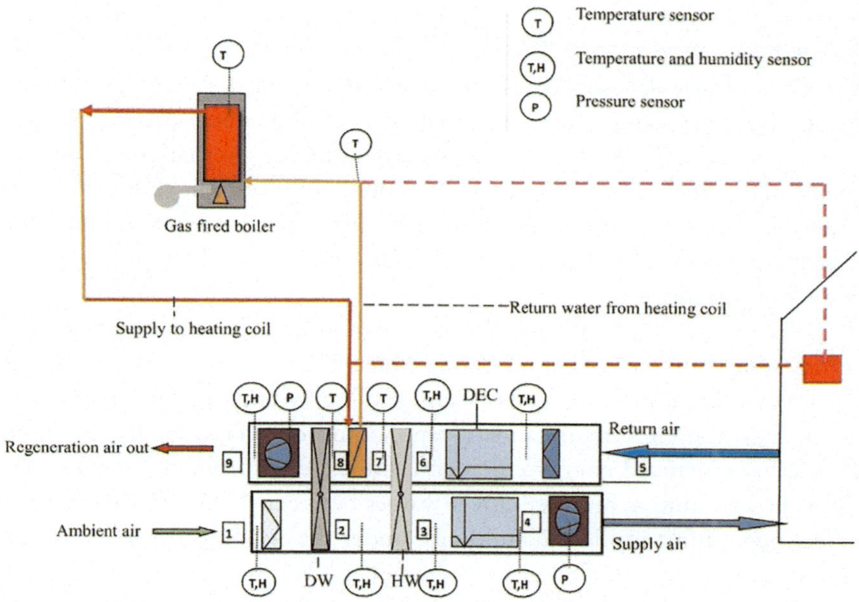

Fig. 14.7 Conventional desiccant air-conditioning (DAC) system.

the direct evaporative cooler from stage 3 to 4. Where it exchanges, it heats up and also becomes moist. Finally, the process air is supplied to the building. In the same way, the return air at stage 5 on the regeneration side passes through the traditional direct evaporative cooler before entering at stage 6. After that, from stage 6 to 7, there is a sensible heat exchange between the two air streams. Air is passed through the heating coils at steps 7 and 8 to reach the required regeneration temperature. The desiccant wheel's stored moisture is removed by the air stream during the regeneration process, which then transfers the humid air to stage 9 for further processing.

In hot and humid conditions, the performance of these systems (DAC) was investigated and compared with the output of a traditional air conditioning system. In these conditions, the proposed system demonstrated a seasonal COP of 5.92, which is over 91% higher than that of the traditional AC system (Venegas and Qu, 2026). Chen et al. (2024) experimentally investigated the performance of a dew-point evaporative cooler in comparison with a solid desiccant-coated heat exchanger–dehumidifier (SDHED). The findings indicate that the solid desiccant-coated heat exchanger performance is greatly influenced by the temperature of the cooling water and the outside environment. On the other hand, the dew point evaporative cooling depends significantly on the supply air to working air ratios. Both developed systems (such as a solid desiccant-coated heat exchanger and a dew-point evaporative cooling system) provide an efficient, eco-friendly, and carbon-emissions-reducing alternative to the conventional vapor-compression air conditioning system. However, the transient behavior of both systems is constrained by rapid temperature changes, slow heat and mass transfer rates, and the need for precise cycle-time optimization.

Zhang et al. (2025b) carried out an experimental study of a two-stage DAC system under hot, humid conditions. Through the use of low-grade or renewable heat sources, desiccant cooling systems have demonstrated a much greater COP than standard vapor compression refrigeration systems, providing enhanced energy efficiency and less environmental impact. The proposed technique produces COP values between 11.3 and 18.4 without affecting comfort inside the buildings, under the condition of 36.22°C and a humidity ratio of 0.028 kg/kg. Sultan et al. (2018b) evaluated the performance of a desiccant cooling system with different hydrophilic organic polymer sorbents. The findings indicate that, despite low mass flow rates providing more dried air at the process air side, high mass flow rates have significant dehumidification at the regeneration air side. Moreover, desiccant dehumidification increases with increasing regeneration temperature, although this effect is not always directly correlated with the overall system COP. Similarly, Güzelel et al. (2022) conducted a simulation-based study of a desiccant cooling system under environmental conditions in Turkey. The findings demonstrate that, with a desiccant cooling system, it is feasible to achieve the target comfort levels throughout the cooling season for building applications. The average monthly thermal COP was highest in October (0.78) and lowest in July (0.22), and the seasonal thermal COP was 0.43.

14.3.4.2 Solar desiccant-integrated MC cooler (SDI-MC)

The two subsystems that combine to make up the most popular, efficient design, such as the solar desiccant-integrated MC coolers (SDI-MC) system, are a desiccant-assisted Maisotsenko cooling system and a solar-assisted water heating system (SWHS), as shown in Fig. 14.8. This system consists of various components such as a solid desiccant dehumidifier, solar thermal collectors, and an efficient evaporative cooler based on the MC (Chaudhary et al., 2024). These components are intended to handle latent loads with a solid desiccant, provide sensible cooling with an MC cooler, and restore

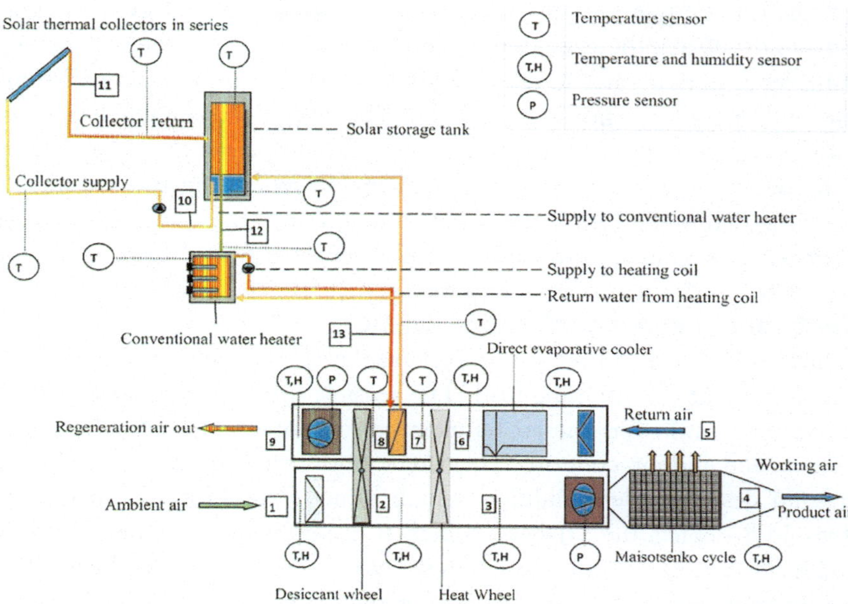

Fig. 14.8 Schematic of experimental setup with sensing and measuring instrument. *(Reproduced from: Chaudhary, G.Q., Ali, M., Sheikh, N.A., Ul, H., Gilani, S.I., Khushnood, S., 2018. Integration of solar assisted solid desiccant cooling system with efficient evaporative cooling technique for separate load handling. Appl. Therm. Eng 140, 696–706. doi:10.1016/j.applthermaleng.2018.05.081.)*

the desiccant wheel by thermal collection at the end of the cycle (Sultan and Miyazaki, 2022). The further explanation of these systems is below.

Solar-assisted water heating system

The SWHS is used to harness solar energy to regenerate the desiccant wheel. The solar thermal collectors are used to fulfill the required thermal energy demand for desiccant wheel regeneration. A stratified storage tank is used for water preheating along with water storage and hot water circulation. By maintaining temperature gradients and maximizing heat recovery, such a tank enhances the overall system's thermal efficiency. When the temperature differential between the collector outlet and the tank bottom exceeds 10°C, all day long, the solar subsystem is in constant operation; otherwise, it turns off. The water in the collector loop is circulated by the pump at a predetermined mass flow rate. It is necessary to have an additional backup heat source because solar energy is intermittent. For this reason, SWHS also has a backup conventional water heater. In addition, a control mechanism is put into place to ensure that the solar system operates efficiently. When the water temperature rises above 60°C, the solar storage tank supplies hot water to the traditional water heater; in other situations, the traditional water heater is sufficient. The heaters activate when the solar storage tank's hot water supply temperature falls below the desired level. The desiccant system is supplied with hot water during the summer or anytime dehumidification is required. In addition, the building's fan coil unit receives a direct supply of this hot water during the winter.

Desiccant integrated MC cooler (DI-MC)

The sensible load and the latent load are the two different components of the overall air conditioning load. The sensible load is handled by an MC) cooler, which lowers the air temperature, improving overall system efficiency and indoor comfort. The latent (moisture) load is controlled by a solid desiccant dehumidifier, which removes humidity from the air. Fig. 14.8 depicts the DI-MC, which is a combination of heating coils, a direct evaporative cooler based on the MC, a solid desiccant wheel, a heat recovery wheel, and an indirect evaporative cooler.

After entering the DI-MC at stage 1, the ambient air is filtered, goes through the desiccant wheel, and exits at stage 2. The air is dehumidified and heated throughout this process. It then goes through a heat wheel for temperature decrease and heat recovery in stages 2 and 3. Based on the MC, the process air goes through the indirect evaporative cooler for further sensible cooling. The process air is separated into two streams at this stage, which are working air and product air. Approximately 40% of the supply air is required for evaporation, and the product air exits the system at stage 4 and enters the room. After that, on the regeneration side, the return air at stage 5 exits at stage 6 after passing through the traditional direct evaporative cooler. Then, from stage 6 to 7, sensible heat exchange occurs between the two air streams. The air from heating coils at stages 7–8 is then passed to reach the required regeneration temperature of 70°C. Ultimately, in stage 9, the regeneration air leaves after removing the desiccant wheel's moisture.

Azam et al. (2023) performed research on desiccant cooling systems integrated with thermal collectors (such as evacuated tube collectors) in subtropical areas based on theMC. The findings showed that evacuated tube solar thermal collectors had an average yearly efficiency of about 35.80%, confirming their suitability for sustainable air-conditioning applications. In addition to improving energy efficiency, the integration of thermal collectors with the desiccant system contributes to decreasing the amount of energy used for air conditioning. Similarly, in subtropical climates, Chaudhary et al. (2026) studied the energy performance and comparative thermodynamics of two different building-integrated solid desiccant cooling system configurations. These two cooling systems are (1) a solar-integrated solid desiccant dehumidifier paired with a novel indirect regenerative evaporative cooler (SDI-IREC) and (2) a solid desiccant dehumidifier with direct evaporative cooling (SDI-DEC). The COP of the SDI-IREC system was 50% higher than that of the SDI-DEC system. Bhat et al. (2023) utilized simulations and an experimental study to evaluate the thermal performance of standalone and MC-based desiccant air cooling systems. The main factors influencing the thermal performance of the MC and standalone desiccant systems have been reported to be the regeneration temperature, humidity, and inlet regeneration air temperature. The results demonstrated that the regeneration conditions are positively correlated with the temperatures of the primary and regeneration outlets, but negatively correlated with the relative humidity of the primary and regeneration outlets.

Another study by Song and Sobhani (2020) investigated the energy and energy efficiency of an integrated desiccant cooling system with thermal collectors based on the Maisotsenko cooler in hot and humid climates. Thermal collectors generated up to 0.77 kW of electricity in October due to low cooling demand, while their electrical efficiency is at its lowest in September, at 13.6%. Moreover, the result also revealed that, in August, the maximum energy and thermal energy efficiency were 68.24% and 5.4%, respectively. Similarly, Iyer et al. (2024) investigated the performance characteristics of a combined desiccant and MC cooling system, with a focus on the effects of cycle time and adsorption-evaporation synergy in counter and crossflow configurations. In the crossflow scenario, the adsorption effects are more noticeable, while the evaporation effects are more noticeable in the counterflow setup. Moreover, in comparison to the crossflow arrangement, the counterflow configuration has a better cycle time performance because it can dehumidify the incoming air for a longer amount of time and cool down from the high regeneration temperatures much more quickly. Another configuration is the integration of a solar-assisted desiccant system with an absorption system. This configuration also consists of two subsystems, i.e., SWHS and a desiccant-integrated gas-fired vapor absorption system. The SWHS is the same as for SDI-MC; however, the desiccant-integrated vapor absorption system (DI-VAS) works in a different manner.

14.3.4.3 Desiccant-integrated vapor absorption system

The DI-VAS consists of two integrated systems to implement the concept of separate load handling. A solid desiccant dehumidifier is used to handle the latent loads, while an absorption system is used for the sensible loads. The heat wheel in between these systems serves the purpose of heat recovery (Abdelgaied et al., 2023a). The air-conditioning process is described in Fig. 14.9.

The ambient air enters the DI-VAS system at stage 1, where it is first filtered before passing through the desiccant wheel, and then exits at stage 2. Throughout this process, the air undergoes dehumidification, resulting in a noticeable increase in its temperature. Afterward, from stage 2 to 3, it passes through a heat wheel for heat recovery and temperature reduction. Afterward, the dehumidified air passes through the cooling coils for sensible cooling from stage 3 to 4. Finally, the chilled air is supplied to the building. Also on the regeneration side, the return air at stage 5 exits at stage 6 after passing through the traditional direct evaporative cooler. The two air streams then experience sensible heat exchange from 6 to 7. The air is

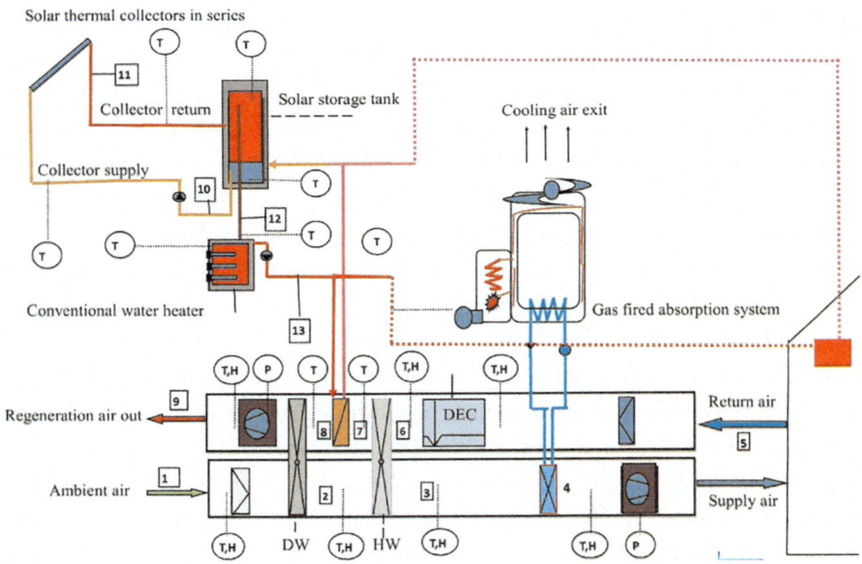

Fig. 14.9 Solar desiccant-assisted vapor absorption system. *(Reproduced with permission from: Habib, M.F., Ali, M., Sheikh, N.A., Badar, A.W., Mehmood, S., 2020. Building thermal load management through integration of solar assisted absorption and desiccant air conditioning systems: a model-based simulation-optimization approach. J. Build. Eng. 30. doi:10.1016/j.jobe.2020.101279.).*

then sent through the heating coils at stages 7–8 to reach the required regeneration temperature. Eventually, at stage 9, the regeneration air removes the moisture from the desiccant wheel.

A DI-VAS's primary advantage is its ability to effectively cool and dehumidify while making use of waste or low-grade heat sources. It is appropriate for application in environmentally friendly building applications since it can be connected with solar thermal collectors (Sharma et al., 2020). Regarding this, Abdelgaied et al. (2023b) used a novel DAC configuration combined with an evacuated tube collector to heat the regeneration air. They also utilized an auxiliary electric heater to elevate the temperature of the regeneration air to the required level for efficient desiccant reactivation. At a regeneration temperature of 90°C, the desiccant air conditioner's novel arrangement has more efficient energy performance, with thermal and electrical coefficients of performance reaching 1.96 and 3.14, respectively. Moreover, the proposed novel desiccant air conditioner configuration has a life cycle cost that is 33% less than that of the traditional air conditioning system. Similarly,

The performance of these systems for building applications was studied by Wang et al. (2019) through an investigation that examined mass diffusion and thermal transport during vapor absorption into hygroscopic liquid desiccant droplets. They found that the temperature and relative humidity affect the growth rate and ultimate size of desiccant droplets, which are moved by variations in vapor pressure and grow by absorbing surrounding water. Moreover, the salt inside the droplet is increasingly concentrated as it absorbs water, which decreases both its capacity to draw in further moisture and the amount of heat that is emitted during the process. Ali et al. (2022) used a gas-fired air-cooled NH_3-H_2O absorption chiller to experimentally investigate an absorption system for handling the thermal load of a separate building. A conventional standalone DAC system equipped with a direct evaporative cooler is employed as a reference for comparison. The performance of the integrated absorption-solid desiccant system is then evaluated under various operating conditions, including changes in air temperature, humidity, and regeneration temperature. The integrated system's COP_{th} was 50%–55% greater than that of the comparable system and nearly equal to that of a double-effect absorption chiller. In summary, combining an absorption cooling system with a solar-assisted desiccant system shows a significant amount of potential for increasing overall energy efficiency and lowering traditional energy usage.

14.4 Application in a controlled environment

Integration of solar thermal collectors with desiccant cooling systems provides an extremely effective and environmentally friendly as well as economically sustainable solution to maintain precise environmental conditions. Applications where maintaining temperature and humidity is crucial, including pharmaceutical storage, museums, libraries, and high-comfort HVAC for humid climates, domestic and commercial, are especially well-suited for this hybrid technology (Salikandi et al., 2021). When compared to traditional vapor-compression systems, the dehumidification process is operated by solar energy, which drastically reduces electricity consumption, maintenance costs, and the carbon impact.

14.4.1 Commercial and residential buildings

Desiccant cooling systems combined with solar thermal collectors provide a very effective climate control alternative for both commercial and residential buildings, especially in humid areas. These systems are excellent at efficiently and economically controlling the high latent loads from required ventilation in large commercial buildings such as offices, hotels, and hospitals. It also uses solar energy to dehumidify fresh air and significantly lowers the requirement for electricity. This method greatly reduces summer energy costs while improving comfort in households by eliminating humid, moist environments without the need for overcooling (Cuce et al., 2025). Moreover, this hybrid strategy offers significant energy savings and a reduced carbon footprint throughout the building spectrum, while also improving indoor air quality and comfort for users (Naik et al., 2020).

14.4.2 Agricultural product drying and storage

These systems are used in agricultural postharvest management to keep storage silos dry and to dry grains, fruits, and herbs. The high level of moisture content in agricultural goods is one of the primary causes of postharvest losses (Sultan and Miyazaki, 2022). Fruits and vegetables go through respiration, transpiration, and ripening processes after being harvested. Moreover, they have a shorter shelf life at higher storage temperatures, but a longer shelf life at lower storage temperatures is shown in Fig. 14.10. The best storage conditions, such as temperature and relative humidity, can reduce these agricultural product deterioration processes. Therefore, DAC systems provide an appropriate temperature and relative

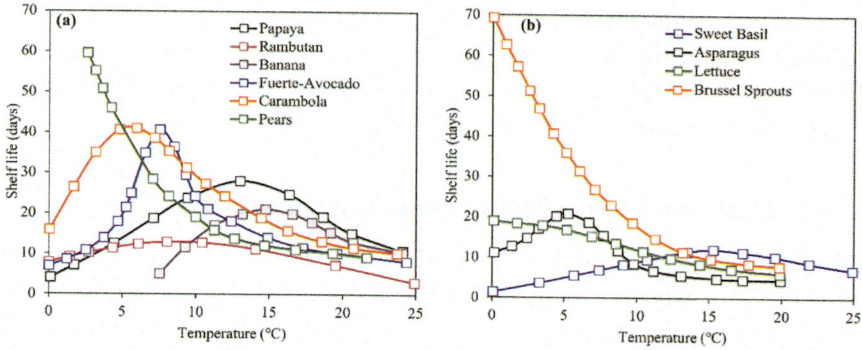

Fig. 14.10 Shelf life is affected by storage temperature in (A) fruit and (B) vegetables. *(Reproduced with permission from: Paull, R.E., 1999. Effect of temperature and relative humidity on fresh commodity quality. Postharvest Biol. Technol 15 (3), 263–277. https://doi.org/10.1016/S0925-5214(98)00090-8.)*

humidity; conditions are both environmentally friendly and energy-efficient (Mahmood et al., 2020).

14.4.3 Pharmaceutical and healthcare facilities

The quality and safety of the medications produced are impacted by moisture control, which is a crucial factor in pharmaceutical manufacturing. Therefore, this integration is essential for protecting vaccines and medications that are easily damaged by moisture in pharmaceutically sterilized labs and storage spaces. Desiccant dehumidifiers provide superior precision in regulating humidity levels, ensuring that medications remain in optimal conditions (Huang et al., 2024). Pharmaceutical plants can function more efficiently and produce more when moisture-related problems are minimized. Regenerating the desiccant with solar energy ensures continuous compliance with complex regulatory requirements and prevents microbial development to preserve sterile conditions. Despite this, these desiccant dehumidifiers can be customized to meet your pharmaceutical plant's unique needs.

14.4.4 Museums and libraries

This technology provides the perfect climate control solution for museums and libraries to preserve valuable cultural material by autonomously controlling humidity and temperature. The expansion, contraction, and mold growth caused by changes in humidity provide a serious threat to organic objects such as paintings, manuscripts, and textiles. With the help of the solar-desiccant system, which creates a completely stable preservation environment driven mainly by renewable energy, this moisture is gently and effectively removed without damaging cycles of overcooling and reheating, keeping priceless things economically and sustainably.

14.5 Challenges and future research directions

Numerous studies have been carried out on the development of highly effective and efficient integrated thermal collector systems coupled with solid desiccant units, as previously discussed. Researchers have made advancements in improving the desiccant cooling system performance by optimizing system arrangement and incorporating thermal collectors for efficient heat utilization. However, with the significant improvements in cooling performance, energy consumption, and system efficiency, the following more investigation is still required:

- Future research should focus on developing novel solid desiccant materials with better efficiency and reduced generation temperature requirements. According to the research, silica gel was frequently used for the majority of investigations as the desiccant system's adsorbent material. The impact of various desiccant materials on system performance was not extensively studied. The equipment space needed and initial cost can be significantly decreased if a single desiccant cooling system using a novel desiccant material can provide the same dehumidification performance as a multistage desiccant unit arrangement.
- Future research should focus on identifying and assessing the most effective configurations for integrating desiccant cooling systems with solar thermal collectors. The objective is to minimize thermal losses throughout the system and maximize energy usage by effectively using solar heat for desiccant regeneration. Utilizing real-time sunlight availability and system load fluctuations develop advanced control algorithms to regulate air temperature and humidity in regulated conditions.
- To improve their performance, optimize design parameters, and assess their potential for large-scale and practical applications, further investigation is required on MC-based desiccant cooling systems. To improve water evaporation, more research should be done on new materials with high surface areas for water evaporation and related water distribution techniques.
- Some researchers have considered numerically using nanofluids as the working fluid instead of water to increase the rate of heat transfer;

however, as this concept has not yet been validated experimentally, as per the author's information, further research could be done to better understand it.

- A possible waste is the direct release of the warm, humid outlet air from the regeneration process onto the environment. Future research might concentrate on recycling the exhaust air's waste heat and moisture. By simply changing the configuration, the waste heat could be recycled to warm the regeneration air, and the water consumption could be decreased by recovering liquid water from water vapor.

14.6 Conclusions

Particularly in areas with higher temperatures, the summer months show the highest demand for electrical energy consumption. Using renewable energy to meet the cooling demand could be a useful strategy for reducing the peak demand for fossil fuel-generated electricity. Therefore, integrating solar thermal collectors with desiccant cooling systems is a promising technique to provide energy-efficient and sustainable air conditioning, especially in controlled environments that require precise temperature and humidity control. Low-grade solar thermal energy can be efficiently used for regeneration in both liquid and solid desiccant systems, thereby lowering greenhouse gas emissions and consumption of electricity. Moreover, the desiccant cooling system displays significant application potential in desiccant-based cooling techniques due to its excellent dehumidification capability. Future developments in solar collector technology, desiccant materials, and advanced control systems will improve system performance, dependability, and applicability even further. The next generation of clean and effective environmental management solutions is expected to depend on solar-assisted desiccant cooling systems, as carbon-neutral technologies and the integration of renewable energy sources become more and more important.

References

Abbas, F., et al., 2025. Comprehensive review on evaporative cooling and desiccant dehumidification technologies for agricultural greenhouses. AgriEngineering 7 (7), 222. doi:10.3390/agriengineering7070222.

Abdelgaied, M., Saber, M.A., Bassuoni, M.M., Khaira, A.M., 2023a. Adsorption air conditioning: a comprehensive review in desiccant materials, system progress, and recent studies on different configurations of hybrid solid desiccant air conditioning systems. Environ. Sci. Pollut. Res. 30 (11), 28344–28372. doi:10.1007/s11356-023-25209-z.

Abdelgaied, M., Saber, M.A., Bassuoni, M.M., Khaira, A.M., 2023b. Performance improvement of solar-assisted air-conditioning systems by using an innovative configuration of a desiccant dehumidifier and thermal recovery unit. J. Therm. Anal. Calorim 148 (14), 7003–7018. doi:10.1007/s10973-023-12215-z.
Alam, M.A., Irshad, K., Rehman, S., Masood, F., Zayed, M.E., 2025. Recent developments in renewable energy assisted cold thermal energy storage air conditioning systems. J. Energy Storage 127, 117124. doi:10.1016/j.est.2025.117124.
Ali, M., Habib, M.F., Sheikh, N.A., Akhter, J., Ul H. Gilani, S.I., 2022. Experimental investigation of an integrated absorption- solid desiccant air conditioning system. Appl. Therm. Eng 203. doi:10.1016/j.applthermaleng.2021.117912.
Ali, S., Jalil, J., Shuraiji, A., 2025. Performance enhancement of evacuated tube solar collectors: a review. Eng. Technol. J. 1–20. doi:10.30684/etj.2025.155236.1853.
Alsagri, A.S., Alrobaian, A.A., Almohaimeed, S.A., 2020. Concentrating solar collectors in absorption and adsorption cooling cycles: an overview. Energy Convers. Manag. 223, 113420. doi:10.1016/j.enconman.2020.113420.
Azam, M.W., Chaudhary, G.Q., Sajjad, U., Abbas, N., Yan, W.M., 2023. Performance investigation of solar assisted desiccant integrated Maisotsenko cycle cooler in subtropical climate conditions. Case Stud. Therm. Eng. 44, 102864. doi:10.1016/j.csite.2023.102864.
Azam, M.W., Chaudhary, G.Q., Simprini, G., He, S., 2025. Integration of renewable energies in warehouses. In: Frikha, M.A., Rahmouni, M., Soufiane, B.O. (Eds.), Innovations in Green and Energy-Efficient Warehousing. IGI Global Scientific Publishing, pp. 263–308. doi:10.4018/979-8-3373-3176-8.ch011.
Baniyounes, A.M., Ghadi, Y.Y., Rasul, M.G., Khan, M.M.K., 2013. An overview of solar assisted air conditioning in queensland's subtropical regions, Australia. Renew. Sustain. Energy Rev. 26, 781–804. doi:10.1016/j.rser.2013.05.053.
Bhat, S.A., et al., 2023. Experiments and modeling on thermal performance evaluation of standalone and M-cycle based desiccant air-conditioning systems. Energy Rep. 11, 1445–1454. doi:10.1016/j.egyr.2024.01.019.
Boutagount, S., et al., 2025. Innovative materials for energy storage systems and photovoltaic solar technologies: a review. Nano-Struct. Nano-Objects 44, 101562. doi:10.1016/j.nanoso.2025.101562.
Bozorgi, M., Tasnim, S.H., Mahmud, S., 2024a. Enhancing indoor thermal comfort and sustainability: a solar-driven desiccant cooling and adsorption chiller system with environmental impact assessment. Sol. Energy 271. doi:10.1016/j.solener.2024.112440.
Bozorgi, M., Tasnim, S.H., Mahmud, S., 2024b. Experimental and numerical study on the integration of solar-driven desiccant and thermoelectric Systems for Sustainable Thermal Comfort. Appl. Energy 376. doi:10.1016/j.apenergy.2024.124253.
Buker, M.S., Riffat, S.B., 2015. Building integrated solar thermal collectors - A review. Renew. Sustain. Energy Rev. 51, 327–346. doi:10.1016/j.rser.2015.06.009.
Chaudhary, G.Q., et al., 2024. Transient analysis of an efficient solar assisted air-conditioning system for subtropical climate with various solar thermal collectors. Energy Convers. Manag. X 23. doi:10.1016/j.ecmx.2024.100634.
Chaudhary, G.Q., et al., 2026. Analysis of building-integrated solar desiccant air cooling systems considering the dynamic sensible and latent cooling loads. Int. J. Refrig. 181, 111–125. doi:10.1016/j.ijrefrig.2025.10.007.
Chen, W., Cui, X., Huang, Z.F., Shao, Y.L., Chua, K.J., 2024. Developing an integrated solid desiccant dehumidifier and dew-point evaporative cooler for green air conditioning. J. Build. Eng. 96. doi:10.1016/j.jobe.2024.110404.
Cuce, P.M., Yilmaz, Y.N., Cuce, E., 2025. A comprehensive review on rotary desiccant wheel systems: the future of smart building climate control. J. Therm. Anal. Calorim 150 (13), 9827–9850. doi:10.1007/s10973-025-14362-x.

Deng, J., Wang, R.Z., Han, G.Y., 2011. A review of thermally activated cooling technologies for combined cooling, heating and power systems. Prog. Energy Combust. Sci. 37 (2), 172–203. doi:10.1016/j.pecs.2010.05.003.

Dincer, I., Rosen, M.A., 2007. Exergy analysis of psychrometric processes. Exergy 76–90. doi:10.1016/B978-008044529-8.50009-4.

Duong, H.C., et al., 2023. Membrane distillation-liquid desiccant air-conditioning for thermal comfort in buildings. In: Hanif, M.A., Nadeem, F., Tariq, R., Rashid, U. (Eds.), Current Developments in Biotechnology and Bioengineering: Membrane Technology for Sustainable Water and Energy Management. Elsevier, pp. 387–409. doi:10.1016/B978-0-443-19180-0.00010-9.

Ensafjoee, K., Daghigh, R., 2023. An evacuated tube solar bubble column liquid desiccant regenerator. Appl. Therm. Eng 235. doi:10.1016/j.applthermaleng.2023.121331.

Essa, M.E.-S.M., et al., 2025. Developments in solar-driven desalination: technologies, photovoltaic integration, and processes. Energy Convers. Manag. X 25, 100861. doi:10.1016/j.ecmx.2024.100861.

Ge, T.S., Dai, Y.J., Wang, R.Z., 2014. Review on Solar Powered Rotary Desiccant Wheel Cooling System. *Elsevier Ltd.* doi:10.1016/j.rser.2014.07.121.

Güzelel, Y.E., Olmuş, U., Büyükalaca, O., 2022. Simulation of a desiccant air-conditioning system integrated with dew-point indirect evaporative cooler for a school building. Appl. Therm. Eng 217. doi:10.1016/j.applthermaleng.2022.119233.

Hanif, M.A., Nadeem, F., Tariq, R., Rashid, U., 2022. Solar thermal energy and photovoltaic systems. In: Hanif, M.A., Nadeem, F., Tariq, R., Rashid, U. (Eds.) Renewable and Alternative Energy Resources. Elsevier, pp. 171–261. doi:10.1016/b978-0-12-818150-8.00007-1.

Hua, L.J., Jiang, Y., Ge, T.S., Wang, R.Z., 2018. Experimental investigation on a novel heat pump system based on desiccant coated heat exchangers. Energy 142, 96–107. doi:10.1016/j.energy.2017.07.141.

Huang, L., Guan, B., Qi, M., Liu, Y., Yang, H., 2024. Performance investigation of a hybrid liquid-desiccant air conditioning system in a pharmaceutical warehouse: a case study and refined strategy. Front. Built Environ 10. doi:10.3389/fbuil.2024.1468537.

Hussain, I., et al., 2022. Evaluating the parameters affecting the direct and indirect evaporative cooling systems. Eng. Anal. Bound. Elem 145, 211–223. doi:10.1016/j.enganabound.2022.09.016.

Ibrahim, N.I., et al., 2026. State-of-the-art of the digital twin concept in HVAC+R systems for thermal management of buildings. Renew. Sustain. Energy Rev. 229, 116601. doi:10.1016/J.RSER.2025.116601.

International Energy Agency, n.d. The Future of Cooling Opportunities for Energy-Efficient Air Conditioning Together Secure Sustainable. Available from: https://www.iea.org/t&c/

Iyer, P.K., Ganguly, A., Maiya, M.P., 2024. Investigations on adsorption—evaporation characteristics and cycle time effects of integrated desiccant coated M-cycle cooler. Therm. Sci. Eng. Prog. 51. doi:10.1016/j.tsep.2024.102594.

Kadam, T.V., Pakhare, S.Y., Godse, A.B., 2025. 'Performance analysis of solar thermal collectors: a comprehensive review,'. Meas. Energy 7, 100059. doi:10.1016/j.meaene.2025.100059.

Khalid, A., Mahmood, M., Asif, M., Muneer, T., 2009. Solar assisted, pre-cooled hybrid desiccant cooling system for Pakistan. Renew. Energy 34 (1), 151–157. doi:10.1016/j.renene.2008.02.031.

Lee, J.H., Kim, B., Chu, H., Lim, H., Yu, K.H., Jeong, J.W., 2025. Empirical analysis of efficient winter operation in reversible heat Pump-Driven Liquid-Desiccant systems. Energy Build. 329. doi:10.1016/j.enbuild.2024.115262.

Lee, Y., Park, S., Kang, S., 2021. Performance analysis of a solid desiccant cooling system for a residential air conditioning system. Appl. Therm. Eng 182. doi:10.1016/j.applthermaleng.2020.116091.

Liang, C., Li, X., 2025. Efficient liquid desiccant dehumidification system using a natural cooling source for solution regeneration. Build. Environ 286. doi:10.1016/j.buildenv.2025.113696.

Mahmood, M.H., Sultan, M., Miyazaki, T., 2020. Experimental evaluation of desiccant dehumidification and air-conditioning system for energy-efficient storage of dried fruits. Build. Serv. Eng. Res. Technol. 41 (4), 454–465. doi:10.1177/0143624419893660.

Naik, B.K., et al., 2020. Investigating solid and liquid desiccant dehumidification options for room air-conditioning and drying applications. Sustainability 12 (24), 1–22. doi:10.3390/su122410582.

Patel, A.R., Modi, K.V., 2025. Impact of distinct input variables on the performance indices of components of liquid desiccant-assisted forced convection indirect type solar drying system—a critical review. Environ. Sci. Pollut. Res. 32 (43), 24554–24632. doi:10.1007/s11356-025-37026-7.

Rekhade, R., Bose, S., Datta, S.P., Matuszczak, M., Pandelidis, D., Jagirdar, M., 2025. State-of-the-art on liquid desiccant dehumidification systems: components, performance, challenges and integration with evaporative cooling systems. J. Build. Eng. 111, 113175. doi:10.1016/J.JOBE.2025.113175.

Robbins, T., Garimella, S., 2020. An autonomous solar driven adsorption cooling system. Sol. Energy 211, 1318–1324. doi:10.1016/j.solener.2020.10.068.

Salikandi, M., Ranjbar, B., Shirkhan, E., Priya, S.S., Thirunavukkarasu, I., Sudhakar, K., 2021. Recent trends in liquid desiccant materials and cooling systems: application, performance and regeneration characteristics. J. Build. Eng. 33, 101579. doi:10.1016/j.jobe.2020.101579.

Shamim, J.A., et al., 2025. Advances in Desiccant Wheels for Dehumidification, VOC Mitigation, and CO_2 Removal for Energy-Efficient IAQ Management. Elsevier Ltd.. doi:10.1016/j.ijheatmasstransfer.2025.126906.

Sharma, D.K., Sharma, D., Ali, A.H.H., 2020. A state of the art on solar-powered vapor absorption cooling systems integrated with thermal energy storage. Environ. Sci. Pollut. Res. 27 (1), 158–189. doi:10.1007/s11356-019-06941-x.

Shehram, M., Hamidi, M.N., Wahab, A.A.A., Desa, M.K.M., 2025. Comprehensive review of hybrid solar cooling systems for buildings: integrating PV and thermal energy storage in phase change materials. J. Therm. Anal. Calorim 150 (7), 4873–4906. doi:10.1007/s10973-025-14119-6.

Song, J., Sobhani, B., 2020. Energy and exergy performance of an integrated desiccant cooling system with photovoltaic/thermal using phase change material and maisotsenko cooler. J. Energy Storage 32. doi:10.1016/j.est.2020.101698.

Sultan, M., El-Sharkawy, I.I., Miyazaki, T., Saha, B.B., Koyama, S., 2015. An overview of solid desiccant dehumidification and air conditioning systems. Renew. Sustain. Energy Rev. 46, 16–29. doi:10.1016/j.rser.2015.02.038.

Sultan, M., Miyazaki, T., Koyama, S., 2018a. Optimization of adsorption isotherm types for desiccant air-conditioning applications. Renew. Energy 121, 441–450. doi:10.1016/j.renene.2018.01.045.

Sultan, M., Miyazaki, T., Koyama, S., Khan, Z.M., 2018b. Performance evaluation of hydrophilic organic polymer sorbents for desiccant air-conditioning applications. Adsorpt. Sci. Technol. 36 (1–2), 311–326. doi:10.1177/0263617417692338.

Sultan, M., Miyazaki, T., 2022. In: Muhammad, S., Miyazaki, T. (Eds.), Energy-Efficient Systems for Agricultural Applications. Springer, Cham. doi:10.1007/978-3-030-86394-4.

Trombe, B.F., Foex, M., n.d. The production of cold by means of solar radiation. Sol. Energy. doi:10.1016/0038-092X(57)90055-5.

United Nations Environment Programme, 2021. Global Status Report for Buildings and Construction. Global Alliance for Building and Construction. Available from: https://www.globalabc.org.

Usman, M., et al., 2025. Performance analysis of a novel solar-assisted desiccant wheel-based heating and humidification system. MTME. doi:10.3390/materproc2025023020.

Venegas, T.P., Qu, M., 2026. Assessing seasonal system performance: design and simulation of a heat pump-coupled internally-cooled membrane-based liquid desiccant air-conditioning system. Int. J. Refrig. 181, 142–153. doi:10.1016/j.ijrefrig.2025.10.011.

Verma, A.K., Yadav, L., 2025. Experimental study of desiccant wheel regeneration using a series parabolic trough collector having a concentric copper tube in evacuated tube solar air heater for medium solar intensity. J. Build. Eng. 111. doi:10.1016/j.jobe.2025.113154.

Wang, Z., Orejon, D., Sefiane, K., Takata, Y., 2019. Coupled thermal transport and mass diffusion during vapor absorption into hygroscopic liquid desiccant droplets. Int. J. Heat Mass Transf 134, 1014–1023. doi:10.1016/j.ijheatmasstransfer.2019.01.084.

Xu, Y., Chen, L., 2024. A Comprehensive Review of Low Dew-point Desiccant Wheel System: Mechanisms, Configuration, and Optimization. *Elsevier Ltd.* doi:10.1016/j.ijrefrig.2024.09.006.

Yang, K.S., Hamid, K., Wu, S.K., Sajjad, U., Wang, C.C., 2021. Experimental analysis of a heat pump dryer with an external desiccant wheel dryer. Processes 9 (7). doi:10.3390/pr9071216.

Zeoli, A., Pacak, A., Gendebien, S., Chorowski, M., Xie, X., Lemort, V., 2025. Feasibility analysis of desiccant evaporative cooling technologies in various climate conditions: present and future potential. Energy 327. doi:10.1016/j.energy.2025.136257.

Zhang, Y., Chen, Y., Luo, J., Yang, H., 2025a. Improving air conditioning efficiency using solar-assisted liquid desiccant systems: experimental and simulation analysis of two desiccants. Renew. Energy 253. doi:10.1016/j.renene.2025.123572.

Zhang, Y., Chen, Y., Yang, H., Zhang, H., Leung, C.W., 2025b. Experimental performance investigation on a desiccant-assisted two-stage evaporative cooling system in hot and humid areas. Appl. Energy 377. doi:10.1016/j.apenergy.2024.124704.

Zhao, L.H., Wang, R.Z., Ge, T.S., 2021. Desiccant Coated Heat Exchanger and Its Applications. *Elsevier Ltd.* doi:10.1016/j.ijrefrig.2021.06.008.

CHAPTER 15

Future outlook and deployment pathways of hybrid cooling and dehumidification systems for controlled environments

Uzair Sajjad[a], Ghulam Abbas Gohar[b], Hafiz Muhammad Ali[a,b], and Muhammad Sultan[c]
[a]Interdisciplinary Research Center for Sustainable Energy Systems (IRC-SES), King Fahd University of Petroleum and Minerals, Dhahran, Saudi Arabia
[b]Department of Mechanical Engineering, King Fahd University of Petroleum and Minerals, Dhahran, Saudi Arabia
[c]Department of Agricultural Engineering, Bahauddin Zakariya University, Multan, Pakistan

15.1 Overview and context

The literature demonstrates that hybrid cooling and dehumidification systems are a feasible and a progressively needed direction toward the realization of the energy-efficient thermal and humidity control of the controlled environment. These systems have been developed to overcome most of the limitations of traditional air-conditioning technologies, especially in hot, arid, semi-arid, subtropical, and tropical climates, by integrating evaporative cooling, desiccant-based dehumidification, membrane energy recovery, renewable energy integration, and intelligent modeling strategies. Evaporative cooling, desiccant air-conditioning, and membrane energy recovery ventilators (ERVs) are individual technologies that have already reached maturity; their hybrid integration and climate-specific optimization are still active areas of research and development. This chapter is a synthesis of the main findings that have been presented throughout the book and a description of future challenges, opportunities and research directions that are necessary to transition sustainable hybrid cooling systems from experimental and pilot-scale applications toward widespread deployment.

Sustainable Hybrid Cooling and Dehumidification Strategies for Controlled Environments
https://doi.org/10.1016/B978-0-443-48555-8.00008-6

15.2 Future directions in evaporative and hybrid cooling technologies

This section demonstrates the significance of airflow control, heat and mass transfer improvement, and indirect evaporative cooling designs to achieve the energy-efficient control of temperature. The future investigations in the field should aim at design optimization under actual climatic variability, such as varying operating conditions and seasonal performance evaluation. The improved thermal effectiveness with minimum pressure losses and water usage can be realized by further development of novel architecture of indirect evaporative cooling, assisted by high-fidelity computational fluid dynamics (CFD) modeling. Also, it is important to increase the focus on the hybrid operating strategies in which the evaporative cooling will be smartly coupled to the mechanical vapor compression to maintain stable performance under adverse climate conditions.

15.3 Advances in desiccant-based cooling and dehumidification systems

Advances in desiccant-based cooling and dehumidification systems show their significance in the mitigation of latent cooling loads, especially in greenhouses and controlled environments. Comparative studies of standalone desiccant systems and Maisotsenko-cycle-assisted systems show a high likelihood of additional performance improvement due to their optimization of the regeneration process and system integration. The next step should be creating high-performance desiccant materials that have an improved moisture adsorption capacity, thermal stability, and durability. Simultaneously, system-level research efforts are necessary to test low-grade heat use, such as solar thermal energy and waste heat sources, under realistic working conditions. Desiccant dehumidification (combined with indirect evaporation cooling) is a viable line of research towards the realization of balanced temperature-humidity conditions in arid and semi-arid conditions. In parallel, data-driven techniques have been proven effective in design optimization of desiccant-based hybrid cooling systems.

15.4 Energy recovery and membrane-based system enhancement

The literature on ERV systems confirms that the performance of the hybrid cooling systems can be greatly improved by the use of membrane-based energy recovery ventilators (MERVs), which is capable of extracting both sensible and latent energy in the exhaust air streams. Further development in this field will rely on the development of better membrane materials, optimization of channel geometry, and channel designs that are resistant to fouling. Future research should be aimed at scaling CFD-optimized membrane designs to actual systems and assessing their long-term performance under varying climatic and operational conditions. Also, the combination of MERVs with evaporative cooling and desiccant systems has the potential of minimizing additional ventilation energy loss, especially in subtropical and tropical climates, further.

15.5 Role of renewable energy and passive cooling strategies

Renewable-assisted cooling systems, such as solar chimney-assisted passive cooling, solar thermal collector integration, and solar-powered desiccant systems, are an important aspect of sustainable hybrid cooling systems. As has been illustrated throughout the book, these solutions can significantly cut reliance on grid electricity and increase resilience in the systems. Future studies ought to focus on system-level optimization of renewable-hybrid coupling, taking into consideration the intermittency, thermal storage needs, and control strategies. Design tools, e.g., building energy simulation platforms, can be of significant importance in the assessment of long-term performance and economic viability in various climatic regions.

15.6 Intelligent modeling, control, and optimization

As proposed in this book, the data-driven methods such as ANNs show the increased significance of intelligent system analysis. More likely, the future hybrid cooling systems will be based on real-time monitoring, adaptive control, and predictive modeling to handle interactions among many subsystems. More effort needs to be put into the integration of AI-based models and physical system simulations to allow robust and explainable decision-making systems. The emergence of hybrid modeling technologies that are used to ensure the integration of first-principles knowledge and data-driven intelligence is a promising field that will enhance system reliability and operational efficiency.

15.7 Deployment challenges and practical considerations

Although hybrid cooling and dehumidification systems have been proven to be useful, several challenges must be addressed to enable widespread implementation. These are the complexity of the system, cost of initial investment, maintenance, and water consumption, as well as limited users' awareness. Design guidelines and standardized measures of performance, which are climate-specific are required to aid technology transfer and commercialization.

Long-term experimental validation should be done, especially in greenhouse agriculture and postharvest preservation, to give confidence to the stakeholders and policymakers. Also, there is a need for training and capacity building to facilitate the operation and maintenance of the hybrid systems in developing regions.

15.8 Concluding outlook

Hybrid cooling and dehumidification is a strategically valuable solution to the realization of energy-effective and climate-resilient thermal treatment in controlled environments. The hybrid strategies outlined in this book highlight that significant reductions in energy consumption and environmental footprint can be attained by having intelligent system design, hybridization, and renewable energy integration. To clarify the transition from research to deployment, Table 15.1 summarizes the system-level technology readiness levels (TRLs), key technical gaps, and priority research and deployment directions for the hybrid cooling and dehumidification pathways discussed in this book (see Fig. 15.1 as well).

Table 15.1 suggests that advances in materials, system integration, and modeling as well as control strategies are expected to significantly influence the scalability and the long-term success of these technologies. By addressing the identified challenges and building upon the insights presented in this volume, hybrid cooling systems can make a significant contribution to sustainable development, food security, and climate adaptation in regions most vulnerable to rising cooling demands.

Table 15.1 Future outlook of hybrid cooling and dehumidification systems: Readiness, gaps, and priority actions.

Technology pathway	Current system-level TRL	Key technical gaps	Near-term research priority	Medium-term deployment focus
Evaporative and hybrid cooling systems	4–6	Climate dependency; water use	Design optimization under variable climatic conditions	Pilot-scale validation in arid and semi-arid regions
Desiccant-based cooling and dehumidification	5–6	Regeneration efficiency; material durability	Advanced desiccant materials and low-grade heat utilization	Greenhouse and mixed-climate deployment
Membrane-based energy recovery systems (system-integrated)	5–7	Fouling resistance; long-term operational stability	Improved membrane materials and optimized channel design	Integration with ventilation-intensive buildings
Renewable-assisted hybrid cooling	5–7	Intermittency; thermal storage integration	System-level renewable-cooling coupling	Off-grid and low-carbon cooling applications
Intelligent modeling and control (CFD, ANN)	3–5	Model generalization; interpretability	Hybrid physics-based and AI modeling frameworks	Real-time monitoring and adaptive control
Integrated hybrid cooling systems	4–6	System complexity; cost	Simplified integration and control strategies	Demonstration projects and standardization

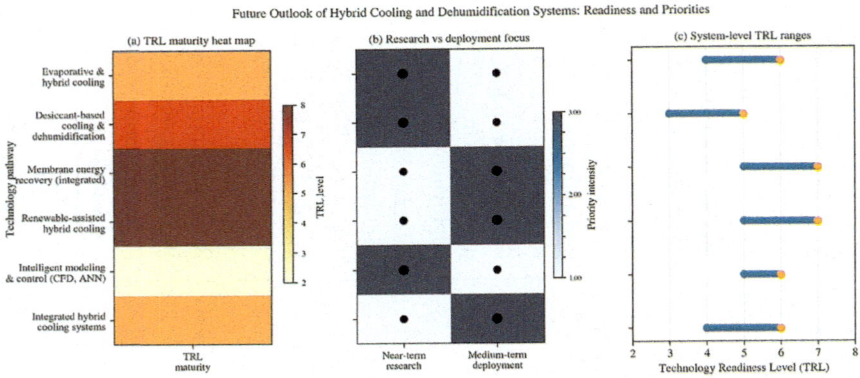

Fig. 15.1 A–C. System-level TRL landscape of hybrid cooling and dehumidification systems.

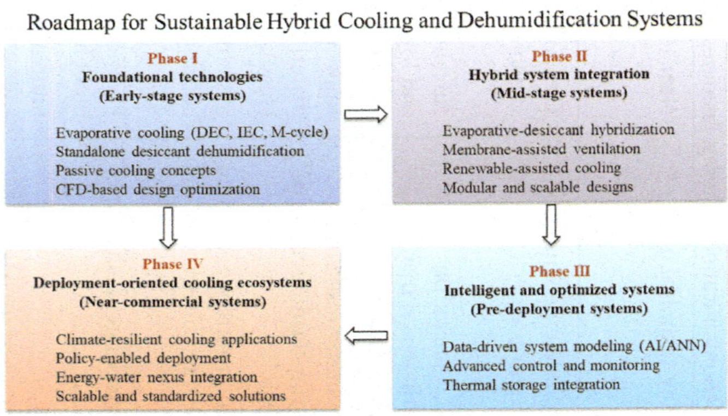

Fig. 15.2 Phased roadmap for sustainable hybrid cooling and dehumidification systems.

Fig. 15.2 complements the TRL analysis by showing how hybrid cooling and dehumidification systems can be developed from fundamental technologies toward scalable, climate-resilient deployment.

AI disclosure

During the preparation of this work, the author(s) used Paper Pal as an AI tool solely to assist with language editing. The authors are fully responsible for the content of this chapter. After using this tool/service, the author(s) reviewed and edited the content as needed and take(s) full responsibility for the content of the publication.

Index

Note: Page numbers followed by "*f*" indicate figures, "*t*" indicate tables.

Printed and bound by CPI Group (UK) Ltd, Croydon, CR0 4YY
14/09/2026
02216410-0004